AN INTRODUCTION
TO THE PHYSICS OF
INTERSTELLAR DUST

Series in Astronomy and Astrophysics

The Series in Astronomy and Astrophysics includes books on all aspects of theoretical and experimental astronomy and astrophysics. Books in the series range in level from textbooks and handbooks to more advanced expositions of current research.

Series Editors:

M Birkinshaw, University of Bristol, UK

J Silk, University of Oxford, UK

Other books in the series

Numerical Methods in Astrophysics: An Introduction
P Bodenheimer, G P Laughlin, M Różyczka, H W Yorke

Very High Energy Gamma-Ray Astronomy
T C Weekes

The Physics of Interstellar Dust
E Krügel

Dust in the Galactic Environment, 2nd Edition
D C B Whittet

Dark Sky, Dark Matter
J M Overduin and P S Wesson

An Introduction to the Science of Cosmology
D J Raine and E G Thomas

The Origin and Evolution of the Solar System
M M Woolfson

The Physics of the Interstellar Medium, 2nd Edition
J E Dyson and D A Williams

Optical Astronomical Spectroscopy
C R Kitchin

Dust and Chemistry in Astronomy
T J Millar and D A Williams (eds)

Stellar Astrophysics
R J Tayler (ed)

Series in Astronomy and Astrophysics

AN INTRODUCTION TO THE PHYSICS OF INTERSTELLAR DUST

Endrik Krügel
Max-Planck Institute for Radioastronomy
Bonn, Germany

CRC Press
Taylor & Francis Group
Boca Raton London New York

CRC Press is an imprint of the
Taylor & Francis Group, an **informa** business

A TAYLOR & FRANCIS BOOK

CRC Press
Taylor & Francis Group
6000 Broken Sound Parkway NW, Suite 300
Boca Raton, FL 33487-2742

First issued in paperback 2019

© 2008 by Taylor & Francis Group, LLC
CRC Press is an imprint of Taylor & Francis Group, an Informa business

No claim to original U.S. Government works

ISBN-13: 978-1-58488-707-2 (hbk)
ISBN-13: 978-0-367-38886-7 (pbk)

Library of Congress Cataloging-in-Publication Data

Krügel, Endrik.
    An introduction to the physics of interstellar dust / Endrik Krügel.
       p. cm. -- (Series in astronomy and astrophysics)
    Includes bibliographical references and index.
    ISBN 978-1-58488-707-2 (alk. paper)
    1. Cosmic dust. 2. Interstellar matter. I. Title. II. Series.

QB791.K78 2007
523.1'125--dc22   2007007710

Visit the Taylor & Francis Web site at
http://www.taylorandfrancis.com

and the CRC Press Web site at
http://www.crcpress.com

Foreword

This book is a derivate of *The Physics of Interstellar Dust* which appeared four years ago at The Institute of Physics. It met with indulgence by the community and when asked by my new publisher, Taylor & Francis, to prepare a new edition, less voluminous and more affordable to students, I gladly and gratefully agreed.

A substantial part of the material had therefore to be abridged or left out. The chapters on the Milky Way, star formation and grain alignment are omitted altogether, the one on polarization is drastically cut. On the other hand, several sections have been expanded (on the charge, motion, optical constants, infrared resonances and surface chemistry of grains) and new topics found their entry (X-ray absoption, discrete dipole approximation, meteorites, AGB stars).

The second major topic of this book, besides the physics of dust, but linked to it, is the transfer of radiation through a dusty medium. When an astronomer is confronted with dust, it is usually only an annoyance that distorts the flow of photons from the object of interest in direction, frequency and polarization. To interpret observations, one therefore has to understand, in some approximate way, how light propagates. Therefore, the chapter on radiative transfer has been enlarged, too.

Overall, the new text is shorter and condensed, but also supplemented, it has been corrected and hopefully in places improved.

It is unlikely that anyone without an astronomical background will pick up this book. It adresses the astronomy student who certainly knows that our Milky Way is one out of billions of galaxies in the universe, that it has the shape of a flat disk, contains billions of stars and is pervaded by a dilute fluid: the interstellar medium of which dust is a part.

Dust, although it accounts for only one percent of the mass of the interstellar matter, absorbs one third of the starlight in the Milky Way and reemits it in the infrared. It weakens the visual light and thus limits our optical view: in the plane of our galaxy, we cannot look much farther than a few kiloparsec, only perpendicular to it can we fathom the vastness of the universe. Dust strongly influences the physical and chemical conditions of the interstellar medium, determines the way in which stars form and is the basis for planet formation. Its overwhelming importance is indisputable.

The other constituent of the interstellar medium is the gas. It is tightly related to the dust, but not treated explicitly here. To anyone who is not familar with the gas component and wishes to learn more about it and the dust-gas connection, I recommend the recent introductions to the interstellar medium by J Lequeux (*The Interstellar Medium*) and by A Tielens (*The Physics and Chemistry of the Interstellar Medium*). Another suitable complement to the present text, on the topic of dust itself, is D Whittet's *Dust in the Galactic Environment* where observational aspects are stressed.

I am much obliged to Peter Schilke and Nikolaj Voshchinnikov for critically reading parts of the manuscript, to Werner Tscharnuter for his helpful comments on the chapter on *Radiative transfer*, and to Alexandr Tutukov for general advice.

Contents

1 The dielectric permeability   **1**
- 1.1   How the electromagnetic field acts on dust   1
  - 1.1.1   Electric field and magnetic induction   1
  - 1.1.2   Electric polarization of the medium   2
  - 1.1.3   Magnetic polarization of the medium   6
  - 1.1.4   Free charges and polarization charges   7
  - 1.1.5   The field equations   9
  - 1.1.6   Waves in a dielectric medium   9
  - 1.1.7   Energy dissipation of a grain in a variable field   12
- 1.2   The harmonic oscillator   13
  - 1.2.1   The Lorentz model   14
  - 1.2.2   Dissipation of energy   17
  - 1.2.3   Dispersion relation of the dielectric permeability . . .   18
  - 1.2.4   The harmonic oscillator and light   20
  - 1.2.5   Radiation damping   23
  - 1.2.6   The cross section of an harmonic oscillator   24
- 1.3   Waves in a conducting medium   25
  - 1.3.1   The dielectric permeability of a conductor   25
  - 1.3.2   Conductivity and the Drude profile   27

2 How to evaluate grain cross sections   **29**
- 2.1   Defining cross sections   29
  - 2.1.1   Cross section for scattering, absorption and extinction   29
  - 2.1.2   Phase function and cross section for radiation pressure   31
  - 2.1.3   Efficiencies, mass and volume coefficients   32
- 2.2   The optical theorem   32
  - 2.2.1   The intensity of forward scattered light   33
  - 2.2.2   The refractive index of a dusty medium   35
- 2.3   Mie theory for a sphere   37
  - 2.3.1   The formalism   37
  - 2.3.2   Scattered and absorbed power   38
- 2.4   Polarization and scattering   39
  - 2.4.1   The amplitude scattering matrix   40
  - 2.4.2   Angle-dependence of scattering   41
  - 2.4.3   The polarization ellipse   41
  - 2.4.4   Stokes parameters   42

2.5	The discrete dipole approximation		45
2.6	The Kramers-Kronig relations		47
	2.6.1	The KK relation for the dielectric permeability	47
	2.6.2	Three corollaries of the KK relation	47
2.7	Composite grains		50
	2.7.1	Effective medium theories	50
	2.7.2	The influence of grain size, ice and porosity	54

3 Very small and very big particles   **59**
3.1	Tiny spheres		59
	3.1.1	Approximating the efficiencies	59
	3.1.2	Polarization and angle-dependent scattering	64
	3.1.3	Small-size effects beyond Mie theory	64
3.2	Tiny ellipsoids		65
	3.2.1	Cross section and shape factor of pancakes and cigars	66
	3.2.2	Randomly oriented ellipsoids	67
3.3	The fields inside a dielectric particle		70
	3.3.1	Internal field and depolarization field	70
	3.3.2	Depolarization field and surface charges	70
	3.3.3	The local field at an atom	71
	3.3.4	The relation of Clausius-Mossotti	72
3.4	Very large particles		73
	3.4.1	Babinet's theorem	73
	3.4.2	Reflection and transmission at a plane surface	75
	3.4.3	Huygens' principle	77
3.5	Grains of small refractive index		80
	3.5.1	Rayleigh Gans particles	80
	3.5.2	X-ray scattering	81
	3.5.3	X-ray absorption	82

4 Case studies of Mie calculus   **85**
4.1	Efficiencies of bare spheres		85
	4.1.1	Scattering and absorption	85
	4.1.2	Efficiency vs. cross section and volume coefficient	90
4.2	Scattering by bare spheres		92
	4.2.1	The intensity pattern of scattered light	92
	4.2.2	The polarization of scattered light	93
4.3	Linear polarization through extinction		96
4.4	Coated spheres		97
4.5	Surface modes in small grains		99
	4.5.1	Small graphite spheres	99
	4.5.2	Ellipsoids and metals	100

5 Structure and composition of dust **101**
  5.1 Crystal structure 101
      5.1.1 Translational symmetry 101
      5.1.2 Lattice types 102
      5.1.3 The reciprocal lattice 105
  5.2 Binding in crystals 107
      5.2.1 Covalent and ionic bonding 107
      5.2.2 Metals 108
      5.2.3 van der Waals forces and hydrogen bridges 110
  5.3 Carbonaceous grains and silicate grains 111
      5.3.1 Origin of the two major dust constituents 111
      5.3.2 The bonding in carbon 112
      5.3.3 Carbon compounds 113
      5.3.4 Silicates 117
      5.3.5 The origin of the elements found in dust grains 120
  5.4 Optical constants of dust materials 121
  5.5 Grain sizes 128
      5.5.1 The MRN size distribution 128
      5.5.2 Collisional fragmentation 129

6 Dust radiation **131**
  6.1 Kirchhoff's law 131
      6.1.1 The emissivity of dust 131
      6.1.2 Thermal emission of grains 132
      6.1.3 Absorption and emission in thermal equilibrium 133
  6.2 The temperature of big grains 134
      6.2.1 The energy equation 134
      6.2.2 Temperature estimates 135
      6.2.3 Relation between grain size and grain temperature . . 137
      6.2.4 Dust temperatures from observations 138
  6.3 The emission of big grains 140
      6.3.1 Constant temperature and low optical depth 140
      6.3.2 Total emission and cooling rate of a grain 143
  6.4 Calorific properties of solids 143
      6.4.1 Traveling waves in a crystal 145
      6.4.2 Internal energy of a grain 148
      6.4.3 The Debye temperature 149
      6.4.4 Specific heat 150
      6.4.5 Two-dimensional lattices 151
  6.5 Temperature fluctuations of very small grains 152
      6.5.1 The probability density $P(T)$ 153
      6.5.2 The transition matrix 153
      6.5.3 The stochastic time evolution of grain temperature . . 155
  6.6 The emission spectrum of very small grains 156
      6.6.1 Moderate fluctuations 156

6.6.2   Strong fluctuations 158
6.6.3   Temperature fluctuations and flux ratios 159

7 Dust and its environment   161
7.1   Grain charge   161
    7.1.1   Charge equilibrium in the absence of a UV field 161
    7.1.2   The photoelectric effect 163
7.2   Grain motion   166
    7.2.1   Random walk 167
    7.2.2   The drag on a grain subjected to an outer force 167
    7.2.3   Brownian motion of a grain 170
7.3   Dust in the solar system   172
    7.3.1   Interplanetary dust 172
    7.3.2   The Poynting-Robertson effect 173
    7.3.3   Electromagnetic forces on grains: Dust from Io 174
    7.3.4   Shooting stars and less belligerent meteoroids 176
7.4   Grain destruction 181
    7.4.1   Mass balance of gas and dust in the Milky Way 181
    7.4.2   Destruction processes 183
7.5   Grain formation 184
    7.5.1   Evaporation temperature and vapor pressure 185
    7.5.2   Vapor pressure of small grains 187
    7.5.3   Critical saturation 189
    7.5.4   Time-dependent homogeneous nucleation 190
    7.5.5   Steady-state nucleation 191
    7.5.6   Solutions to time-dependent homogeneous nucleation . 195

8 Grain surfaces   201
8.1   Gas accretion on grains   201
    8.1.1   Physical adsorption and chemisorption 202
    8.1.2   The sticking probability 205
8.2   Mobility of atoms on grain surfaces 206
    8.2.1   Thermal hopping 207
    8.2.2   Evaporation 208
    8.2.3   Tunneling 208
    8.2.4   Photodesorption 209
8.3   Grain surface chemistry 210
    8.3.1   Chemical reactions in the gas 210
    8.3.2   Chemical reactions on dust 211
    8.3.3   The formation of H_2 in diffuse clouds 214
8.4   Ice mantles   215

9 PAHs and spectral features of dust **219**
  9.1 Polycyclic Aromatic Hydrocarbons 219
    9.1.1 Microcanonic emission of PAHs 220
    9.1.2 An example: anthracene 221
    9.1.3 Photo-excitation of PAHs 224
    9.1.4 Cutoff wavelength for electronic excitation 225
    9.1.5 Photo-destruction and ionization 226
    9.1.6 Cross sections and line profiles of PAHs 227
  9.2 ERE and DIBs 229
  9.3 The silicate bands at 10μm and 18μm 230
    9.3.1 The strength of the resonances 230
    9.3.2 How the bands change with temperature and grain
           size 231
  9.4 Crystalline silicates 233
    9.4.1 Where they are found and how they form 233
    9.4.2 Thermal expansion of grains 234
    9.4.3 The frequency shift of a resonance in grain heating . . 236
  9.5 The feature at 3.4μm 237

10 Interstellar reddening and dust models **239**
  10.1 Reddening by interstellar grains 239
    10.1.1 Stellar photometry 239
    10.1.2 The interstellar extinction curve 241
    10.1.3 Two-color-diagrams 245
    10.1.4 Spectral indices 246
    10.1.5 The mass absorption coefficient 246
  10.2 Dust models 249
    10.2.1 Description of the model components 250
    10.2.2 Extinction and scattering of the dust model 252
    10.2.3 Extinction and absorption mass coefficients 254

11 Radiative transport **259**
  11.1 Basic transfer relations 259
    11.1.1 Definition of intensity, mean intensity and flux 259
    11.1.2 The general transfer equation 262
    11.1.3 Transfer equation in spherical and slab symmetry . . . 264
    11.1.4 Frequency averages 266
    11.1.5 Analytical solutions to the transfer equation 267
  11.2 Spherical clouds 268
    11.2.1 Integral equations for the intensity 269
    11.2.2 Practical hints 270
  11.3 Passive disks 272
    11.3.1 Radiative transfer in a plane parallel layer 272
    11.3.2 Disks of high optical thickness 274
    11.3.3 The grazing angle 275

11.4 Galactic nuclei 278
    11.4.1 Hot spots in a spherical stellar cluster 278
    11.4.2 Low and high luminosity stars 279
11.5 The pursuit of random photons 281
    11.5.1 The strategy 281
    11.5.2 Grains with temperature fluctuations 284
    11.5.3 Anisotropic scattering 286
    11.5.4 Practical considerations 287

12 Spectral energy distribution of dusty objects   **289**
12.1 Early stages of star formation 289
    12.1.1 Globules 289
    12.1.2 Isothermal gravitationally-bound clumps 291
    12.1.3 The density structure of a protostar 292
12.2 Accretion disks 296
    12.2.1 Flat disks 296
    12.2.2 Inflated disks 299
12.3 Reflection nebulae 302
12.4 Starburst nuclei 304
12.5 Mass loss giants 307
    12.5.1 Flow equations 307
    12.5.2 Solutions to the flow equations 310
12.6 The effective extinction curve 314
    12.6.1 The effective optical thickness 314
    12.6.2 Monte Carlo simulations 316

A Various dust related physics   **319**
A.1 Boltzmann statistics 319
    A.1.1 The probability of an arbitrary energy distribution . . 319
    A.1.2 Partition function and population of energy cells . . . 321
    A.1.3 The mean energy of harmonic oscillators 323
    A.1.4 The Maxwellian velocity distribution 323
A.2 Quantum statistics 325
    A.2.1 The unit cell h^3 of phase space 325
    A.2.2 Bosons and fermions 326
A.3 Thermodynamics 328
    A.3.1 The ergodic hypothesis 328
    A.3.2 Definition of entropy and temperature 330
    A.3.3 The canonical distribution 331
    A.3.4 Thermodynamic relations 332
    A.3.5 Equilibrium conditions of the state functions 335
A.4 Blackbody radiation 337
    A.4.1 The Planck function 337
    A.4.2 Low and high frequency limit 338
    A.4.3 The laws of Wien and Stefan-Boltzmann 339

A.5 The classical Hamiltonian 340
    A.5.1 Normal coordinates 341
A.6 The Hamiltonian in quantum mechanics 342
    A.6.1 The time-dependent Schrödinger equation 342
    A.6.2 Stationary solutions 343
    A.6.3 The dipole moment of a transition 343
    A.6.4 The quantized harmonic oscillator 344
A.7 The Einstein coefficients A and B 346
    A.7.1 Induced and spontaneous transitions 346
A.8 Potential wells and tunneling 348
    A.8.1 Wave function of a particle in a constant potential . . 348
    A.8.2 Potential walls and Fermi energy 349
    A.8.3 Rectangular potential barriers 350
    A.8.4 The double potential well 354

B Miscellaneous **357**
B.1 Mathematical notations 357
B.2 Mathematical formulae 358
    B.2.1 Sums and integrals 358
    B.2.2 The bell curve 358
    B.2.3 Polynomials 359
    B.2.4 Vector analysis 360
    B.2.5 The time average of an harmonically varying field . . . 361
B.3 Cosmic constants 362
B.4 Problem set 363
B.5 List of symbols 374

Bibliography **377**

Index **381**

1

The dielectric permeability

We begin by acquainting ourselves with the polarization of matter. The fundamental quantity describing how an interstellar grain responds to an electromagnetic wave is the dielectric permeability which relates the polarization of matter to the applied field. We recall the basic equations of electrodynamics and outline how plane waves travel in an infinite non-conducting (dielectric) medium and in a plasma. We summarize the properties of harmonic oscillators, including the absorption, scattering and emission of light by individual dipoles. Approximating a solid body by an ensemble of such dipoles (identical harmonic oscillators), we learn how its dielectric permeability changes with frequency. This study is carried out for

- a dielectric medium where the electron clouds oscillate about the atomic nuclei

- and a metal where the electrons are free.

1.1 How the electromagnetic field acts on dust

At the root of all phenomena of classical electrodynamics, such as the interaction of light with interstellar dust, are Maxwell's formulae. They can be written in different ways and the symbols, their names and meaning are not universal, far from it. Before we exploit Maxwell's equations, we therefore first define the quantities which describe the electromagnetic field.

1.1.1 Electric field and magnetic induction

A charge q traveling with velocity $\mathbf{v}$ in a fixed electric field $\mathbf{E}$ and a fixed magnetic field of flux density $\mathbf{B}$ experiences a force

$$\mathbf{F} = q\left[\mathbf{E} + \frac{1}{c}\mathbf{v} \times \mathbf{B}\right] , \qquad (1.1)$$

called the *Lorentz* force; the cross $\times$ denotes the vector product. $\mathbf{B}$ is also called magnetic induction. Equation (1.1) shows what happens mechanically to a charge in an electromagnetic field and we use it to define $\mathbf{E}$ and $\mathbf{B}$.

The force $\mathbf{F}$ has an electric part, $q\mathbf{E}$, which pulls a positive charge in the direction of $\mathbf{E}$, and a magnetic component, $(q/c)\,\mathbf{v}\times\mathbf{B}$, working perpendicular to $\mathbf{v}$ and $\mathbf{B}$. In case the moving charges are electrons in an atom driven by an electromagnetic wave, their velocities are small. A typical value is the velocity of the electron in the ground state of the hydrogen atom for which classically v/c equals the fine structure constant

$$\alpha \;=\; \frac{e^2}{\hbar c} \;\simeq\; \frac{1}{137} \;\ll\; 1 \;.$$

Protons move still more slowly because of their greater inertia. For an electromagnetic wave in vacuum, $|\mathbf{E}| = |\mathbf{B}|$ and therefore the term $(\mathbf{v}\times\mathbf{B})/c$ in formula (1.1) is much smaller than $\mathbf{E}$ and usually irrelevant for the motion of the charge.

1.1.2 Electric polarization of the medium

1.1.2.1 Dielectric permeability and electric susceptibility

In an electrically neutral medium, the integral of the local charge density $\rho(\mathbf{x})$ at locus $\mathbf{x}$ over any volume V vanishes,

$$\int_V \rho(\mathbf{x})\,dV \;=\; 0 \;.$$

However, a neutral body of volume V may have a dipole moment $\mathbf{p}$ (small letter) given by

$$\mathbf{p} \;=\; \int_V \mathbf{x}\,\rho(\mathbf{x})\,dV \;. \tag{1.2}$$

The value of the integral is independent of the choice of the coordinate system. The dipole moment per unit volume is called polarization $\mathbf{P}$ (capital letter),

$$\mathbf{P} \;=\; \frac{\mathbf{p}}{V} \;. \tag{1.3}$$

It may be interpreted as the number density N of molecules times the average electric dipole moment $\mathbf{p}_{\text{mol}}$ per molecule,

$$\mathbf{P} \;=\; N\,\mathbf{p}_{\text{mol}} \;.$$

$\mathbf{E}$ and $\mathbf{P}$ being defined, we introduce as an additional quantity the displacement $\mathbf{D}$. It adds nothing new and is just another way of expressing the polarization of matter, $\mathbf{P}$, through an electric field $\mathbf{E}$,

$$\mathbf{D} \;=\; \mathbf{E} \;+\; 4\pi\,\mathbf{P} \;. \tag{1.4}$$

The local polarization $\mathbf{P}$ and the local average electric field $\mathbf{E}$ in the dielectric medium are much weaker than the fields on the atomic level and constitute just a small perturbation. Therefore, **P**, **D** and **E** are proportional to each other and we can write

$$\mathbf{D} = \varepsilon\,\mathbf{E} \tag{1.5}$$

$$\mathbf{P} = \chi\,\mathbf{E} \tag{1.6}$$

with

$$\chi = \frac{\varepsilon - 1}{4\pi}\,. \tag{1.7}$$

The proportionality factors ε and χ are material constants, they mean more or less the same; ε is called *dielectric permeability* or *dielectric constant* or *permittivity*, χ bears the name *electric susceptibility*. In the trivial case of vacuum, the polarization **P** vanishes, $\mathbf{E} = \mathbf{D}$ and $\varepsilon = 1$. In a dielectric medium, however, a constant field **E** induces a dipole moment so that $\varepsilon > 1$ and $\mathbf{P} \neq 0$.

We have altogether three quantities describing the electric field: **D**, **E** and **P**, although two would suffice. For example, we could replace **D** in all equations by $\varepsilon\mathbf{E}$. Equation (1.5) is the first of two constitutive relations complementing the set of Maxwell's formulae.

1.1.2.2 The electric polarizability

Another quantity we will need is the *electric polarizability* α_e. If we place a small grain of volume V into a constant electric field **E**, it acquires a dipole moment

$$\mathbf{p} = \alpha_e\,V\mathbf{E}\,. \tag{1.8}$$

With respect to atoms or point-like dipoles, the polarizabilty is usually defined by $\mathbf{p} = \alpha_e\mathbf{E}$. Because of the dipole moment, the field in the vicinity of the grain becomes distorted. It has no longer the value **E**, but some other value, say $\mathbf{E}^e$ (e for external). As one recedes from the grain, $\mathbf{E}^e$ approaches **E** asymptotically. There is also a field $\mathbf{E}^i$ inside the grain. It is different from the constant outer field **E**. The relation between them is described in detail in section 3.3. The polarization **P** (capital letter) depends linearly both on $\mathbf{E}^i$ (from (1.6)) and on **E** (from (1.8)), the proportionality factors being α_e and χ, respectively,

$$\mathbf{P} = \chi\,\mathbf{E}^i = \alpha_e\,\mathbf{E}\,. \tag{1.9}$$

In the general case, α_e and χ are tensors, and the dipole moment and the fields do not have the same direction.

1.1.2.3 The dependence of the dielectric permeability on frequency

Unless specified otherwise, we always assume (on a macroscopic level with dimensions much greater than an atom) homogeneity so that the dielectric permeability ε does not depend on position in the grain. The relation (1.5),

$\mathbf{D} = \varepsilon\mathbf{E}$, is linear, which means that ε is independent of the field strength. However, it may depend on the direction of the fields because, on a microscopic level, a grain is made up of atoms and thus not homogeneous. Equation (1.5) must then be generalized to

$$D_i = \sum_i \varepsilon_{ik} E_i + D_{0i} .$$

The constant term D_{0i} implies a frozen-in dipole moment even in the absence of an outer field. This happens only in special cases known as pyroelectricity; for mono-crystals with a cubic structure (see section 5.1), D_{0i} is always zero. The tensor ε_{ik} is symmetric, $\varepsilon_{ik} = \varepsilon_{ki}$, and can always be brought into a diagonal form by the right choice of coordinate system. It then has, at most, three independent components. If the three diagonal elements are equal, ε reduces to a scalar and the grain is said to be isotropic. Crystals of the cubic system (section 5.1) have this pleasant property, but in interstellar grains we will also encounter anisotropy.

The dielectric permeability ε changes with frequency

$$\omega = 2\pi \nu ,$$

and so will the optical behavior of interstellar dust. The functional form of $\varepsilon(\omega)$ is called the dispersion relation. So the term dielectric constant, which is the other frequently used name for $\varepsilon(\omega)$, is misleading when taken at its face value.

In electrostatics, only one parameter is needed to specify the polarization and χ or ε are real. But in an alternating field, one needs for each frequency ω two independent numbers, which out of mathematical convenience are written as *one* complex variable,

$$\chi(\omega) = \chi_1(\omega) + i\,\chi_2(\omega) \tag{1.10}$$
$$\varepsilon(\omega) = \varepsilon_1(\omega) + i\,\varepsilon_2(\omega) . \tag{1.11}$$

As the field changes, E and P are not in phase because the electrons always need a little while to adjust. The polarization P therefore lags behind by some time $\Delta t \ll \omega^{-1}$ corresponding to a small angle

$$\varphi = \omega\Delta t = \tan(\chi_2/\chi_1) \simeq \chi_2/\chi_1 \ll 1 . \tag{1.12}$$

The interval Δt must be positive for reasons of causality: sensible people squeal only *after* they have been pinched. Therefore, the imaginary part χ_2 is positive and determines, for a given χ_1, the time lag. While adapting to the field, the electrons have to overcome internal friction, which implies dissipational losses. Some of the field energy is inevitably drained into the dielectric medium. If χ_2 and Δt were negative, the field would draw energy

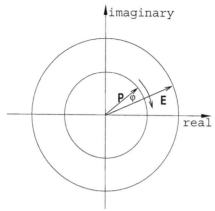

FIGURE 1.1 The electric field vector, **E**, in the medium is presented as a complex quantity of length E, which rotates at frequency ω. It induces a complex polarization **P** in the material of length P. When damping is weak, **P** lags behind **E** by a small angle φ. The real part of the complex electric susceptibility is equal to the ratio of the radii of the two circles, $\chi_1 = P/E$, whereas the imaginary part gives the phase lag, $\chi_2 = \chi_1\varphi$. Only the real components of **P** and **E** have a physical meaning, the complex representation is for mathematical convenience.

from the dielectric body, decreasing its entropy which contradicts the second law of thermodynamics. Even $\chi_2 = 0$ is impossible as it would mean that E and P are absolutely synchronous and dissipation is completely absent.

If the frequency ω is not small, but arbitrary, the angle $\omega\Delta t$ is also not necessarily small. Nevertheless, we can exclude $\chi_2 \leq 0$ for the same reasons as above. The real part of the susceptibility, χ_1, on the other hand, may become zero as well as negative. The maximum of P is now given by $P_{\max} = \sqrt{\chi_1^2 + \chi_2^2}\, E_0$.

Of course, we could also choose a time dependence $e^{i\omega t}$ of the field, and nothing would change. The sign of i is just a convention, although one that must be followed consistently. When $E = E_0\, e^{i\omega t}$, the field rotates anti-clockwise in figure 1.1 and the susceptibility, instead of (1.10), would be $\chi(\omega) = \chi_1(\omega) - i\,\chi_2(\omega)$, again with positive χ_2.

1.1.2.4 Dielectrics and metals

We will be dealing with two kinds of substances for which Maxwell's equations are often formulated separately:

- *Dielectrics.* They are insulators and no constant current can be sustained within them. Nevertheless, alternating currents produced by a time-variable electric field are possible. In these currents, the charges do not travel far from their equilibrium positions.

- *Metals.* This is a synonym for conductors, and in this sense they are the opposite of dielectrics. When a piece of metal is connected at its ends to the poles of a battery, a steady current flows under the influence of an electric field. When this piece of metal is placed in a static electric field, the charges accumulate at its surface and arrange themselves in such a way that the electric field inside vanishes and then there is no internal current. However, time-varying electric fields and currents are possible.

In the interstellar medium, one finds both dielectric and metallic particles, but the latter are probably far from being perfect conductors.

1.1.3   Magnetic polarization of the medium

In a macroscopic picture, a stationary motion of charges with density ρ and velocity $\mathbf{v}$, corresponding to a current density $\mathbf{J} = \rho\mathbf{v}$, produces a magnetic moment

$$\mathbf{m} = \frac{1}{2c} \int (\mathbf{x} \times \mathbf{J})\, dV \ . \tag{1.13}$$

When the integral extends over a unit volume, it is called magnetization $\mathbf{M}$,

$$\mathbf{M} = \frac{\mathbf{m}}{V} \ . \tag{1.14}$$

For example, a charge q traveling with velocity v in a circular orbit of radius r constitutes a current $I = qv/2\pi r$. The magnetic moment of this moving charge is

$$m = \frac{qvr}{2c} = \frac{A\,I}{c} \ ,$$

where $A = \pi r^2$ signifies the area of the loop. If A is small, the accompanying field is that of a dipole.

Without macroscopic charge motion, the magnetization of matter comes, in a classical picture, from the atomic currents (electron orbits). The magnetic moment per unit volume $\mathbf{M}$ may be interpreted as the number density N of molecules in the substance times the total magnetic moment $\mathbf{m}_{\mathrm{mol}}$ per molecule,

$$\mathbf{M} = N\,\mathbf{m}_{\mathrm{mol}} \ .$$

The magnetic field $\mathbf{H}$ is defined by

$$\mathbf{H} = \mathbf{B} - 4\pi\,\mathbf{M} \ . \tag{1.15}$$

Similar to the situation of the electric field, we could replace in all equations $\mathbf{H}$ by $\mathbf{B}$ and retain instead of three field quantities only two, $\mathbf{B}$ and $\mathbf{M}$. We would prefer the pairs $(\mathbf{E}, \mathbf{P})$ and $(\mathbf{B}, \mathbf{M})$, because the electric field $\mathbf{E}$, the magnetic flux density $\mathbf{B}$ as well as the polarizations $\mathbf{P}$ and $\mathbf{M}$ allow a direct physical interpretation, however, conventions urge us to drag $\mathbf{D}$ and $\mathbf{H}$ along.

1.1.3.1 The magnetic susceptibility

The second constitutive equation concerns the magnetic field $\mathbf{H}$ and the induction $\mathbf{B}$. If the substance is not ferromagnetic, a linear relation holds as in (1.5),

$$\mathbf{B} = \mu\,\mathbf{H}\,. \tag{1.16}$$

μ is the magnetic permeability and generally complex too,

$$\mu = \mu_1 + i\,\mu_2\,.$$

It also shows dispersion. In vacuum, $\mathbf{B} = \mathbf{H}$ and $\mu = 1$. For diamagnetic substances, μ is a little smaller than one; for paramagnetic substances a little bigger, μ is large only for ferromagnets. Similar to (1.7), we define the magnetic susceptibility by

$$\chi = \frac{\mu - 1}{4\pi}\,, \tag{1.17}$$

so that $\mathbf{M} = \chi\mathbf{H}$. If there is any chance of confusion, one must write explicitly χ_{m} or χ_{e} for the magnetic or electric case. In analogy to (1.8), a constant *outer* magnetic field $\mathbf{H}$ induces in a small body of volume V a magnetic dipole moment

$$\mathbf{m} = V\mathbf{M} = \alpha_{\mathrm{m}}\,V\mathbf{H}\,, \tag{1.18}$$

where α_{m} is the magnetic polarizability and $\mathbf{M}$ the magnetization *within* the body. For interstellar grains, which are weakly magnetic substances with $|\mu|$ close to one, the magnetic induction and magnetic field are practically the same inside and outside, and $\chi \simeq \alpha$. If $\mathbf{B}$ denotes the interstellar magnetic field and $\mathbf{M}$ the magnetization of the grain, one usually writes $\mathbf{M} = \chi\mathbf{B}$.

1.1.4 Free charges and polarization charges

We are generally interested in the electromagnetic field averaged over regions containing many atoms. So we usually picture the grain material to be a continuous medium. However, sometimes we have to dive into the atomic world. On a scale of 1Å or less, the medium becomes structured. Atoms appear with positive nuclei and negative electrons spinning and moving in complicated orbits, and the electric and magnetic fields, which are smooth on a macroscopic scale, have huge gradients.

Consider a microscopic volume δV in some piece of matter. It contains protons and electrons and the net charge divided by δV gives the total local charge density ρ_{tot}. For an electrically neutral body, electrons and protons exactly cancel and ρ_{tot} integrated over the body is zero. Often ρ_{tot} is written as the sum of two terms,

$$\rho_{\mathrm{tot}} = \rho_{\mathrm{pol}} + \rho_{\mathrm{free}}\,. \tag{1.19}$$

The first arises from polarization. In a dielectric medium, the electric field polarizes the atoms and thus produces aligned dipoles. If the polarization within the particle is uniform, ρ_{pol} is zero because the positive and negative charges of the neighboring dipoles exactly balance; the only exception is the surface of the particle where charges of one kind accumulate. However, if the polarization $\mathbf{P}$ is spatially non-uniform, ρ_{pol} does not vanish. Then the separation of charges is inhomogeneous and leads to a charge density

$$\rho_{pol} = -\operatorname{div} \mathbf{P} . \tag{1.20}$$

The second term in (1.19), ρ_{free}, comprises all other charges besides ρ_{pol}. For example, when the polarization is constant, $\operatorname{div} \mathbf{P} = 0$ and ρ_{free} stands for *all* charges, positive and negative. In an uncharged body, their sum is zero. When charges are brought in from outside, $\rho_{free} \neq 0$. Within a metal, $\mathbf{P} = 0$, and polarization charges can appear only on the surface.

Moving charges constitute a current. The total current density $\mathbf{J}_{tot}$ may also be split into parts similarly to (1.19),

$$\mathbf{J}_{tot} = \mathbf{J}_{free} + \mathbf{J}_{pol} + \mathbf{J}_{mag} . \tag{1.21}$$

The first term on the right side is associated with the motion of the free charges, the second with the time-varying polarization of the medium, and the third represents the current that gives rise to its magnetization,

$$\mathbf{J}_{free} = \mathbf{v}\rho_{free}, \quad \mathbf{J}_{pol} = \dot{\mathbf{P}}, \quad \mathbf{J}_{mag} = c \cdot \operatorname{rot} \mathbf{M} . \tag{1.22}$$

1.1.4.1 Sign conventions

We have to make a remark on the sign convention. An electron has of course a negative charge, its absolute value being $e = 4.803 \times 10^{-10}$ esu (electrostatic units). After Maxwell's equation (1.27), the electric field of an isolated positive charge (proton) is directed away from it and repels other positive charges according to (1.1). The moment $\mathbf{p}$ of a dipole created by two opposite, but equal charges is after (1.2) directed from the minus to the plus charge and anti-parallel to the electric field along the line connecting the two charges. Hence a dielectric grain in a constant field $\mathbf{E}$ has a polarization $\mathbf{P}$ parallel to $\mathbf{E}$ and surface charges as depicted in figure 3.5 of chapter 3. With the help of this figure, we can explain the minus sign in equation (1.20). Going from left to right, the polarization jumps at the left edge of the grain from zero to its value inside. The gradient there is positive and thus a negative charge appears on the surface. In the case of non-uniform polarization within the grain, at a place where $\operatorname{div} \mathbf{P} > 0$, the electric field pulls a small charge δq out of a tiny volume δV leaving behind an unbalanced negative charge $-\delta q$.

1.1.5 The field equations

We have now defined all quantities that appear in Maxwell's equations, and we write them down, first, for a neutral dielectric medium,

$$\text{div } \mathbf{D} = 0 \tag{1.23}$$

$$\text{div } \mathbf{B} = 0 \tag{1.24}$$

$$\text{rot } \mathbf{E} = -\frac{1}{c}\dot{\mathbf{B}} \tag{1.25}$$

$$\text{rot } \mathbf{H} = \frac{1}{c}\dot{\mathbf{D}}, \tag{1.26}$$

and second, for a medium with free charges and currents,

$$\text{div } \mathbf{D} = 4\pi\,\rho_{\text{free}} \tag{1.27}$$

$$\text{div } \mathbf{B} = 0 \tag{1.28}$$

$$\text{rot } \mathbf{E} = -\frac{1}{c}\dot{\mathbf{B}} \tag{1.29}$$

$$\text{rot } \mathbf{H} = \frac{1}{c}\dot{\mathbf{D}} + \frac{4\pi}{c}\,\mathbf{J}_{\text{free}}. \tag{1.30}$$

A dot above a letter stands for partial time derivative, for instance, $\dot{\mathbf{B}} = \partial\mathbf{B}/\partial t$. Applying the operator div to (1.30) gives the expression for charge conservation,

$$\dot{\rho}_{\text{free}} + \text{div } \mathbf{J}_{\text{free}} = 0. \tag{1.31}$$

$\mathbf{E}$ denotes the electric and $\mathbf{H}$ the magnetic field, $\mathbf{D}$ the electric displacement, $\mathbf{B}$ the magnetic induction, $\mathbf{J}$ the current density and c, of course, the velocity of light. Our choice of mathematical symbols is summarized in Appendix B.1 together with some common relations of vector analysis.

1.1.6 Waves in a dielectric medium

1.1.6.1 The wave equation

In case of harmonic fields, which have a sinusoidal time dependence proportional to $e^{-i\omega t}$, Maxwell's equations (1.25) and (1.26) become simpler,

$$\text{rot } \mathbf{E} = i\frac{\omega\mu}{c}\mathbf{H} \tag{1.32}$$

$$\text{rot } \mathbf{H} = -i\frac{\omega\varepsilon}{c}\mathbf{E}. \tag{1.33}$$

Applying the operator rot to (1.32) and (1.33), we arrive with the formula (B.23) of vector analysis at

$$\Delta\mathbf{E} + \frac{\omega^2}{c^2}\mu\varepsilon\,\mathbf{E} = 0, \qquad \Delta\mathbf{H} + \frac{\omega^2}{c^2}\mu\varepsilon\,\mathbf{H} = 0 \tag{1.34}$$

or

$$\Delta \mathbf{E} - \frac{\mu\varepsilon}{c^2} \ddot{\mathbf{E}} = 0, \qquad \Delta \mathbf{H} - \frac{\mu\varepsilon}{c^2} \ddot{\mathbf{H}} = 0 . \tag{1.35}$$

These are the wave equations which describe the change of the electromagnetic field in space and time. In an infinite medium, one solution to (1.34) is a plane harmonic wave,

$$\mathbf{E}(\mathbf{r}, t) = \mathbf{E_0} \cdot e^{i(\mathbf{k}\cdot\mathbf{r} - \omega t)} \tag{1.36}$$

$$\mathbf{H}(\mathbf{r}, t) = \mathbf{H_0} \cdot e^{i(\mathbf{k}\cdot\mathbf{r} - \omega t)} , \tag{1.37}$$

where $\mathbf{r} = (x, y, z)$ and $\mathbf{k}$ is a vector with $\mathbf{k}^2 = \omega^2 \mu\varepsilon/c^2$. The characteristics of a plane harmonic wave are an $e^{i\mathbf{k}\cdot\mathbf{r}}$ space variation and an $e^{-i\omega t}$ time variation. All waves in interstellar space that interact with interstellar matter are plane because the sources from which they arise are very distant.

1.1.6.2   The wavenumber

The vector $\mathbf{k} = (k_x, k_y, k_z)$ in (1.36) is called wavenumber and in the most general case is complex, but we only consider the standard case in which $\mathbf{k}$ is real and has the length

$$k = |\mathbf{k}| = \frac{\omega}{c} \sqrt{\varepsilon\mu} . \tag{1.38}$$

Inserting the field of a plane wave given by (1.36), (1.37) into equations (1.32) and (1.33) yields

$$\frac{\omega\mu}{c} \mathbf{H} = \mathbf{k} \times \mathbf{E} \qquad \text{and} \qquad \frac{\omega\varepsilon}{c} \mathbf{E} = -\mathbf{k} \times \mathbf{H} \tag{1.39}$$

and after scalar multiplication with $\mathbf{k}$

$$\mathbf{k} \cdot \mathbf{E} = \mathbf{k} \cdot \mathbf{H} = 0 . \tag{1.40}$$

As $\mathbf{k}$ is real, the imaginary parts ε_2 and μ_2, which are responsible for the dissipation of energy, must be zero. This is strictly possible only in vacuum. Any other medium is never fully transparent and an electromagnetic wave always suffers some losses. As $\mathbf{k}$ specifies the direction of wave propagation, it follows from (1.39) and (1.40) that the vectors of the electric and magnetic field are perpendicular to each other and to $\mathbf{k}$. The wave is plane and travels with undiminished amplitude at a phase velocity

$$v_{\mathrm{ph}} = \frac{\omega}{k} = \frac{c}{\sqrt{\varepsilon\mu}} . \tag{1.41}$$

The real amplitude of the magnetic and electric field, E_0 and H_0, are related through

$$H_0 = \sqrt{\frac{\varepsilon}{\mu}} E_0 . \tag{1.42}$$

In particular, in vacuum

$$v_{\mathrm{ph}} = c, \qquad H_0 = E_0, \qquad k = \frac{\omega}{c} = \frac{2\pi}{\lambda} .$$

1.1.6.3 Flux and momentum of the electromagnetic field

An electromagnetic wave carries energy. The flux, which is the energy per unit time and area, is given by the Poynting vector $\mathbf{S}$. For *real* fields, its *momentary* value is

$$\mathbf{S} = \frac{c}{4\pi}\,\mathbf{E} \times \mathbf{H} \,. \tag{1.43}$$

In vacuum, when E_0 denotes the amplitude of the electric field vector, the *time average* is

$$\langle S \rangle = \frac{c}{8\pi}\,E_0^2 \,. \tag{1.44}$$

When we use *complex* fields, the time average of the Poynting vector is a real quantity and equals after (B.31)

$$\langle \mathbf{S} \rangle = \frac{c}{8\pi}\,\mathrm{Re}\,\{\mathbf{E} \times \mathbf{H}^*\} \,. \tag{1.45}$$

$\mathbf{H}^*$ is the complex conjugate of $\mathbf{H}$. The wave also transports momentum, through a unit area at a rate $\mathbf{S}/c$. In the corpuscular picture, an individual photon travels at the speed of light, has an energy $h\nu$, a momentum $h\nu/c$ and an angular momentum $\hbar$.

1.1.6.4 The optical constant or refractive index

For the propagation of light, the electromagnetic properties of matter may be described either by the wave vector after (1.38) or by the optical constant (also called complex refractive index)

$$m = \sqrt{\varepsilon\mu} \,. \tag{1.46}$$

It is clear from the definition that m does not contain any new information, in fact less than the material constants ε and μ together. The name *optical constant* is unfortunate as m is not constant, but variable with frequency. It is a complex dimensionless number,

$$m(\omega) = n(\omega) + i\,k(\omega) \,, \tag{1.47}$$

with real part n and imaginary part k. In the common case of a nonmagnetic medium, where

$$\mu = 1 \,,$$

the real and imaginary parts of the optical constant follow from ε_1 and ε_2,

$$n = \frac{1}{\sqrt{2}}\sqrt{\sqrt{\varepsilon_1^2 + \varepsilon_2^2} + \varepsilon_1} \tag{1.48}$$

$$k = \frac{1}{\sqrt{2}}\sqrt{\sqrt{\varepsilon_1^2 + \varepsilon_2^2} - \varepsilon_1} \tag{1.49}$$

and vice versa,

$$\varepsilon_1 = n^2 - k^2 \tag{1.50}$$
$$\varepsilon_2 = 2nk . \tag{1.51}$$

(1.46) has two solutions and we pick the one with positive n and non-negative k.*

1.1.7 Energy dissipation of a grain in a variable field

The total energy of an *electrostatic field* $\mathbf{E}$ in vacuum, produced by a fixed distribution of charges, is

$$U = \frac{1}{8\pi} \int \mathbf{E} \cdot \mathbf{E} \, dV ,$$

where the integration extends over all space. This formula follows readily from the potential energy between the charges, which is due to their Coulomb attraction. It suggests an energy density

$$u = \frac{1}{8\pi} \mathbf{E} \cdot \mathbf{E} . \tag{1.52}$$

If the space between the charges is filled with a dielectric,

$$u = \frac{1}{8\pi} \mathbf{E} \cdot \mathbf{D} . \tag{1.53}$$

When we compare (1.52) with (1.53), we find that the energy density of the dielectric medium contains an extra term $\frac{1}{2}\mathbf{E}\,\mathbf{P}$ which accounts for the work needed to produce in the medium a polarization $\mathbf{P}$. Therefore, if we place into a constant field in vacuum some small dielectric object of volume V, the total field energy changes by an amount

$$w = -\tfrac{1}{2} V \mathbf{P} \cdot \mathbf{E} . \tag{1.54}$$

It becomes smaller because some work is expended on the polarization of the grain. If the outer field slowly oscillates, $\mathbf{E} = \mathbf{E}_0 \, e^{-i\omega t}$, so does the polarization and for small energy changes

$$dw = -\tfrac{1}{2} V \, d(\mathbf{P} \cdot \mathbf{E}) = -V\mathbf{P} \cdot d\mathbf{E} .$$

The time derivative $\dot{w}$, which we denote by W, gives the dissipated power and its mean is

$$\langle W \rangle = -V \langle \mathbf{P}\dot{\mathbf{E}} \rangle = \tfrac{1}{2} V\omega \, \mathrm{Im}\{\alpha_e\} \, |\mathbf{E}_0|^2 > 0 . \tag{1.55}$$

---

*In order to abide by the customary nomenclature, we use the same letter for the imaginary part of the optical constant (italic type k) and for the wavenumber (Roman type k).

It is positive and proportional to frequency ω and volume V, which makes sense. The bracket $\langle \ldots \rangle$ denotes the time average, and we have used $\mathbf{P} = \alpha_e \, \mathbf{E}$ after (1.8). Because of the mathematical relation (B.31), the real part of the electric polarizability $\text{Re}\{\alpha_e\}$ disappears in the product $\langle \mathbf{P}\dot{\mathbf{E}} \rangle$ confirming that only the imaginary part $\text{Im}\{\alpha_e\}$ is responsible for heating the grain.

Formally similar expressions, but based on different physics, hold for magnetism. We first look for the formula of the magnetic energy density. Whereas static electric fields are produced by fixed charges and the total energy of the *electric* field is found by bringing the charges to infinity (which determines their potential energy), static magnetic fields are produced by constant currents. These currents exert also a force $\mathbf{F}$ on a charge, but from (1.1) do no work because $\mathbf{F} \cdot \mathbf{v} = (q/c) \, \mathbf{v} \cdot (\mathbf{v} \times \mathbf{B}) = 0$. Therefore, in the case of a magnetic field, one has to evaluate its energy by *changing* $\mathbf{H}$. This produces according to (1.25) an electric field $\mathbf{E}$ and thus a loss rate $\mathbf{E} \cdot \mathbf{J}$. After some elementary vector analysis, one gets for the total energy of the magnetostatic field

$$ U \;=\; \frac{1}{8\pi} \int \mathbf{H} \cdot \mathbf{B} \, dV \; , $$

which implies an energy density of the magnetic field

$$ u \;=\; \frac{1}{8\pi} \, \mathbf{H} \cdot \mathbf{B} \; . \tag{1.56} $$

In complete analogy to the electric field, a time variable magnetic field, $\mathbf{H} = \mathbf{H}_0 \, e^{-i\omega t}$, leads in a small body of volume V and magnetic polarizability α_m (see 1.18) to a heat dissipation rate

$$ \langle W \rangle \;=\; \tfrac{1}{2} \, V\omega \, \text{Im}\{\alpha_m\} \, |\mathbf{H}_0|^2 \; . \tag{1.57} $$

(1.55) and (1.57) are the basic equations for understanding the absorption of radiation by interstellar grains.

1.2   The harmonic oscillator

In the early history of atomic theory, H.A. Lorentz applied the harmonic oscillator model to the motion of an electron in an atom. Despite its simplicity and the fact that electrons "move" in reality in complicated paths, the Lorentz model is quite successful in the quantitative description of many phenomena that occur on the atomic level and reveal themselves in the macroscopic world. The oscillator concept is very fruitful, although a precise idea of what we mean by it is often missing. Usually we have in mind the electrons in an atom, or the oppositely charged atomic nuclei of a crystal or some other kind of dipoles, and we assume that a grain is built up of such oscillators. Using this concept, we derive below the dispersion relation $\varepsilon = \varepsilon(\omega)$ of the dielectric permeability around a resonance at frequency ω_0.

1.2.1 The Lorentz model

We imagine the following idealized situation: An electron of mass m_e and charge e is attached to a spring of force constant κ. A harmonic wave with electric field

$$E = E_0 e^{-i\omega t}$$

exerts a force $F = eE$ which causes the electron to move, say, in x-direction. Its motion is governed by the equation

$$m_e \ddot{x} + b\dot{x} + \kappa x = F . \qquad (1.58)$$

On the left side, there is *a)* an inertia term $m_e\ddot{x}$; *b)* a frictional force, $-b\dot{x}$, which is proportional to velocity and leads to damping of the system unless it is powered from outside; *c)* a restoring force, $-\kappa x$, that grows linearly with the displacement from the equilibrium position. Putting

$$\omega_0^2 = \frac{\kappa}{m_e}, \qquad \gamma = \frac{b}{m_e} ,$$

we get

$$\ddot{x} + \gamma\dot{x} + \omega_0^2 x = \frac{eE}{m_e} . \qquad (1.59)$$

The oscillator has only three properties:

  – charge-to-mass ratio e/m_e
  – damping constant γ
  – resonant frequency ω_0. If the electron is unbound, $\omega_0 = 0$.

In the following, we discuss only the one-dimensional oscillator, where the electron moves along the x-axis, but one may readily generalize to three dimensions. It is assumed that the electric field is spatially constant over the displacement x of the electron. This is correct as long as the velocity v of the electron is small, $v \ll c$, because then $x \simeq v/\omega$ can be neglected in comparison to the wavelength c/ω of the field. Indeed, the velocity of an electron in an atom is typically of order $v/c \sim 1\%$.

1.2.1.1 Free oscillations

In the simplest case, when there is no friction ($\gamma = 0$) and no perturbation from outside ($E = 0$), (1.59) reduces to

$$\ddot{x} + \omega_0^2 x = 0 \qquad (1.60)$$

and the electron oscillates harmonically forever at its natural frequency ω_0. If there is friction ($\gamma \neq 0$), but no external force,

$$\ddot{x} + \gamma\dot{x} + \omega_0^2 x = 0 , \tag{1.61}$$

the solution is called a *transient*. It has the general form

$$x = e^{-\gamma t/2} \cdot (A e^{-i\omega_\gamma t} + B e^{i\omega_\gamma t}) , \tag{1.62}$$

where A and B are complex constants and

$$\omega_\gamma = \sqrt{\omega_0^2 - \gamma^2/4} .$$

For x to be real, it is required that B is the complex conjugate of A,

$$B = A^* ,$$

which means $\mathrm{Re}\{A\} = \mathrm{Re}\{B\}$ and $\mathrm{Im}\{A\} = -\mathrm{Im}\{B\}$. If $\gamma^2 < 4\omega_0^2$, i.e., when damping is weak, ω_γ is real. The electron then oscillates at a frequency ω_γ somewhat smaller than the natural frequency ω_0, with an exponentially decaying amplitude. If friction is strong, $\gamma^2 > 4\omega_0^2$ and ω_γ is imaginary. The amplitude then subsides exponentially without any oscillations. Critical damping occurs for

$$\gamma = 2\omega_0 . \tag{1.63}$$

1.2.1.2 The general solution to the oscillator equation

In the most general case of equation (1.59), there is friction plus an external field that acts on the electron. We use complex variables and sometimes mark, for clarity, the complexity of a quantity explicitly by a bar, for instance $\bar{x}, \bar{E}, \bar{P}$. As before, we consider an harmonic field $\bar{E} = E_0 e^{-i\omega t}$ and assume that E_0 is real, which is always possible by adjusting the time. To solve equation (1.59), written here as

$$\ddot{\bar{x}} + \gamma\dot{\bar{x}} + \omega_0^2\bar{x} = \frac{e}{m_e} E_0 e^{-i\omega t} , \tag{1.64}$$

we note that one obtains the general solution of this inhomogeneous equation by adding to the general solution (1.62) of the associated homogeneous equation (1.61) one particular solution of (1.64), for example

$$\bar{x} = \bar{x}_0 e^{-i\omega t} . \tag{1.65}$$

In such a sum, the transient (1.62) eventually dies out and only (1.65) remains. The electron then oscillates at frequency ω, and not ω_0. Equation (1.65) describes the steady state solution into which any particular solution, satisfying certain initial values for x and $\dot{x}$ at some earlier time t_0, evolves. For the complex amplitude of the displacement of the electron one finds

$$\bar{x}_0 = \frac{eE_0}{m_e(\omega_0^2 - \omega^2 - i\omega\gamma)} . \tag{1.66}$$

Putting $x_0 = |\bar{x}_0|$, the real amplitude of the electron's displacement becomes

$$x_0 = \frac{eE_0}{m_e\sqrt{(\omega_0^2 - \omega^2)^2 + \omega^2\gamma^2}} . \tag{1.67}$$

If damping (γ) is small and the incoming wave vibrates with the natural frequency of the electron ($\omega = \omega_0$), the amplitude x_0 is proportional to γ^{-1} and can become very large.

A static field E_0 induces in the oscillator a permanent dipole moment

$$p = ex_0 = \frac{e^2 E_0}{m_e \omega_0^2} .$$

One can apply this equation to atoms to obtain a rough estimate of their electric polarizability $\alpha_e = e^2/m_e\omega_0^2$ if one chooses for the characteristic frequency ω_0 a value such that $\hbar\omega_0$ equals the ionization potential of the atom. One then comes up with $\alpha_e \sim 10^{-24}$ cm^3.

1.2.1.3   The phase shift

Equation (1.65) tells us that although the electron has a natural frequency ω_0, it moves in the forced oscillation with the frequency ω of the external field. Without friction ($\gamma=0$), the electron and the field are synchronous. With friction, the complex variables for position and field, $\bar{x}$ and $\bar{E}$, are out of phase because of the imaginary term $i\omega\gamma$ in (1.66). The electron always lags behind the field by a positive angle

$$\Phi = \begin{cases} \tan^{-1}(x) & : \quad \text{if} \quad \omega < \omega_0 \\ \tan^{-1}(x) + \pi & : \quad \text{if} \quad \omega > \omega_0 \end{cases} \qquad \text{where} \quad x = \frac{\omega\gamma}{\omega_0^2 - \omega^2} .$$

- At low frequencies ($\omega \ll \omega_0$), the lag is small.

- Around the resonance frequency, the phase shift changes continuously and amounts to 90° at $\omega = \omega_0$.

- For $\omega \gg \omega_0$, it approaches 180°. Then the electron moves opposite to the direction in which it is being pushed by the external force eE. This is not a miracle, but reflects the *steady state* response of the electron. Initially, when the field E was switched on, the acceleration vector of the electron pointed, of course, in the same direction as the electric field.

1.2.2 Dissipation of energy

1.2.2.1 Dissipation in a forced oscillation

The total energy of the oscillator is the sum of kinetic energy T plus potential energy V,

$$T + V = \tfrac{1}{2}m_e \left(\dot{x}^2 + \omega_0^2 x^2 \right) . \tag{1.68}$$

The total energy declines when there is friction. If we think of a grain to be composed of many oscillators (atoms), friction results from collisions of the electrons with the lattice and leads to heating. The damping constant γ, which has the dimension s^{-1}, is then interpreted as the collisional frequency. Because the mechanical power W, which is converted in a forced oscillation into heat, equals force times velocity,

$$W = F\dot{x} ,$$

multiplication of formula (1.58) with $\dot{x}$ gives

$$W = F\dot{x} = \frac{d}{dt} \left(\tfrac{1}{2}m_e \dot{x}^2 + \tfrac{1}{2}m_e \omega_0^2 x^2 \right) + \gamma m_e \dot{x}^2 .$$

In the brackets stands the total energy. As it is constant in the forced oscillation, its time derivative vanishes and only the term $\gamma m_e \dot{x}^2$ remains. With $\dot{x} = -i\omega x$, the *time-averaged* dissipation rate is (see B.31)

$$W = \tfrac{1}{2}\gamma m_e \omega^2 x_0^2 = \frac{\gamma e^2 E_0^2}{2 m_e} \cdot \frac{\omega^2}{(\omega_0^2 - \omega^2)^2 + \omega^2 \gamma^2} . \tag{1.69}$$

- When the frequency is very high or very low, the heating rate W goes to zero because the velocities are small.

- But near the resonance frequency ($\omega \simeq \omega_0$), and especially when damping is weak, the power W becomes large.

1.2.2.2 Dissipation in a free oscillation

If the system is not driven by an external force and if damping is small, the electron swings almost freely near its natural frequency ω_0, while its amplitude gradually declines like $e^{-\gamma t/2}$. If the motion of the electron is described by

$$x(t) = \begin{cases} x_0 e^{-\gamma t/2} e^{-i\omega_0 t} & : \quad \text{for} \quad t \geq 0 \\ 0 & : \quad \text{for} \quad t < 0 \end{cases} , \tag{1.70}$$

the total energy E (unfortunately, the same letter as for the electric field) at time $t = 0$ is

$$E_0 = \tfrac{1}{2}m_e \omega_0^2 x_0^2 ,$$

and afterwards it falls due to dissipational losses like

$$E = E_0 e^{-\gamma t} . \tag{1.71}$$

The energy of the system drops by a factor e in a time $\tau = 1/\gamma$ or after $\omega_0/2\pi\gamma$ cycles; the initial loss rate is

$$W = \tfrac{1}{2}\gamma m_e \omega_0^2 x_0^2 .$$

The Fourier transform of $x(t)$ in equation (1.70) is by definition

$$f(\omega) = \int_{-\infty}^{\infty} x(t) e^{i\omega t} dt \tag{1.72}$$

or in view of the reciprocity relation

$$x(t) = \frac{1}{2\pi} \int_{-\infty}^{\infty} f(\omega) e^{-i\omega t} d\omega . \tag{1.73}$$

The two functions $x(t)$ and $f(\omega)$ form the Fourier transform pair. The integral (1.73) can be regarded as an infinite expansion of the motion of the electron, $x(t)$, into harmonic functions $e^{-i\omega t}$ of amplitude $f(\omega)$ and with continuously varying frequencies ω. For the amplitudes in the Fourier expansion of the free oscillator of (1.70) we find

$$f(\omega) = x_0 \int_0^{\infty} e^{-\gamma t/2} e^{-i(\omega_0 - \omega)t} dt = \frac{x_0}{i(\omega_0 - \omega) + \gamma/2} . \tag{1.74}$$

In the decomposition of $x(t)$ into harmonics, only frequencies of substantial amplitude $f(\omega)$ are relevant. As we assume weak damping, they all cluster around ω_0. The loss rate corresponding to each frequency component is then proportional to the square of the amplitude,

$$W(\omega) \propto \gamma |f(\omega)|^2 \propto \frac{\gamma}{(\omega - \omega_0)^2 + (\gamma/2)^2} .$$

Equation (1.74) is the basis for the Lorentz profile (see section 1.2.6).

1.2.3   Dispersion relation of the dielectric permeability

Because the restoring force, $-\kappa x$, in (1.58) is due to electrostatic attraction between the electron and a proton, we may consider the oscillating electron as an alternating dipole of strength $\bar{x}e$. A real grain contains many such oscillators (electrons) that are driven by the incoming wave up and down. If they swing in phase and their volume density equals N, the dipole moment per unit volume becomes after (1.66)

$$\bar{P} = N\bar{x}e = \frac{Ne^2}{m_e} \cdot \frac{\bar{E}}{\omega_0^2 - \omega^2 - i\omega\gamma} = \frac{\varepsilon - 1}{4\pi} \bar{E} . \tag{1.75}$$

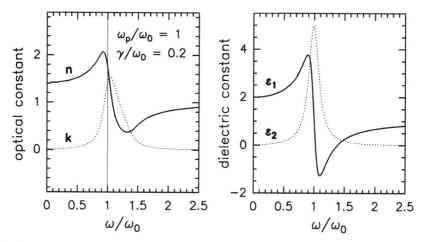

FIGURE 1.2 The dispersion relation specifies how the dielectric permeability $\varepsilon = \varepsilon_1 + i\varepsilon_2$, or equivalently the optical constant $m = n + ik$, changes with frequency. Here for a harmonic oscillator after (1.77). The vertical line helps to locate the maxima with respect to the resonance frequency.

The equal sign on the far right comes from (1.6). The bar designates complex quantities. Equation (1.75) represents the so called dispersion relation of the complex dielectric permeability and specifies how ε varies with frequency. If we use the plasma frequency

$$\omega_p = \sqrt{\frac{4\pi N e^2}{m_e}} \,, \tag{1.76}$$

where N is the number density of *free* electrons, we get

$$\varepsilon = \varepsilon_1 + i\varepsilon_2 = 1 + \frac{\omega_p^2(\omega_0^2 - \omega^2)}{(\omega_0^2 - \omega^2)^2 + \gamma^2\omega^2} + i\,\frac{\omega_p^2\gamma\omega}{(\omega_0^2 - \omega^2)^2 + \gamma^2\omega^2} \cdot \tag{1.77}$$

Figure 1.2 presents an example of the dispersion relation $\varepsilon(\omega)$ calculated with $\gamma/\omega_0 = 0.2$ and $\omega_p = \omega_0$. Despite this particular choice, it shows the characteristic behavior of the dielectric permeability at a resonance. Of interest are also the limiting values in very rapidly and very slowly changing fields. As we can compute m from ε after (1.48) and (1.49), figure 1.2 gives also the dispersion relation for n and k.

- In a constant field E, all quantities are real. The induced dipole moment per unit volume and the static permeability are

$$P = \frac{N e^2 E}{m_e \omega_0^2}$$

$$\varepsilon(0) = 1 + \frac{\omega_p^2}{\omega_0^2} .$$

- In a slowly varying field ($\omega \ll \omega_0$),

$$\varepsilon = 1 + \frac{\omega_p^2}{\omega_0^2} + i \frac{\omega_p^2 \gamma}{\omega_0^4} \omega . \tag{1.78}$$

The real part ε_1 approaches the electrostatic value $\varepsilon(0)$, whereas ε_2 becomes proportional to frequency ω times the damping constant γ and is thus small. Correspondingly, n goes towards a constant value and k falls to zero.

- Around the resonance frequency, ε_2 is quite symmetric and has a prominent peak. But note in figure 1.2 the significant shifts of the extrema in n, k and ε_1 with respect to ω_0. At exactly the resonance frequency, $\varepsilon_1(\omega_0) = 1$ and $\varepsilon_2(\omega_0) = \omega_p^2 / \gamma \omega_0$.

- The imaginary part ε_2 goes to zero far away from the resonance on either side, but stays always positive. ε_1 increases from its static value as ω nears ω_0, then sharply drops and rises again. It may become zero or negative, as displayed in figure 1.2.

- At high frequencies ($\omega \gg \omega_0$) and far from the resonance, there is very little polarization as the electrons cannot follow the field due to their inertia and ε can be approximated by

$$\varepsilon = 1 - \frac{\omega_p^2}{\omega^2} + i \frac{\omega_p^2 \gamma}{\omega^3} . \tag{1.79}$$

ε is then an essentially real quantity asymptotically approaching unity. This is a necessary condition for the fulfillment of the Kramers-Kronig relations (see section 2.6). Refraction disappears ($n \to 1$), and the material becomes transparent because there are no dissipational losses ($k, \varepsilon_2 \to 0$).

- At *very high* frequencies, when the wavelength is reduced to the size of an atom, the concept of a continuous medium breaks down and modifications are necessary.

1.2.4   The harmonic oscillator and light

We continue to discuss the optical constant from the viewpoint that matter is made of harmonic oscillators. We derive the emission of an accelerated charge, compute how an oscillator is damped by its own radiation and evaluate the cross section for absorption and scattering of a single oscillator.

1.2.4.1 Retarded potentials of a moving charge

The retarded potentials created by a moving charge of density $\rho(\mathbf{r})$ are given by

$$\phi(\mathbf{r}_2, t) = \int \frac{\rho(\mathbf{r}_1, t')}{|\mathbf{r}_2 - \mathbf{r}_1|} dV \tag{1.80}$$

$$\mathbf{A}(\mathbf{r}_2, t) = \int \frac{\mathbf{J}(\mathbf{r}_1, t')}{c\,|\mathbf{r}_2 - \mathbf{r}_1|} dV . \tag{1.81}$$

Because in Maxwell's equations div $\mathbf{B} = 0$, there exists a vector potential $\mathbf{A}$ such that

$$\mathbf{B} = \mathrm{rot}\ \mathbf{A} \tag{1.82}$$

and because rot $(\mathbf{E} + \dot{\mathbf{A}}/c) = 0$, there exists a scalar potential ϕ with

$$\mathbf{E} + \frac{1}{c}\dot{\mathbf{A}} = -\nabla\phi . \tag{1.83}$$

These potentials are gauged by imposing on them the *Lorentz condition*,

$$\mathrm{div}\ \mathbf{A} + \frac{1}{c}\dot{\phi} = 0 . \tag{1.84}$$

$\mathbf{A}$ and ϕ are called retarded potentials as they refer to the *present time t*, but are determined by the configuration at an earlier epoch t'. The delay corresponds to the time it takes light to travel from $\mathbf{r}_1$ to $\mathbf{r}_2$,

$$t' = t - \frac{|\mathbf{r}_2 - \mathbf{r}_1|}{c} . \tag{1.85}$$

If the charges are near the center of the coordinate system and spread out over a region of size $d \sim |\mathbf{r}_1|$ that is small compared to the distance $r = |\mathbf{r}_2 - \mathbf{r}_1|$ between charge and observer,

$$d \ll r ,$$

we may take $|\mathbf{r}_2 - \mathbf{r}_1|$ out from under the integrals. Moreover, $|\mathbf{r}_2 - \mathbf{r}_1| \simeq |\mathbf{r}_2| - \mathbf{r}_1 \cdot \mathbf{e}$, where $\mathbf{e}$ is the unit vector pointing from the atom to the observer, therefore

$$t' \simeq t - (r - \mathbf{r}_1 \cdot \mathbf{e})/c . \tag{1.86}$$

If d is also small compared to the wavelength,

$$d \ll \lambda ,$$

there will be no phase shift among the waves emitted from different parts of the source and the emission is coherent, so we can write

$$t' \simeq t - r/c$$

and (1.81) becomes

$$A(r_2, t) = \frac{1}{cr} \int J(r_1, t - r/c) \, dV .$$ (1.87)

This expression applies, for example, to atoms where a charge is oscillating at frequency ν and the linear dimension of the system $d \sim \nu^{-1}v = \lambda v/c$, where v is the non-relativistic charge velocity.

1.2.4.2   Emission of an harmonic oscillator

Under such simplifications, one gets the *dipole field*. The vector potential A follows from a volume integral over the current density J at the earlier epoch $t' = t - r/c$. If there is only one small charge q of space density $\rho(r)$ that is oscillating about an opposite charge at rest, equation (1.87) yields

$$A(r_2, t) = \frac{1}{cr} \dot{p}(r_1, t - r/c)$$ (1.88)

because the current density is $J = \rho v$ and the wiggling charge q constitutes a dipole $p = q r_1$ whose derivative is given by the integral over the current density. If there are many charges q_i with velocities v_i, the result is the same with $p = \sum q_i r_i$.

Far away from the charge, the wave is plane and

$$H = e \times E, \qquad E = H \times e ,$$

and $\nabla \phi$ in (1.83) is then negligible because ϕ represents the potential of the charge and the electric field of a charge falls off like $1/r^2$, whereas the field of a light source falls off more slowly like $1/r$. We thus have $E = -\dot{A}/c$ and

$$H = \frac{1}{c^2 r} \ddot{p} \times e = \frac{\omega^2}{c^2 r} e \times p$$ (1.89)

$$E = \frac{\omega^2}{c^2 r} (e \times p) \times e ,$$ (1.90)

where we assumed that the dipole varies harmonically at frequency ω. Having determined the fields E and H via the dipole moment p, we can compute the flux dW carried in a solid angle $d\Omega = \sin\theta \, d\theta d\phi$ into the direction e which forms an angle θ with the dipole moment p. This flux is given by the Poynting vector of (1.43), so

$$dW = \frac{|\ddot{p}|^2}{4\pi c^3} \sin^2\theta \, d\Omega .$$ (1.91)

Emission is zero in the direction of the motion of the charge, and maximum perpendicular to it. Integration over all directions yields for the total *momentary power* radiated by a dipole,

$$W = \frac{2}{3c^3} |\ddot{p}|^2 .$$ (1.92)

If the dipole oscillates harmonically proportional to $\cos\omega t$, the *time-averaged power* equals half the maximum value of W in (1.92).

1.2.5 Radiation damping

An accelerating electron radiates and the emitted light reacts on the electron because of conservation of energy and momentum. Although theory (quantum electrodynamics) and experiment agree extremely well, there is currently no strict, self-consistent description of the feedback. We describe below the feedback in classical terms.

When a force F accelerates a free and otherwise undamped electron, we write the equation of motion in the form

$$F + F_{\text{rad}} = m_e \dot{u} .$$

The additional force F_{rad} accounts for the retardation caused by the radiative loss; $u = \dot{x}$ is the velocity. F_{rad} acts oppositely to F. Averaged over some time interval Δt, we assume that the emitted power due to the radiative deceleration (see 1.92) is equal to the work that the force F_{rad} has done on the electron,

$$-\int_{\Delta t} F_{\text{rad}}\, u\, dt = \frac{2e^2}{3c^3} \int_{\Delta t} \dot{u}^2\, dt .$$

When we integrate the right side by parts and assume a periodic motion where $u\dot{u} = 0$ at the beginning and at the end of the time interval Δt, we find

$$\int_{\Delta t} \left(F_{\text{rad}} - \frac{2e^2}{3c^3}\ddot{u} \right) u\, dt = 0 .$$

So the force becomes

$$F_{\text{rad}} = m_e \tau \ddot{u} \tag{1.93}$$

with $\tau = 2e^2/3m_e c^3$. When F_{rad} is added in the equation of motion of the harmonic oscillator, we get for (1.59)

$$\omega_0^2 x + \gamma u + \dot{u} - \tau \ddot{u} = \frac{eE_0}{m_e} e^{-i\omega t} .$$

The steady state solution is again $\bar{x} = \bar{x}_0\, e^{-i\omega t}$ (a bar denotes complexity) and the complex amplitude becomes

$$\bar{x}_0 = \frac{e\bar{E}_0}{m_e} \frac{1}{\omega_0^2 - \omega^2 - i\omega(\gamma + \tau\omega^2)} .$$

This is the same expression for $\bar{x}_0$ as in (1.66) if one replaces there the dissipation constant γ by

$$\gamma + \tau\omega^2 = \gamma + \gamma_{\text{rad}} .$$

The term

$$\gamma_{\text{rad}} = \frac{2e^2}{3m_e c^3}\, \omega^2 \tag{1.94}$$

is the damping constant due to radiation alone and follows from equating the time-averaged losses $\frac{1}{2}\gamma_{\text{rad}} m_e \omega^2 x_0^2$ of the harmonic oscillator after (1.69) to the time-averaged radiated power $p_0^2 \omega^4/3c^3$ of (1.92).

1.2.6 The cross section of an harmonic oscillator

1.2.6.1 Scattering

An oscillating electron with dipole moment

$$\bar{p} = \bar{p}_0\,e^{-i\omega t} = e\,\bar{x}_0\,e^{-i\omega t}\,,$$

scatters after (1.67) and (1.92) the power

$$W^{\text{sca}} = \frac{2}{3c^3}\,\omega^4 p_0^2 = \frac{2}{3c^3}\,\frac{e^4 E_0^2}{m_e^2}\,\frac{\omega^4}{(\omega_0^2 - \omega^2)^2 + \omega^2\gamma^2} \qquad (1.95)$$

where we put $p_0 = |\bar{p}_0|$. So its cross section σ^{sca}, defined as the scattered power divided by the incident flux $S = (c/4\pi)E_0^2$ of (1.43), becomes

$$\sigma^{\text{sca}}(\omega) = \frac{W^{\text{sca}}}{S} = \sigma_{\text{T}}\,\frac{\omega^4}{(\omega_0^2 - \omega^2)^2 + \omega^2\gamma^2} \qquad (1.96)$$

where

$$\sigma_{\text{T}} = \frac{8\pi r_0^2}{3} = 6.65 \times 10^{-25} \quad \text{cm}^2 \qquad (1.97)$$

is the *Thomson* scattering cross section of a single electron; it is frequency-independent. r_0 is the classical electron radius and follows from equaling the rest mass energy $m_e c^2$ to the electrostatic energy e^2/r_0,

$$r_0 = \frac{e^2}{m_e c^2} = 2.82 \times 10^{-13} \quad \text{cm}\,. \qquad (1.98)$$

As there is only radiative damping (scattering is assumed to be conservative), we have to set $\gamma = \gamma_{\text{rad}}$ after (1.94). From equation (1.96) we find the following approximations for the scattering cross section at the resonance, and at low and high frequencies,

$$\sigma^{\text{sca}}(\omega) = \begin{cases} \sigma_{\text{T}}\left(\dfrac{\omega}{\omega_0}\right)^4 & : \quad \text{if } \omega \ll \omega_0 \\[3mm] \dfrac{\sigma_{\text{T}}}{4}\dfrac{\omega_0^2}{(\omega - \omega_0)^2 + (\gamma/2)^2} & : \quad \text{if } \omega \simeq \omega_0 \\[3mm] \sigma_{\text{T}} & : \quad \text{if } \omega \gg \omega_0\,. \end{cases} \qquad (1.99)$$

When the frequency is high, the electron is essentially free and the cross section constant and equal to σ_{T}. When the frequency is small, σ^{sca} falls with the fourth power of ω (Rayleigh scattering). Near the resonance, we have put $\omega^2 - \omega_0^2 \simeq 2\omega_0(\omega - \omega_0)$.

1.2.6.2 Absorption

In a similar way, we get from (1.69) for the absorption cross section of the harmonic oscillator

$$\sigma^{\text{abs}}(\omega) \;=\; \frac{W^{\text{abs}}}{S} \;=\; \frac{4\pi e^2}{m_e c}\,\frac{\omega^2\gamma}{(\omega_0^2 - \omega^2)^2 + \omega^2\gamma^2} \tag{1.100}$$

with the approximations

$$\sigma^{\text{abs}}(\omega) \;=\; \frac{\pi e^2}{m_e c} \begin{cases} 4\gamma\omega_0^{-4}\,\omega^2 & : \quad \text{if } \;\omega \ll \omega_0 \\[2mm] \dfrac{\gamma}{(\omega - \omega_0)^2 + (\gamma/2)^2} & : \quad \text{if } \;\omega \simeq \omega_0 \\[2mm] 4\gamma\,\omega^{-2} & : \quad \text{if } \;\omega \gg \omega_0 \;. \end{cases} \tag{1.101}$$

A frequency dependence according to the middle line of (1.101) or (1.99) results in a Lorentz profile. It has the characteristic feature that the intensity over an emission or absorption line changes proportional to $[(\omega - \omega_0)^2 + (\gamma/2)^2]^{-1}$. As we have seen in the Fourier analysis of the motion of a free oscillator after (1.74), at the root of such a profile is the exponential decline $e^{-\gamma t/2}$ of the amplitude of the electron.

1.3 Waves in a conducting medium

We modify the expression for the dielectric permeability or the complex wave number for the case that some of the electrons in the material are not bound to individual atoms, but are free so that they can conduct a current. The medium is then a plasma and the dispersion relation is known as Drude profile.

1.3.1 The dielectric permeability of a conductor

A conductor contains free charges that can support a constant current, as we know from everyday experience with electricity. Ohm's law asserts a proportionality between the electric field $\mathbf{E}$ and the current density $\mathbf{J}$ (omitting the suffix "free"),

$$\mathbf{J} \;=\; \sigma\,\mathbf{E} \;. \tag{1.102}$$

σ is called conductivity and is usually defined for a quasi-stationary electric field; in vacuum, $\sigma = 0$. When a conductor is placed in a *static field*, no current can flow because the charges arrange themselves on its surface in such a way that the electric field inside cancels. In a *dielectric medium*, on the other hand, a static electric field is, of course, possible. However, an electric field may exist in a metal, either because it is quickly variable and the charges do not have time to reach their equilibrium positions, or because it is produced by a changing magnetic field according to (1.25).

We treat conductors because some interstellar grains have metallic properties and because the interstellar medium is a plasma. Let us start with the Maxwell equations (1.29) and (1.30) which are appropriate for conductors,

$$\text{rot } \mathbf{E} = -\frac{1}{c} \dot{\mathbf{B}} , \qquad \text{rot } \mathbf{H} = \frac{1}{c} \dot{\mathbf{D}} + \frac{4\pi}{c} \mathbf{J} .$$

When we insert into these formulae harmonic fields, we get for a homogeneous medium with the help of (1.102):

$$\text{rot } \mathbf{E} = i\frac{\omega\mu}{c} \mathbf{H}$$

$$\text{rot } \mathbf{H} = -i\frac{\omega}{c} \left(\varepsilon + i\frac{4\pi\sigma}{\omega} \right) \mathbf{E} . \qquad (1.103)$$

We may look at (1.103) as an expansion of rot $\mathbf{H}$ into powers of ω. The term with the conductivity σ dominates at low frequencies, more precisely, when

$$|\varepsilon| \ll \frac{4\pi\sigma}{\omega} .$$

Then equation (1.103) reduces to

$$\text{rot } \mathbf{H} = \frac{4\pi\sigma}{c} \mathbf{E} , \qquad (1.104)$$

which describes the magnetic field produced by a quasi-stationary current. Formally, we can retain the field equation

$$\text{rot } \mathbf{H} = \frac{\varepsilon}{c} \dot{\mathbf{E}} ,$$

as formulated in (1.33) for a dielectric medium, also for a conducting medium if ε denotes the sum of the permittivity of (1.5), written below as ε_d to indicate that it refers to a dielectric, plus a term relating to the conductivity,

$$\varepsilon = \varepsilon_d + i\frac{4\pi\sigma}{\omega} . \qquad (1.105)$$

The complex wavenumber (1.38) is in a conducting medium

$$k^2 = \frac{\omega^2\mu}{c^2} \left(\varepsilon_d + i\frac{4\pi\sigma}{\omega} \right) \qquad (1.106)$$

and the optical constant m follows from ε of (1.105) via $m = \sqrt{\varepsilon\mu}$.

At low frequencies, the dielectric permeability of a metal is approximately (see equation (1.113) for an exact expression)

$$\varepsilon(\omega) = i\,\frac{4\pi\sigma}{\omega}\;.$$

ε is then purely imaginary ($\varepsilon_1 = 0$), much greater than one ($|\varepsilon| \gg 1$) and has a singularity at $\omega = 0$. The optical constants n and k are therefore after (1.48), (1.49) large too and roughly equal,

$$n \simeq k \to \infty \qquad \text{for} \qquad \omega \to 0\;. \tag{1.107}$$

The difference between dielectrics and metals vanishes when the electromagnetic field changes so rapidly that the electrons make elongations small compared to the atomic radius. Then the restoring force, $-\kappa x$, in (1.58) is negligible and all electrons are essentially free. This happens around $\omega \sim 10^{17}\,\mathrm{s}^{-1}$.

1.3.2 Conductivity and the Drude profile

We repeat the analysis of the harmonic oscillator in section 1.2 for a plasma. The electrons are then free; they experience no restoring force after an elongation and therefore $\omega_0{=}0$ in (1.59). With the same ansatz $\bar{x} = \bar{x}_0 e^{-i\omega t}$ as before, we find the velocity of a free electron in an harmonic electric field,

$$\bar{v} = \frac{e}{m_e(\gamma - i\omega)}\,\bar{E}\;.$$

The conductivity follows from the current density $\mathbf{J} = \sigma\,\mathbf{E} = Ne\mathbf{v}$. If there are N free electrons per cm^3,

$$\sigma = \frac{Ne^2}{m_e(\gamma - i\omega)} = \frac{\omega_p^2}{4\pi(\gamma - i\omega)}\;, \tag{1.108}$$

where $\omega_p = \sqrt{4\pi Ne^2/m_e}$ is the plasma frequency of (1.76). So generally speaking, the conductivity is a complex function and depends on ω,

$$\sigma(\omega) = \sigma_1(\omega) + i\,\sigma_2(\omega) = \frac{\omega_p^2}{4\pi}\frac{\gamma}{\gamma^2 + \omega^2} + i\,\frac{\omega_p^2}{4\pi}\frac{\omega}{\gamma^2 + \omega^2}\;. \tag{1.109}$$

- At long wavelengths ($\omega \ll \gamma$), σ becomes frequency-independent. It is then an essentially real quantity approaching the direct-current limit

$$\sigma(\omega = 0) = \frac{\omega_p^2}{4\pi\gamma}\;, \tag{1.110}$$

which appears in Ohm's law. The conductivity depends on the density of free electrons through ω_p and on the collision time through γ.

- If the frequency is high or damping weak ($\gamma \ll \omega$), the conductivity is purely imaginary and inversely proportional to frequency. There is then no dissipation of energy in the current.

When we put $\varepsilon_d = 0$, $\mu = 1$, and substitute for the conductivity σ the expression from (1.108) into (1.106), we find for the wavenumber of a plasma

$$k^2 = \frac{\omega^2}{c^2}\left(1 - \frac{\omega_p^2}{\gamma^2 + \omega^2} + i\frac{\gamma}{\omega}\frac{\omega_p^2}{\gamma^2 + \omega^2}\right). \qquad (1.111)$$

Without damping,

$$k^2 = \frac{\omega^2}{c^2}\left(1 - \frac{\omega_p^2}{\omega^2}\right). \qquad (1.112)$$

In this case, when ω is greater than the plasma frequency, the real part n of the optical constant is after (1.48) smaller than 1, so the phase velocity is greater than the velocity of light, $v_{ph} > c$. For $\omega = \omega_p$, n becomes zero. A wave with $\omega \leq \omega_p$ cannot penetrate into the medium and is totally reflected. The bending of short-wavelength-band radio waves in the ionosphere or the reflection on metals (see 3.39) are illustrations.

The dielectric permeability $\varepsilon(\omega)$ corresponding to (1.111) is known as the *Drude profile* (see 1.38),

$$\varepsilon(\omega) = 1 - \frac{\omega_p^2}{\gamma^2 + \omega^2} + i\frac{\gamma}{\omega}\frac{\omega_p^2}{\gamma^2 + \omega^2}. \qquad (1.113)$$

To get some feeling for the numbers associated with the conductivity, here are two examples:

- *Copper* is a pure metal with a free electron density $N \simeq 8 \times 10^{22}\,\mathrm{cm^{-3}}$, a plasma frequency $\omega_p = 1.6 \times 10^{16}\,\mathrm{s^{-1}}$ and a damping constant $\gamma \simeq 3 \times 10^{13}\,\mathrm{s^{-1}}$. It has a high direct-current conductivity $\sigma \sim 7 \times 10^{17}\,\mathrm{s^{-1}}$. Therefore copper cables are ideal to carry electricity.

- *Graphite*, which is found in interstellar space, has $N \simeq 1.4 \times 10^{20}\,\mathrm{cm^{-3}}$ and $\gamma \simeq 5 \times 10^{12}\,\mathrm{s^{-1}}$, so the direct-current conductivity becomes $\sigma \simeq 7 \times 10^{15}\,\mathrm{s^{-1}}$; it is two orders of magnitude smaller than for copper. Graphite is a reasonable conductor only when the electrons move in the basal plane.

2

How to evaluate grain cross sections

In section 2.1, we define cross sections, the most important quantities describing the interaction between light and interstellar grains. Section 2.2 deals with the optical theorem which relates the intensity of light that is scattered by a particle into exactly the forward direction to its extinction cross section. In section 2.3 to 2.5, we learn how to compute the scattering and absorption coefficients of particles. Section 2.6 is concerned with a strange, but important property of the material constants that appear in Maxwell's equations, like ε or μ. They are complex quantities and Kramers and Kronig discovered a dependence between the real and imaginary part. In the final section, we approximate the material constants of matter that is a mixture of different substances.

2.1   Defining cross sections

2.1.1   Cross section for scattering, absorption and extinction

For a single particle, the scattering cross section is defined as follows: Consider a plane monochromatic electromagnetic wave at frequency ν and with flux F_0. The flux is the energy carried per unit time through a unit area and given in (1.44) as the absolute value of the Poynting vector. When the wave hits the particle, some light is scattered into the direction specified by the angles (θ, ϕ) as depicted in figure 2.1. The flux of this scattered light, $F(\theta, \phi)$, when it is received at a large distance r from the particle, is obviously proportional to F_0/r^2; we therefore write

$$F(\theta, \phi) = \frac{F_0}{k^2 r^2} \cdot \mathcal{L}(\theta, \phi) \,. \tag{2.1}$$

The function $\mathcal{L}(\theta, \phi)$ does not depend on r nor on F_0. We have included in the denominator the wavenumber

$$k = \frac{2\pi}{\lambda}$$

to make $\mathcal{L}(\theta, \phi)$ dimensionless; the wavelength λ is then the natural length unit to measure the distance r.

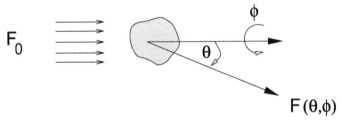

FIGURE 2.1 A grain scatters light from a plane wave with flux F_0 into the direction (θ, ϕ). In this direction, the scattered flux is $F(\theta, \phi)$.

- The cross section for scattering of the particle, C^{sca}, follows from the condition that $F_0 C^{\mathrm{sca}}$ be the total energy scattered into *all* directions per unit time. Consequently,

$$C^{\mathrm{sca}} \;=\; \frac{1}{k^2} \int_{4\pi} \mathcal{L}(\theta, \phi)\, d\Omega \;=\; \frac{1}{k^2} \int_0^{2\pi} d\phi \int_0^{\pi} d\theta\, \mathcal{L}(\theta, \phi)\, \sin\theta$$

with the element of solid angle

$$d\Omega \;=\; \sin\theta\, d\theta\, d\phi \,. \tag{2.2}$$

The scattering cross section C^{sca} has the dimension of an area. It is assumed that the frequency of radiation is not changed in the scattering process.

- Besides scattering, a particle inevitably absorbs some light. The corresponding cross section C^{abs} is defined by the condition that $F_0 C^{\mathrm{abs}}$ equals the energy absorbed by the particle per unit time.

- The sum of absorption plus scattering is called extinction. The extinction cross section,

$$C^{\mathrm{ext}} \;=\; C^{\mathrm{abs}} + C^{\mathrm{sca}} \,, \tag{2.3}$$

determines the *total* amount of energy removed from the impinging beam of light.

- The albedo is defined as the ratio of scattering over extinction,

$$A \;=\; \frac{C^{\mathrm{sca}}}{C^{\mathrm{ext}}} \,. \tag{2.4}$$

It lies between 0 and 1; an object with a high albedo scatters a lot of light.

All these various cross sections generally do not depend on the radiation field, the temperature of the dust material or the density of the interstellar medium. In this respect, it is much simpler to calculate cross sections of grains than of gas atoms.

2.1.2 Phase function and cross section for radiation pressure

In equation (2.1), $\mathcal{L}(\theta, \phi)$ specifies how the intensity of the scattered radiation changes with direction. We form a new function $\tilde{f}(\theta, \phi)$, which is proportional to $\mathcal{L}(\theta, \phi)$, but normalized so that the integral of $\tilde{f}$ over all directions equals 4π,

$$\int_{4\pi} \tilde{f}(\theta, \phi) \, d\Omega = 4\pi , \qquad (2.5)$$

and call it phase function. An isotropic scatterer has $\tilde{f}=1$.

For spheres, but also for a large ensemble of arbitrarily shaped and oriented grains, there is for reasons of symmetry no dependence on the angle ϕ, only on θ. It is then convenient to have a phase function f which has $\cos\theta$ as the argument, so we put $f(\cos\theta) = \tilde{f}(\theta)$. Again $f = 1$ for isotropic scattering, and the normalization condition is

$$1 = \tfrac{1}{2} \int_{-1}^{+1} f(x) \, dx . \qquad (2.6)$$

When one does not know or does not need the full information contained in $f(\cos\theta)$, one sometimes uses just one number to characterize the scattering pattern. This number is the asymmetry factor g, the mean of $\cos\theta$ over all directions weighted by the phase function $f(\cos\theta)$,

$$g = <\cos\theta> = \tfrac{1}{2} \int_{-1}^{+1} f(x) \, x \, dx . \qquad (2.7)$$

- It is easy to verify that g lies between -1 and 1.

- When scattering is isotropic, and thus independent of direction, $g = 0$.

- When there is mainly forward scattering, g is positive, otherwise it is negative. In the limit of pure forward scattering, $g = 1$; for pure backscattering one would have $g = -1$.

Knowing the asymmetry factor g, the phase function $f(\cos\theta)$ is often approximated by [Hen41]

$$f(\cos\theta) = \frac{1 - g^2}{(1 + g^2 - 2g\cos\theta)^{3/2}} \qquad (2.8)$$

which satisfies (2.6) and (2.7).

Electromagnetic radiation exerts also a pressure on a grain. A photon that is absorbed deposits its full momentum $h\nu/c$. If it is scattered at an angle θ (see figure 2.1), the grain receives (in the forward direction) only the fraction $(1 - \cos\theta)$. Therefore, the cross section for radiation pressure, C^{rp}, can be written as

$$C^{\mathrm{rp}} = C^{\mathrm{ext}} - g\, C^{\mathrm{sca}} . \qquad (2.9)$$

As g can be negative, C^{rp} may be greater than C^{ext}. To obtain the momentum transmitted per second to the grain by a flux F, we have to divide by the velocity of light c, so the transmitted momentum is

$$\frac{F\,C^{\mathrm{rp}}}{c}\ . \tag{2.10}$$

2.1.3   Efficiencies, mass and volume coefficients

The definition of the cross section of a single particle can be extended to $1\,\mathrm{g}$ of interstellar matter or $1\,\mathrm{g}$ of dust, or $1\,\mathrm{cm}^3$ of space. We use the following nomenclature (omitting the dependence on frequency):

- *Cross section of a single particle:* C

- *Efficiency*, Q, defined as

$$Q = \frac{C}{\sigma_{\mathrm{geo}}}\ . \tag{2.11}$$

  where σ_{geo} is the projected geometrical surface area of the grain. There are efficiencies for absorption, scattering, extinction and radiation pressure, again

$$Q^{\mathrm{ext}} = Q^{\mathrm{abs}} + Q^{\mathrm{sca}} \tag{2.12}$$
$$Q^{\mathrm{rp}} = Q^{\mathrm{ext}} - g\,Q^{\mathrm{sca}}\ . \tag{2.13}$$

  With the exception of spheres, σ_{geo} as well as the Cs and Qs change with the direction of the incoming light.

- *Mass coefficient*, K, is the cross section per unit mass. It refers either to $1\,\mathrm{g}$ of dust or to $1\,\mathrm{g}$ of interstellar matter. The latter quantity is some hundred times smaller.

- *Volume coefficient* is the cross section per unit volume and also denoted by the letter K. It refers either to $1\,\mathrm{cm}^3$ in space (which typically contains $10^{-23}\,\mathrm{g}$ of interstellar matter and $10^{-25}\,\mathrm{g}$ of dust) or to $1\,\mathrm{cm}^3$ of dust material with a mass of about $1\,\mathrm{g}$.

2.2   The optical theorem

When a beam of light falls on a particle, some light is absorbed by the grain, heating it, and some is scattered. The optical theorem asserts that the reduction of intensity in the forward direction fully determines the particle's extinction cross section.

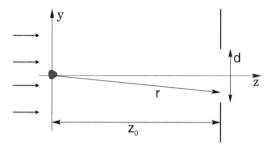

FIGURE 2.2 Light enters from the left, is scattered by a particle and falls at a large distance r through a disk of diameter d. The x-axis is perpendicular to the (y, z)-plane.

2.2.1 The intensity of forward scattered light

Consider a plane electromagnetic wave of wavelength λ propagating in vacuum in z-direction with electric field (figure 2.2)

$$E_{\mathrm{i}} = e^{i(kz-\omega t)} . \tag{2.14}$$

For easier writing, we neglect the vector character of the field and assume a unit amplitude. When the wave encounters a grain located at the origin of the coordinate system, some light is absorbed and the rest is scattered into all directions (θ, ϕ). At a distance r which is large when measured in wavelength units ($kr = 2\pi r/\lambda \gg 1$), the scattered field can be presented as

$$E_{\mathrm{s}} = S(\theta, \phi) \frac{e^{i(kr-\omega t)}}{-ikr} . \tag{2.15}$$

The information about the amplitude of the scattered wave lies in the complex function $S(\theta, \phi)$; the exponential term contains the phase. Because $|E_{\mathrm{s}}|^2$ is proportional to the scattered intensity, comparison of (2.15) with (2.1) tells us that $|S(\theta, \phi)|^2$ corresponds to $\mathcal{L}(\theta, \phi)$. As before, conservation of energy requires $E_{\mathrm{s}} \propto 1/r$. The wavenumber $k = 2\pi/\lambda$ is introduced to make $S(\theta, \phi)$ dimensionless,; the factor $-i$ in the denominator is just a convention.

Let us now determine the flux F through a disk of area A far behind the grain. The disk has a diameter d, so $d^2 \sim A$. Its center coordinates are $(x = 0, y = 0, z_0)$. It lies in the (x, y)-plane and is thus oriented perpendicular to the z-axis. All points (x, y) in the disk fulfill the inequalities

$$|x| \ll z_0, \qquad |y| \ll z_0 ,$$

and their distance to the particle is approximately

$$r \simeq z_0 + \frac{x^2 + y^2}{2z_0} .$$

The radiation that goes through the disk consists of the incident light (E_{i}) and the light scattered by the particle (E_{s}). The two fields interfere, and to obtain the flux through the disk, they have to be added

$$F = \int_A |E_i + E_s|^2 \, dA . \tag{2.16}$$

Dividing (2.15) by (2.14), we get

$$E_s = E_i \cdot S(\theta, \phi) \frac{e^{ik(r-z)}}{-ikr} \tag{2.17}$$

and for the sum of the incident and scattered field

$$E_i + E_s = E_i \left\{ 1 - \frac{S(0)}{ikz_0} \exp\left(ik \frac{x^2 + y^2}{2z_0} \right) \right\} . \tag{2.18}$$

Here we put

$$S(0) = S(\theta \simeq 0, \phi)$$

because $d \ll z_0$, so the disk as viewed from the grain subtends a very small angle. Therefore, in the above expression for $|E_i + E_s|^2$, we may neglect terms with z_0^{-2}. In this way, we find

$$|E_i + E_s|^2 = |E_i|^2 - \frac{2|E_i|^2}{kz_0} \operatorname{Re}\left\{ \frac{S(0)}{i} \exp\left(ik \frac{x^2 + y^2}{2z_0} \right) \right\} .$$

To extract the flux F from (2.16), we have to evaluate the integral (see B.7)

$$\int_{-\infty}^{\infty} e^{\frac{ik(x^2 + y^2)}{2z_0}} \, dx \, dy = \left[\int_{-\infty}^{\infty} e^{\frac{ikx^2}{2z_0}} \, dx \right]^2 = i \frac{2\pi z_0}{k} . \tag{2.19}$$

Of course, the disk does not really extend to infinity. But the integral still gives more or less the correct value as long as the disk diameter is much greater than $\sqrt{z_0 \lambda} = \sqrt{2\pi z_0/k}$; i.e., it is required that

$$\lambda \ll d \simeq \sqrt{A} \ll z_0 .$$

We therefore obtain the flux

$$F = |E_i|^2 \left(A - \frac{4\pi}{k^2} \operatorname{Re}\{S(0)\} \right) .$$

Now without an obstacle, the flux through the disk would obviously be $A|E_i|^2$ and thus greater. The light that has been removed by the particle determines its cross section for extinction,

$$C^{\text{ext}} = \frac{\lambda^2}{\pi} \operatorname{Re}\{S(0)\} . \tag{2.20}$$

This is the grand extinction formula, also known as the optical theorem. It is baffling because it asserts that C^{ext}, which includes absorption plus scattering

FIGURE 2.3 Light passes through a plane-parallel slab of thickness l with one face lying in the (x, y)-plane. The slab is uniformly filled with identically scattering particles. We determine the field at point P with coordinates $(0, 0, z)$.

into *all directions*, is fixed by the scattering amplitude in the *forward direction* alone. The purely mathematical derivation of (2.20) may not be satisfying. But obviously C^{ext} must depend on $S(0)$: The extinction cross section of an obstacle specifies how much light is removed from a beam if one observes from a large distance, no matter how and where to it is removed (by absorption or scattering). If we let the disk of figure 2.2 serve as a detector and place it far away from the particle, it receives of course only the forward scattered light and therefore $S(0)$ contains the information about C^{ext}.

2.2.2 The refractive index of a dusty medium

One may also assign a refractive index to a dusty medium, like a cloud of grains. Let a plane wave traveling in z-direction pass through a slab as depicted in figure 2.3. The slab is of thickness l and uniformly filled with identically scattering grains of number density N. When we compute the field at point P on the z-axis resulting from interference of all waves scattered by the grains, we have to sum over all particles. This leads to an integral in the (x, y)-plane like the one in (2.19) of the preceding subsection where we considered only one particle. But there is now another integration necessary in z-direction extending from 0 to l. Altogether the field at P is

$$E_{\text{i}} + E_{\text{s}} = E_{\text{i}} \left\{ 1 - N \frac{S(0)}{ikz} \int_0^l dz \, \frac{1}{z} \int_{-\infty}^{\infty} dx \, dy \, e^{\frac{ik(x^2+y^2)}{2z}} \right\}$$

$$= E_{\text{i}} \left(1 - S(0) \frac{\lambda^2}{2\pi} Nl \right) . \tag{2.21}$$

The main contribution to the double integral comes again from a region of area $z_0 \lambda$ (see figure 2.2 for definition of z_0). We see from (2.21) that the total field at P is different from the incident field E_{i} and it is obtained by multiplying E_i with the factor

$$1 \; - \; S(0) \, \frac{\lambda^2}{2\pi} \, Nl \; . \tag{2.22}$$

Not only do the grains reduce the intensity of radiation in the forward direction, but they also change the phase of the wave because the function $S(0)$ is complex. We can therefore assign to the slab an optical constant (with a bar on top)

$$\overline{m} \; = \; \overline{n} \, + \, i\,\overline{k} \; .$$

The field within the slab then varies according to (1.38) like $e^{i(k\overline{m}z - \omega t)}$. If the slab were empty containing only vacuum, it would vary like $e^{i(kz - \omega t)}$. Therefore, the presence of the grains causes a *change* in the field at P by a factor $e^{ikl(\overline{m}-1)}$. When $\overline{m}$ is close to one ($|\overline{m} - 1| \ll 1$), which is certainly true for any interstellar dust cloud,

$$e^{ikl(\overline{m}-1)} \; \simeq \; 1 \, + \, ik\,l\,(\overline{m} - 1) \; . \tag{2.23}$$

Setting (2.22) equal to (2.23) yields

$$\overline{n} - 1 \; = \; -\frac{2\pi N}{k^3} \, \mathrm{Im}\left\{S(0)\right\} \tag{2.24}$$

$$\overline{k} \; = \; \frac{2\pi N}{k^3} \, \mathrm{Re}\left\{S(0)\right\} \; . \tag{2.25}$$

The refractive index of the slab, $\overline{m} = \overline{n} + i\,\overline{k}$, refers to a medium that consists of vacuum plus uniformly distributed particles. The way $\overline{m}$ is defined, it has the property that if the particles are pure scatterers, $\overline{k}$ is nevertheless positive. This follows from comparing (2.20) and (2.25) because the extinction coefficient C^{ext} of a single grain does not vanish. A positive $\overline{k}$ implies some kind of dissipation, which seems unphysical as no light is absorbed. One may therefore wonder whether $\overline{n}$ and $\overline{k}$ obey the Kramers-Kronig relation of section 2.6. However, they do, as one can show by studying the frequency dependence of $S(0)$.

One finds the refractive index $\overline{m} = \overline{n} + i\,\overline{k}$ of a cloud filled with identical spheres of size parameter $x = 2\pi a/\lambda$ and refractive index $m = n + ik$ from equation (2.24) and (2.25) where $S(0)$ is given by

$$S(0) \; = \; \tfrac{1}{2} \sum_n (2n + 1) \cdot (a_n + b_n) \; . \tag{2.26}$$

The coefficients a_n, b_n are the same as in (2.37) for the extinction efficiency. $S(0)$ is equal to the matrix elements S_1, S_2 in (2.44) taken in the forward direction: $S(0) = S_1(0) = S_2(0)$.

2.3 Mie theory for a sphere

2.3.1 The formalism

To determine the scattering and absorption coefficients of particles is a complicated problem of classical electrodynamics which was first solved in the general case for spheres by G. Mie [Mie08] and the underlying theory bears his name. A full derivation can be found in the special literature (for instance, in [Boh83], [Hul57], [Ker69]). Summarizing, the scattered electromagnetic field and the field inside the sphere are found by expanding both into an infinite series of independent solutions to the wave equation; the series coefficients follow from the boundary conditions on the particle surface.

To sketch the theory, consider a spherical particle in vacuum illuminated by a linearly polarized monochromatic plane wave of frequency $\nu = \omega/2\pi$. Let $\mathbf{E}_i$ and $\mathbf{H}_i$ describe the incident field. We denote the field within the particle by $\mathbf{E}_1, \mathbf{H}_1$ and outside of it by $\mathbf{E}_2, \mathbf{H}_2$. The field outside is the superposition of the incident and the scattered field (subscript s),

$$\mathbf{E}_2 = \mathbf{E}_i + \mathbf{E}_s \qquad (2.27)$$

$$\mathbf{H}_2 = \mathbf{H}_i + \mathbf{H}_s . \qquad (2.28)$$

To solve the wave equations (1.34), the calculations are greatly simplified by the fact: When ψ is a solution to the *scalar wave equation*

$$\Delta\psi + k^2\psi = 0 , \qquad (2.29)$$

then, for an arbitrary constant vector $\mathbf{c}$, the vector function $\mathbf{M}$ defined by

$$\mathbf{M} = \mathrm{rot}\,(\mathbf{c}\psi) , \qquad (2.30)$$

is a solution to the *vector equation*

$$\Delta\mathbf{M} + k^2\,\mathbf{M} = 0 . \qquad (2.31)$$

Furthermore, the vector function $\mathbf{N}$ given by

$$\mathbf{N} = \frac{1}{k}\,\mathrm{rot}\,\mathbf{M} \qquad (2.32)$$

also obeys the wave equation

$$\Delta\mathbf{N} + k^2\,\mathbf{N} = 0 , \qquad (2.33)$$

and $\mathbf{M}$ and $\mathbf{N}$ are related through

$$\mathrm{rot}\,\mathbf{N} = k\,\mathbf{M} . \qquad (2.34)$$

All material properties are taken into account by the wavenumber k,

$$k^2 = \frac{\omega^2 \mu \varepsilon}{c^2}$$

after (1.38), or after (1.106) in case of conductivity. In this way, the vector wave equation is reduced to a scalar wave equation. One starts with the vector function

$$\mathbf{M} = \text{rot}\,(\mathbf{r}\psi) \;. \tag{2.35}$$

In spherical polar coordinates (r, θ, ϕ), the wave equation for ψ reads

$$\frac{1}{r^2}\frac{\partial}{\partial r}\left(r^2\frac{\partial \psi}{\partial r}\right) + \frac{1}{r^2 \sin\theta}\frac{\partial}{\partial \theta}\left(\sin\theta\frac{\partial \psi}{\partial \theta}\right) + \frac{1}{r^2 \sin^2\theta}\frac{\partial^2 \psi}{\partial \phi^2} + k^2\,\psi = 0 \;. \tag{2.36}$$

It is basically the same equation that is encountered in finding the eigenstates of the hydrogen atom, and here one also makes an ansatz by separating variables

$$\psi(r, \theta, \phi) = R(r)\,T(\theta)\,P(\phi) \;.$$

The rest is algebra, but one must admire the skill with which such equations were solved a hundred years ago. Only after they have been solved can one employ a computer. Nowadays such a machine supplies us with the expansion coefficients, a_n and b_n, which are determined from the boundary conditions of the electromagnetic field on the grain surface. The scattering and extinction efficiencies follow then immediately,

$$Q^{\text{ext}} = \frac{2}{x^2} \cdot \sum_{n=1}^{\infty}(2n+1)\cdot\text{Re}\left\{a_n + b_n\right\} \tag{2.37}$$

$$Q^{\text{sca}} = \frac{2}{x^2} \cdot \sum_{n=1}^{\infty}(2n+1)\cdot\left[|a_n|^2 + |b_n|^2\right] \;. \tag{2.38}$$

The efficiencies depend only on the complex optical constant, m, of the material in the sphere and the size parameter

$$x = \frac{2\pi a}{\lambda} \;.$$

2.3.2   Scattered and absorbed power

If we imagine the particle to be surrounded by a large spherical and totally transparent surface A, the energy absorbed by the grain, W_{a}, is given by the difference between the flux which enters and which leaves the sphere. If $\mathbf{S}$ is the Poynting vector of the electromagnetic field outside the particle,

$$\mathbf{S} = \frac{c}{8\pi}\,\text{Re}\left\{\mathbf{E}_2 \times \mathbf{H}_2^*\right\} \;,$$

where $\mathbf{E}_2$ or $\mathbf{H}_2$ is the sum of the incident and scattered field after (2.27) and (2.28), W_a is determined by the integral

$$W_a = - \int_A \mathbf{S} \cdot \mathbf{e}_r \, dA . \tag{2.39}$$

Here $\mathbf{e}_r$ is the outward normal of the surface. The minus sign ensures that W_a is positive. The Poynting vector $\mathbf{S}$ can be considered to consist of three parts,

$$\mathbf{S} = \mathbf{S}_i + \mathbf{S}_s + \mathbf{S}_{ext}$$

with

$$
\begin{aligned}
\mathbf{S}_i &= (c/8\pi) \operatorname{Re} \{ \mathbf{E}_i \times \mathbf{H}_i^* \} \\
\mathbf{S}_s &= (c/8\pi) \operatorname{Re} \{ \mathbf{E}_s \times \mathbf{H}_s^* \} \\
\mathbf{S}_{ext} &= (c/8\pi) \operatorname{Re} \{ \mathbf{E}_i \times \mathbf{H}_s^* + \mathbf{E}_s \times \mathbf{H}_i^* \}
\end{aligned}
$$

and therefore

$$-W_a = \int_A \mathbf{S}_i \cdot \mathbf{e}_r \, dA + \int_A \mathbf{S}_s \cdot \mathbf{e}_r \, dA + \int_A \mathbf{S}_{ext} \cdot \mathbf{e}_r \, dA .$$

The first integral vanishes because the incident field (subscript i) enters and leaves the sphere without modification. The second integral obviously describes the scattered energy,

$$W_s = \frac{c}{8\pi} \operatorname{Re} \int_0^{2\pi} \int_0^\pi \left(E_{s\theta} H_{s\phi}^* - E_{s\phi} H_{s\theta}^* \right) r^2 \sin\theta \, d\theta d\phi . \tag{2.40}$$

The negative of the third integral, which we denote by $-W_{ext}$, is therefore the sum of absorbed plus scattered energy,

$$W_{ext} = W_a + W_s \tag{2.41}$$

$$= \frac{c}{8\pi} \operatorname{Re} \int_0^{2\pi} \int_0^\pi \left(E_{i\phi} H_{s\theta}^* - E_{i\theta} H_{s\phi}^* - E_{s\theta} H_{i\phi}^* + E_{s\phi} H_{i\theta}^* \right) r^2 \sin\theta \, d\theta d\phi$$

and thus the total energy removed from the beam at a large distance. According to our definition in section 2.1, W_a is related to the absorption coefficient of the particle, C^{abs}, through $W_a = S_i \, C^{abs}$, where S_i is the time-averaged Poynting vector of the incident field. Likewise for the scattering and extinction coefficient, $W_s = S_i \, C^{sca}$ and $W_{ext} = S_i \, C^{ext}$.

2.4   Polarization and scattering

We introduce the amplitude scattering matrix, define Stokes parameters and compute the radiation field scattered into a certain direction.

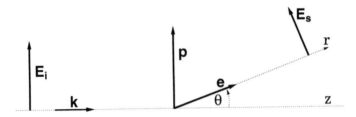

FIGURE 2.4 An incident wave of electric field $\mathbf{E}_i$ traveling in the direction of the wavenumber $\mathbf{k}$ is scattered by a particle. The scattering plane is defined by $\mathbf{k}$ and the scattering direction $\mathbf{e}$. When $\mathbf{E}_i$ excites a dipole $\mathbf{p}$, the electric field of the scattered wave, $\mathbf{E}_s$, lies in the same plane as $\mathbf{p}$ and $\mathbf{e}$ (see 1.90). In the figure, $\mathbf{E}_i$ lies also in the scattering plane; it is parallel to $\mathbf{p}$. Alternatively, when $\mathbf{E}_i$ and $\mathbf{p}$ are perpendicular to the scattering plane (not shown), with $\mathbf{k}$ and $\mathbf{e}$ unchanged, $\mathbf{E}_s$ and $\mathbf{E}_i$ will be parallel.

2.4.1 The amplitude scattering matrix

Consider a plane harmonic wave propagating in the z-direction and a particle at the origin of the coordinate system. Some light is scattered by the particle into a certain direction given by the unit vector $\mathbf{e}$. The z-axis together with $\mathbf{e}$ define what is called the *scattering plane* (see figure 2.4).

The amplitude of the incident electric field $\mathbf{E}_i$ may be decomposed into two vectors, one parallel, the other perpendicular to the scattering plane; their lengths are denoted by $E_{i\parallel}$ and $E_{i\perp}$, respectively. Likewise, one may decompose the scattered electric field $\mathbf{E}_s$, the components being $E_{s\parallel}$ and $E_{s\perp}$.

At a large distance r from the particle, in the far field where $r \gg \lambda$, there is a linear relation between $(E_{s\perp}, E_{s\parallel})$ and $(E_{i\perp}, E_{i\parallel})$ described by the amplitude scattering matrix,

$$\begin{pmatrix} E_{s\parallel} \\ E_{s\perp} \end{pmatrix} = \frac{e^{ik(r-z)}}{-ikr} \begin{pmatrix} S_2 & S_3 \\ S_4 & S_1 \end{pmatrix} \begin{pmatrix} E_{i\parallel} \\ E_{i\perp} \end{pmatrix} . \tag{2.42}$$

The factor before the amplitude scattering matrix is the same as in (2.17) where we considered a scalar field, with only one function S. Here we deal with vectors and there is an amplitude scattering matrix consisting of four elements S_j. They depend of course on the scattering direction, which is specified by the unit vector $\mathbf{e}$ or by two angles, θ and ϕ.

As an example, we work out the scattering matrix of a small grain of unit volume. If its polarizability α is isotropic, its dipole, $\mathbf{p}$, due to the incident wave is

$$\mathbf{p} = \alpha\,\mathbf{E}_i .$$

The scattered electric field is given by equation (1.90),

$$\mathbf{E_s} = \frac{\omega^2 \alpha}{c^2 r} (\mathbf{e} \times \mathbf{E_i}) \times \mathbf{e} .$$

It is always transverse to $\mathbf{e}$ and depends only on $\mathbf{e}$ and $\mathbf{p}$ (or $\mathbf{E_i}$), and not on the direction $\mathbf{k}$ from where the incident wave is coming. One finds that $E_{s\parallel} = E_{i\parallel} \cos\theta$ and $E_{i\perp} = E_{s\perp}$, so for a dipole

$$\begin{pmatrix} S_2 & S_3 \\ S_4 & S_1 \end{pmatrix} = -i\,k^3\alpha \begin{pmatrix} \cos\theta & 0 \\ 0 & 1 \end{pmatrix} . \tag{2.43}$$

2.4.2 Angle-dependence of scattering

In case of a sphere, the amplitude scattering matrix (2.42) acquires also a diagonal structure as element S_3 and S_4 vanish,

$$\begin{pmatrix} E_{s\parallel} \\ E_{s\perp} \end{pmatrix} = \frac{e^{ik(r-z)}}{-ikr} \begin{pmatrix} S_2 & 0 \\ 0 & S_1 \end{pmatrix} \begin{pmatrix} E_{i\parallel} \\ E_{i\perp} \end{pmatrix} . \tag{2.44}$$

S_1, S_2 depend only on $\mu = \cos\theta$, where $\theta = 0$ denotes the forward direction. They can again be expressed with the help of the expansion coefficients a_n, b_n that appear in (2.37) and (2.38).

- For unpolarized incident light ($E_{i\parallel} = E_{i\perp}$), the intensity of the radiation scattered into the direction θ is given by

$$S_{11}(\cos\theta) = \tfrac{1}{2}\left[|S_1|^2 + |S_2|^2\right] . \tag{2.45}$$

  The factor before the matrix in (2.44) has been neglected. The notation S_{11} comes from equation (2.55); in (2.1) the same quantity was denoted $\mathcal{L}$. When integrated over all directions,

$$\int_0^\pi S_{11}(\cos\theta) \sin\theta \, d\theta = \tfrac{1}{2} x^2 Q^{\text{sca}} .$$

- The angle-dependence of the normalized phase function $f(\cos\theta)$ (see 2.6) is related to S_{11} through

$$S_{11}(\cos\theta) = \frac{x^2 Q^{\text{sca}}}{4} f(\cos\theta) . \tag{2.46}$$

2.4.3 The polarization ellipse

Consider a plane harmonic wave of angular frequency ω, wavenumber $\mathbf{k}$ and electric field

$$\mathbf{E(r,}t) = \mathbf{E_0} \cdot e^{i(\mathbf{k}\cdot\mathbf{r}-\omega t)}$$

that travels in z-direction of a Cartesian coordinate system. The amplitude $\mathbf{E_0}$ is in the general case complex,

$$\mathbf{E_0} \;=\; \mathbf{E_1} \,+\, i\,\mathbf{E_2}\;,$$

with real $\mathbf{E_1}$ and $\mathbf{E_2}$. At any fixed z, the real part of the electric vector $\mathbf{E}$ rotates at frequency ω and the tip of the vector describes in the (x,y)-plane, which is perpendicular to the z-axis, an ellipse,

$$\mathrm{Re}\,\{\mathbf{E}\} \;=\; \mathbf{E_1}\,\cos\omega t \,+\, \mathbf{E_2}\,\sin\omega t\;. \tag{2.47}$$

The sense of rotation changes when the plus sign in (2.47) is altered into a minus sign, i.e., when $\mathbf{E_2}$ flips direction by an angle π. There are two special cases:

- *Linear polarization.* For $\mathbf{E_1}=0$ or $\mathbf{E_2}=0$, or when $\mathbf{E_1},\mathbf{E_2}$ are linearly dependent, the ellipse degenerates into a line. By adjusting the time t, it is always possible to make $\mathbf{E_0}$ real. At a fixed location z, the electric vector swings up and down along a straight line whereby its length changes; it vanishes when $\mathrm{Re}\{\mathbf{E}\}$ switches direction. The field can drive a linear oscillator in the (x,y)-plane.

- *Circular polarization.* The vectors are of equal length, $|\mathbf{E_1}|=|\mathbf{E_2}|$, and perpendicular to each other, $\mathbf{E_1}\cdot\mathbf{E_2}=0$. The ellipse is a circle. At a fixed location z, the electric vector never disappears, but rotates retaining its full length. The circularly polarized wave sets a two-dimensional harmonic oscillator in the (x,y)-plane (with equal properties in the x- and y-direction) into a circular motion; it transmits angular momentum.

One can combine two linearly polarized waves to get circular polarization and two circularly polarized waves to obtain linear polarization. The magnetic field has the same polarization as the electric field because $\mathbf{E}$ and $\mathbf{H}$ are in phase and have a constant ratio.

2.4.4   Stokes parameters

The polarization ellipse is completely determined by the length of its major and minor axis, a and b, plus some specification of its orientation in the (x,y)-plane. This could be the angle γ between the major axis and the x-coordinate. Instead of these geometrical quantities (a,b,γ), polarization is usually described by the Stokes parameters I,Q,U and V. They are equivalent to (a,b,γ), but have the practical advantage that they can be measured directly.

When a plane harmonic wave is scattered by a grain, the Stokes parameters of the incident (subscript i) and the scattered wave (subscript s) are linearly related through

$$
\begin{pmatrix} I_\mathrm{s} \\ Q_\mathrm{s} \\ U_\mathrm{s} \\ V_\mathrm{s} \end{pmatrix} = \frac{1}{k^2 r^2} \begin{pmatrix} S_{11} & S_{12} & S_{13} & S_{14} \\ S_{21} & S_{22} & S_{23} & S_{24} \\ S_{31} & S_{32} & S_{33} & S_{34} \\ S_{41} & S_{42} & S_{43} & S_{44} \end{pmatrix} \begin{pmatrix} I_\mathrm{i} \\ Q_\mathrm{i} \\ U_\mathrm{i} \\ V_\mathrm{i} \end{pmatrix} . \tag{2.48}
$$

The Stokes parameters of the scattered light as well as the matrix elements refer to a particular scattering direction (θ, ϕ); r is the distance from the particle. The matrix S_{ij} does not contain more information than the matrix in (2.42) and therefore only seven of its 16 elements are independent, which corresponds to the fact that the four elements S_j in the matrix of (2.42) have four absolute values $|S_j|$ and three phase differences between them.

- Only three of the four Stokes parameters are independent as for fully polarized light

$$
I^2 = Q^2 + U^2 + V^2 . \tag{2.49}
$$

- Unpolarized light has

$$
Q = U = V = 0 ,
$$

and for partially polarized light, one has instead of the strict equality $I^2 = Q^2 + U^2 + V^2$, an inequality

$$
I^2 > Q^2 + U^2 + V^2 .
$$

- For unit intensity $(I=1)$, linearly polarized light has

$$
V = 0, \qquad Q = \cos 2\gamma, \qquad U = \sin 2\gamma
$$

so that

$$
Q^2 + U^2 = 1 .
$$

The degree of linear polarization is defined by

$$
\frac{\sqrt{Q^2 + U^2}}{I} \leq 1 . \tag{2.50}
$$

It can vary between 0 and 1.

- Circular polarization implies

$$
Q = U = 0, \qquad V = \pm 1 ,
$$

the sign determines the sense of rotation of the electric vector. The degree of circular polarization is

$$
-1 \leq \frac{V}{I} \leq 1 . \tag{2.51}
$$

- Even when the incident light is unpolarized, i.e., when

$$Q_i = U_i = V_i = 0 , \qquad (2.52)$$

it becomes partially polarized after scattering if S_{21}, S_{31} or S_{41} are non-zero. Indeed, dropping the factor $1/k^2r^2$ in (2.48), we get

$$I_s = S_{11}I_i, \quad Q_s = S_{21}I_i, \quad U_s = S_{31}I_i, \quad V_s = S_{41}I_i . \qquad (2.53)$$

2.4.4.1　The Stokes parameters of light scattered by a sphere

In case of a sphere, the transformation matrix (2.48) between incident and scattered Stokes parameters simplifies,

$$\begin{pmatrix} I_s \\ Q_s \\ U_s \\ V_s \end{pmatrix} = \frac{1}{k^2r^2} \begin{pmatrix} S_{11} & S_{12} & 0 & 0 \\ S_{12} & S_{11} & 0 & 0 \\ 0 & 0 & S_{33} & S_{34} \\ 0 & 0 & -S_{34} & S_{33} \end{pmatrix} \begin{pmatrix} I_i \\ Q_i \\ U_i \\ V_i \end{pmatrix} . \qquad (2.54)$$

Out of the 16 matrix elements, eight are nontrivial and they have only four significantly different values,

$$\begin{align} S_{11} &= \tfrac{1}{2} \left(|S_1|^2 + |S_2|^2 \right) \qquad (2.55) \\ S_{12} &= \tfrac{1}{2} \left(|S_2|^2 - |S_1|^2 \right) \\ S_{33} &= \tfrac{1}{2} \left(S_2^* S_1 + S_2 S_1^* \right) \\ S_{34} &= \tfrac{i}{2} \left(S_1 S_2^* - S_2 S_1^* \right) . \end{align}$$

Only three of them are independent because

$$S_{11}^2 = S_{12}^2 + S_{33}^2 + S_{34}^2 .$$

- If the incident light is 100% polarized and its electric vector *parallel* to the scattering plane so that $I_i = Q_i$, $U_i = V_i = 0$, we get, dropping the factor $1/k^2r^2$ in (2.54),

$$I_s = (S_{11} + S_{12})I_i, \quad Q_s = I_s, \quad U_s = V_s = 0 .$$

So the scattered light is also 100% polarized parallel to the scattering plane.

- Likewise, if the incident light is 100% polarized *perpendicular* to the scattering plane ($I_i = -Q_i$, $U_i = V_i = 0$), so is the scattered light and

$$I_s = (S_{11} - S_{12})I_i, \quad Q_s = -I_s, \quad U_s = V_s = 0 .$$

- If the incident wave is unpolarized ($Q_i = U_i = V_i = 0$), the scattered light is nevertheless usually polarized; in this case,

$$I_s = S_{11}I_i, \quad Q_s = S_{12}I_i, \quad U_s = V_s = 0 .$$

When one defines the quantity p as the difference of the intensities $|S_1|^2$ and $|S_2|^2$ divided by their sum,

$$p = \frac{|S_1|^2 - |S_2|^2}{|S_1|^2 + |S_2|^2} = -\frac{S_{12}}{S_{11}}, \qquad (2.56)$$

the absolute value $|p|$ is equal to the degree of linear polarization of (2.50). We will not compose a new name for p, but call it also degree of polarization, although it contains via the sign additional information. In the forward direction, $S_1(0) = S_2(0)$ and $p = 0$.

- The sign of p, or of S_{12}, specifies the direction of polarization. Usually, S_{12} is negative and then linear polarization is *perpendicular* to the scattering plane. But S_{12} can after (2.55) also be positive, as happens for big spheres (see figure 4.7). Then polarization is *parallel* to the scattering plane.

In the general case, when the particles are not spherical (anisotropic), which is outside the scope of what we calculate here, none of the matrix elements in (2.48) vanishes. Then the polarization vector can have any inclination towards the scattering plane and p may be non-zero also in the forward direction.

2.5 The discrete dipole approximation

The cross sections of arbitrarily shaped and compounded grains can be found with the discrete dipole approximation, or DDA. This method was invented by [Pur73] and refined by [Dra88]. Its essence is to replace the dust particle by a number of point-like dipoles located at the lattice points of a simple cubic grid (type sc in figure 5.2). If d is the grid constant, one dipole represents the atoms in a volume d^3.

Let $\mathbf{r}_j$, where $j = 1, \ldots, N$, be the Cartesian coordinates of the N dipoles in the grain and α_j their polarizability. If $\mathbf{E}(\mathbf{r}_j)$ is the electric field at $\mathbf{r}_j$, the dipole moment there will be

$$\mathbf{p}_j = \alpha_j \, \mathbf{E}(\mathbf{r}_j) .$$

Let $\mathbf{E}^{\mathrm{inc}}(\mathbf{r}, t) = \mathbf{E}_0 \, e^{i(\mathbf{k} \cdot \mathbf{r} - \omega t)}$ be the field of the incident wave with wavenumber $k = 2\pi/\lambda$. It interacts with the dipoles, and the field $\mathbf{E}(\mathbf{r}_j)$ is the sum of $\mathbf{E}^{\mathrm{inc}}(\mathbf{r}_j)$ plus the scattered fields, $\mathbf{E}_k^{\mathrm{sca}}(\mathbf{r}_j)$, from all other dipoles $k \neq j$,

$$\mathbf{E}(\mathbf{r}_j) = \mathbf{E}^{\mathrm{inc}}(\mathbf{r}_j) + \sum_{k \neq j} \mathbf{E}_k^{\mathrm{sca}}(\mathbf{r}_j) . \qquad (2.57)$$

As the distance between nearest dipoles is smaller than the wavelength, the scattered fields $\mathbf{E}_k^{\mathrm{sca}}$ are also needed in the near zone. The general electric dipole equation which includes the near and the far field (1.90) is given in [Jac99]. When applied to the scattered field $\mathbf{E}_k^{\mathrm{sca}}(\mathbf{r}_j)$ in (2.57), we get

$$\mathbf{E}_k^{\mathrm{sca}}(\mathbf{r}_j) = k^2 \left(\mathbf{e} \times \mathbf{p}_k\right) \times \mathbf{e} \; \frac{e^{ikr}}{r} + \left[3\mathbf{e}(\mathbf{e} \cdot \mathbf{p}_k) - \mathbf{p}_k\right] \left(\frac{1}{r^3} - \frac{ik}{r^2}\right) e^{ikr} \quad (2.58)$$

with $r = |\mathbf{r}_j - \mathbf{r}_k|$ and $\mathbf{e} = (\mathbf{r}_j - \mathbf{r}_k)/r$. Because equation (2.57) is also valid when the point $\mathbf{r}_j$ lies outside the grain, solving it yields all information about scattering, absorption and polarization. In practice, one turns (2.57) into a matrix equation for the dipole moments $\mathbf{p}_j$ and then determines, for example, the extinction coefficient from

$$C^{\mathrm{ext}} = \frac{4\pi k}{|\mathbf{E}^{\mathrm{inc}}|^2} \sum_j \mathrm{Im}\left(\mathbf{E}^{\mathrm{inc}\,*} \cdot \mathbf{p}_j\right)$$

(see also 1.55) or the scattering cross section from the power of the scattered radiation field.

But how does one relate the dielectric susceptibility χ of the grain material to the dipole polarizability α_j? This is done with the help of the Clausius-Mossotti formula (3.35) which we derived in section 3.3 for a cubic grid. As the dipoles are arranged on such a grid, we can apply it, and it then reads $\chi = n\alpha/(1 - 4\pi n\alpha/3)$ where $n = d^{-3}$ is the dipole volume density.

As shown in [Dra88], a consistent treatment necessitates a modification of the relation $\chi = n\alpha/(1 - 4\pi n\alpha/3)$ such that radiation damping is included (see section 1.2.5). Otherwise a pure scatterer ($\chi_2 = 0$) would imply a real polarizability ($\alpha_2 = 0$) and thus no attenuation of the incident wave, i.e., no scattering. However, this refinement is probably not of practical relevance.

In principle, the DDA is an extremely versatile method. It gives satisfactory results if the dipoles are closely spaced such that $|m|d \lesssim 0.1\lambda$, where m is the complex optical constant, and if the total number of dipoles is sufficiently large ($N \gtrsim 1000$). In praxi, however, the DDA often suffers from limitations set by the severe demand on the computing machine. Furthermore, although the DDA can and must be employed when a grain is of irregular shape or structure, such a particle possesses many cross sections, depending on its orientation. In an application to interstellar grains, which are rotating, they should all be calculated and then averaged. The multitude of cross sections must thus be condensed into a few numbers, and they are probably more efficiently derived by assuming simple geometries, like spheres or spheroids, possibly layered or of composite material. It may not even make much sense to compute the exact angle-dependent cross sections, $C(\theta, \varphi)$, of one particular interstellar grain because all others, which also interact with the light that we observe, are different. We shall always need simple-shaped optical analogues of the myriad of complex grains.

2.6 The Kramers-Kronig relations

The formulae of *Kramers* and *Kronig* establish a link between the real and imaginary part of the material constants (like dielectric or magnetic permeability, conductivity. Their deduction is very abstract and the result baffling. The relations have some fundamental consequences, although their practical value is at present moderate.

2.6.1 The KK relation for the dielectric permeability

In any medium, the real and imaginary part of the dielectric permeability, ε_1 and ε_2, are not completely independent of each other, but for any frequency ω_0,

$$\varepsilon_1(\omega_0) - 1 = \frac{2}{\pi} \int_0^\infty \frac{\omega \, \varepsilon_2(\omega)}{\omega^2 - \omega_0^2} \, d\omega \qquad (2.59)$$

$$\varepsilon_2(\omega_0) = -\frac{2\omega_0}{\pi} \int_0^\infty \frac{\varepsilon_1(\omega)}{\omega - \omega_0^2} \, d\omega \ . \qquad (2.60)$$

For the static limit of ε_1 at zero frequency,

$$\varepsilon_1(0) - 1 = \frac{2}{\pi} \int_0^\infty \frac{\varepsilon_2(\omega)}{\omega} \, d\omega \ . \qquad (2.61)$$

Similar relations hold for the electric susceptibility χ, the electric polarizability α_e or the optical constant $m = n + ik$. Whenever the vacuum value of the material constant is one (as for ε), the -1 appears on the left side (see 2.59 and 2.61), when the vacuum value is zero (as for χ), the -1 is missing.

2.6.2 Three corollaries of the KK relation

2.6.2.1 The dependence between ε_1 and ε_2

As any set of physically possible values $\varepsilon_1(\omega)$ and $\varepsilon_2(\omega)$ for any grain material must obey the KK relations (2.59) and (2.60), they serve as a check for the internal consistency of data. Furthermore, it is sufficient to know either $\varepsilon_1(\omega)$ or $\varepsilon_2(\omega)$ over the entire wavelength range to compute the other. Whereas $\varepsilon_1(\omega)$ is not restricted at all, $\varepsilon_2(\omega)$ is associated with the entropy and must be positive everywhere. A data set for $\varepsilon_1(\omega)$ is wrong if it yields at just one frequency a negative value for $\varepsilon_2(\omega)$.

As an example of a dispersion formula that obeys the KK relations we may take equation (1.77) or (1.113). They apply to the harmonic oscillator or a metal, respectively. That they fulfill the KK relations may be verified from general mathematical considerations for the function $\varepsilon(\omega)$, which is the smart way; or by doing explicitly the KK integrals for (1.77), that is the hard way; or numerically, which is the brute way. Even the last method requires some delicacy when handling the Cauchy principal value. Numerical integration is inevitable when $\varepsilon(\omega)$ is available only in tabulated form.

When we look at the dispersion relation (1.77) and realize that ε_1 and ε_2 have the same denominator and contain the same quantities e, m_e, γ, ω and ω_0, equations (2.59) to (2.61) which link ε_1 with ε_2 are not so perplexing any more. On the other hand, the physics associated, for example, with the optical constants n and k, are quite different. n fixes the phase velocity which results from the superposition of all waves involved (see section 3.4.3), whereas k determines the amount of energy dissipation. Here n and k appear as very distinct parameters and any general connection between them comes, at first glance, as a surprise.

2.6.2.2 Dust absorption at very long wavelengths

In a slowly varying electromagnetic field, a dielectric grain of arbitrary shape and composition absorbs after (1.55) the power

$$W = \tfrac{1}{2} V \omega \operatorname{Im}\{\alpha_e\} E_0^2 = C^{\text{abs}}(\omega) \cdot S \,,$$

where V is the volume of the grain, C^{abs} its cross section and $S = (c/8\pi) E_0^2$ the Poynting vector. The particle has to be small, which means that the frequency of the wave must stay below some critical value, say ω_1. We now apply the KK relation to the polarizability α of the grain and split the integral into two,

$$\frac{\pi}{2} \alpha_1(\omega = 0) = \int_0^\infty \frac{\alpha_2(\omega)}{\omega}\, d\omega = \int_0^\infty \frac{\alpha_2(\lambda)}{\lambda}\, d\lambda = \int_0^{\lambda_1} \frac{\alpha_2(\lambda)}{\lambda}\, d\lambda + \int_{\lambda_1}^\infty \frac{\alpha_2(\lambda)}{\lambda}\, d\lambda \,.$$

Because α_2 is positive and the integral over α_2/λ in the total interval $[0, \infty]$ finite, the last integral must also be finite. If we make in the range $\lambda > \lambda_1 = 2\pi c/\omega_1$ the replacement

$$\frac{\alpha_2(\lambda)}{\lambda} = \frac{C^{\text{abs}}(\lambda)}{8\pi^2 V} \,,$$

and write $C^{\text{abs}} = \sigma_{\text{geo}} Q^{\text{abs}}$, where σ_{geo} is the geometrical cross section, we get the following convergence condition

$$\int_0^\infty Q^{\text{abs}}(\lambda)\, d\lambda < \infty \,.$$

Therefore, the absorption efficiency of any grain must at long wavelengths fall off more steeply than λ^{-1}. But this last constraint attains practical importance only when we know the threshold wavelength after which it is valid.

2.6.2.3 Total grain volume

One may apply the KK relation not only to grain material, but also to the interstellar medium as such [Pur69]. Because it is so tenuous, its optical constant $m = n + ik$ is very close to one. A small volume V in the interstellar medium of size d may therefore be considered to represent a Rayleigh-Gans particle. This class of grains, for which $|m - 1| \ll 1$ and $d|m - 1|/\lambda \ll 1$, is discussed in section 3.4. Of course, the particle is inhomogeneous as it consists of a mixture of gas, dust and mostly vacuum, but for its absorption cross section we may nevertheless use (3.27),

$$C^{\text{abs}}(\lambda) = \frac{4\pi V}{\lambda} k(\lambda) = \frac{2\pi V}{\lambda} \varepsilon_2(\lambda) .$$

The static limit (2.61) of the KK relation then gives

$$\varepsilon_1(0) - 1 = \frac{2}{\pi} \int_0^\infty \frac{\varepsilon_2(\lambda)}{\lambda} d\lambda = \frac{1}{\pi^2 V} \int_0^\infty C^{\text{abs}}(\lambda) d\lambda . \tag{2.62}$$

In a *static* electric field $\mathbf{E}$, a volume V of interstellar medium with dielectric permeability $\varepsilon(0)$ has after equations (1.3), (1.6) and (2.62) a dipole moment

$$\mathbf{p} = \frac{\varepsilon(0) - 1}{4\pi} \mathbf{E} V = \frac{\mathbf{E}}{4\pi^3} \int_0^\infty C^{\text{abs}}(\lambda) d\lambda .$$

Alternatively, we may express $\mathbf{p}$ to result from the polarization of the individual grains within V. They have a total volume V_g and at zero frequency a dielectric constant $\varepsilon_g(0)$, so after (3.11)

$$\mathbf{p} = \frac{3}{4\pi} V_g \frac{\varepsilon_g(0) - 1}{\varepsilon_g(0) + 2} \mathbf{E} .$$

When we crudely evaluate the long wavelength limit of the above ratio from figures 5.10 and 5.11, we obtain a value of order one,

$$\frac{\varepsilon_g(0) - 1}{\varepsilon_g(0) + 2} \sim 1 .$$

Along a line of sight of length L, the optical depth for absorption is $\tau(\lambda) = L C^{\text{abs}}$. Therefore, the total grain volume V_{dust} in a column of length L that produces an optical depth $\tau(\lambda)$ and has a cross section of $1 \, \text{cm}^2$ is

$$V_{\text{dust}} \simeq \frac{1}{3\pi^2} \int_0^\infty \tau(\lambda) d\lambda .$$

With the standard interstellar extinction curve $\tau(\lambda)/\tau_V$ of figure 10.1, we get for the above integral in the interval from 0.1 to $10\mu\text{m}$ a value of about 2×10^{-4}. Consequently, the total dust volume in a column of $1 \, \text{cm}^2$ cross section with $A_V = 1 \, \text{mag}$ (or $\tau_V = 1$) is

$$V_{\text{dust}} \simeq 6 \times 10^{-6} \, \text{cm}^{-3} . \tag{2.63}$$

Such an estimate is of principle value, although its precision does not allow discrimination between grain models.

2.7   Composite grains

Interstellar grains are probably not solid blocks made of one kind of material, but are more likely to be inhomogeneous so that the dielectric function within them varies from place to place. There are many ways inhomogeneity can come about. For instance,

- when particles coagulate, the result will be a bigger particle with voids inside. The new big grain has then a fluffy structure and for its description, even if the grains before coagulation were homogeneous and chemically identical, at least one additional dielectric function is needed, namely that of vacuum ($\varepsilon = 1$). If chemically diverse particles stick together, one gets a heterogenous mixture. We note that purely thermal Brownian motion of grains (section 7.2.3) is too small to make encounters between them significant, but omnipresent turbulent velocities of $10\,\mathrm{m/s}$ are sufficient to ensure coagulation in dense clouds.

- A grain becomes inhomogeneous when gas molecules freeze out on its surface; this happens in cold clouds. The ice mantle itself is inhomogeneous because it consists of various condensates attributable to different molecules.

- During the formation or evolution of a grain tiny solid subparticles, like PAHs (chapter 9) or metal atoms, are built into the bulk material and contaminate it chemically.

We expect to find composite grains in dense protostellar cloud cores, in the cooler parts of stellar disks, but also in comets. We figure such particles as fluffy aggregates, probably substantially bigger than normal interstellar grains. They are composed of refractory (resistant to heating) and compact subparticles made of silicates or amorphous carbon. The subparticles are enshrouded by a thick sheet of ice as a result of molecules that have frozen out. The volatile (easy to evaporate) ice balls with their compact cores are loosely bound together and form, as a whole, the porous grain. In this picture, the frosting of molecules in ice layers on the surface of the subparticles precedes the process of coagulation.

2.7.1   Effective medium theories

The cross section of composite particles can be computed exactly in those few cases where the components are homogeneous and the geometrical structure is simple; examples are spherical shells, cylinders with mantles or coated ellipsoids. For real composite particles, where different media are intermixed in a most complicated manner, such computations are out of the question and one may try the discrete dipole approximation (section 2.5).

Another way to estimate the optical behavior of composite particles, is to derive an average dielectric function ε_{av} representing the mixture as a whole. Once determined, ε_{av} is then used in Mie theory, usually assuming a spherical shape for the total composite grain. The starting point is the constitutive relation

$$\langle \mathbf{D} \rangle = \varepsilon_{av} \langle \mathbf{E} \rangle . \tag{2.64}$$

Here $\langle \mathbf{E} \rangle$ is the average internal field and $\langle \mathbf{D} \rangle$ the average displacement defined as

$$\langle \mathbf{E} \rangle = \frac{1}{V} \int \mathbf{E}(\mathbf{x}) \, dV , \qquad \langle \mathbf{D} \rangle = \frac{1}{V} \int \varepsilon(\mathbf{x}) \, \mathbf{E}(\mathbf{x}) \, dV . \tag{2.65}$$

The integration extends over the whole grain volume V. If we envisage the grain to consist of a finite number of homogeneous components (subscript j), each with its own dielectric function ε_j and volume fraction f_j, we can write

$$\langle \mathbf{E} \rangle = \sum_j f_j \, \mathbf{E}_j , \qquad \langle \mathbf{D} \rangle = \sum_j f_j \, \varepsilon_j \, \mathbf{E}_j . \tag{2.66}$$

The $\mathbf{E}_j$ are averages themselves after (2.65) over the subvolume $f_j V$; we just dropped the brackets. The constitutive relation (2.64) is thus replaced by

$$\sum_j \varepsilon_j \, f_j \, \mathbf{E}_j = \varepsilon_{av} \sum_j f_j \, \mathbf{E}_j . \tag{2.67}$$

We envisage the components to be present in the form of many identical subparticles that are much smaller than the wavelength. When such a subparticle is placed into an extended medium with a spatially constant, but time-variable field $\mathbf{E}'$, there is a linear relation between the field in the subparticle and the field in the outer medium (see section 3.1 and 3.2),

$$\mathbf{E}_j = \beta \, \mathbf{E}' . \tag{2.68}$$

When one assumes that such a large scale average field $\mathbf{E}'$ in the grain exists, one can get rid of the local fields $\mathbf{E}_j$ and find the average ε_{av}.

2.7.1.1 The mixing rule of Garnett

To exploit (2.67), we first imagine the grain to consist of a matrix (subscript m) containing inclusions (subscript i). For simplicity, let there be only one kind of inclusions with a total volume fraction f_i so that $f_i + f_m = 1$. For the constant large scale field $\mathbf{E}'$ in (2.68), we take the field in the matrix and obtain

$$\varepsilon_i f_i \beta + \varepsilon_m f_m = \varepsilon_{av} f_i \beta + \varepsilon_{av} f_m .$$

For spherical inclusions, the proportionality factor β in equation (2.68) has the form

$$\beta = \frac{3\,\varepsilon_m}{\varepsilon_i + 2\varepsilon_m} \, ,$$

which is a generalization of equation (3.10) when the medium surrounding the sphere is not vacuum, but has some permeability ε_m. We thus arrive at the *Garnett* formula

$$\varepsilon_{\mathrm{av}} = \varepsilon_m \, \frac{1 + 2\, f_i \, (\varepsilon_i - \varepsilon_m)/(\varepsilon_i + 2\varepsilon_m)}{1 - f_i \, (\varepsilon_i - \varepsilon_m)/(\varepsilon_i + 2\varepsilon_m)} \, . \tag{2.69}$$

The expression is evidently not symmetric with respect to inclusion and matrix. One has to make up one's mind which component to regard as the inclusion that pollutes the matrix. If the concentration of the inclusions, f_i, is small, equation (2.69) simplifies to

$$\varepsilon_{\mathrm{av}} = \varepsilon_m \left(1 + 3f_i \, \frac{\varepsilon_i - \varepsilon_m}{\varepsilon_i + 2\varepsilon_m} \right) \, .$$

The Garnett mixing rule is very similar to the law of Clausius-Mossotti (see 3.35). Indeed, the latter follows almost immediately from (2.69). Clausius-Mossotti gives the dielectric constant of an inhomogeneous medium that consists of a vacuum matrix ($\varepsilon_m = 1$) with embedded spherical inclusions, the latter being atoms of polarizability $\alpha = (3/4\pi)\,(\varepsilon_i - 1)/(\varepsilon_i + 2)$.

2.7.1.2   The mixing rule of Bruggeman

Next we suppose that the grain consists of various components (index j) that distinguish themselves only through their permeability and volume fraction. Therefore, no assumption is made about the average field $\mathbf{E}'$. Inserting (2.68) into (2.67) yields the *Bruggeman* rule

$$0 = \sum_{j} (\varepsilon_j - \varepsilon_{\mathrm{av}}) \, f_j \beta_j \, , \tag{2.70}$$

with $\sum f_j = 1$. Contrary to the Garnett rule, this formula is symmetric in all components j. If they consist of spherical entities,

$$0 = \sum_{j} f_j \, \frac{\varepsilon_j - \varepsilon_{\mathrm{av}}}{\varepsilon_j + 2\varepsilon_{\mathrm{av}}} \, . \tag{2.71}$$

Thus for n components, $\varepsilon_{\mathrm{av}}$ is determined from a complex polynomial of n-th degree. When we imagine the interstellar dust to be a democratic compound of silicate, amorphous carbon, ice and vacuum, the Bruggeman mixing rule is preferred. On the other hand, when ice becomes dirty through contamination by tiny impurities (metal atoms or PAHs) that amount to only a small volume fraction ($f_j \ll f_m$), the Garnett rule is more appropriate.

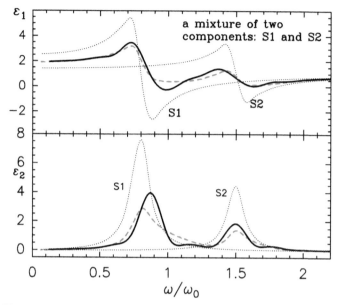

FIGURE 2.5 The dielectric permeability of two substances, S1 and S2, is represented by harmonic oscillators (dotted curves, $\omega_0 = 1$, $\omega_1 = 0.8$, $\omega_2 = 1.5$, $\gamma_1 = 0.16$, $\gamma_2 = 0.15$). The **average** permeability of a mixture of these two substances calculated from the Bruggeman mixing rule is shown by the broken curve, when calculated from the Garnett mixing rule, by the solid line.

An illustration of how the dielectric permeabilities of two components combine to an avarage ε_{av} is shown in figure 2.5 for the two mixing rules and for a very idealized situation. The materials have the optical properties of harmonic oscillators with different resonance frequencies. Both components have equal volume and both averaged permeabilities obey Kramers-Kronig's relations, as do, of course, the dielectric constants of S1 and S2. In the mixture, the resonances are damped, broadened and shifted.

2.7.1.3 Composition of grains in protostellar cores

Likely and astronomically relevant candidates for composite grains, as described above, are the solid particles in cold and dense protostellar cloud cores. We figure them as fluffy aggregates composed of refractory and compact subparticles of silicates or amorphous carbon enshrouded by a thick sheet of ice. Altogether, the grains in protostellar cores have four distinct components:

   – compact silicate subparticles
   – compact carbon subparticles
   – ice sheets
   – vacuum.

Their possible diversity is infinite and their structure can be extremely complex. The volume fractions of the components, f, add up to one,

$$f^{\mathrm{Si}} + f^{\mathrm{C}} + f^{\mathrm{ice}} + f^{\mathrm{vac}} = 1 .$$

To be specific, we adopt for the volume ratio of silicate to carbon material in the grain

$$\frac{f^{\mathrm{Si}}}{f^{\mathrm{C}}} = 1.4 \ldots 2 ,$$

compatible with the cosmic abundances in solids, and for the specific weight of the refractory subparticles and of ice

$$\rho_{\mathrm{ref}} \simeq 2.5\,\mathrm{g\,cm}^{-3} , \qquad \rho_{\mathrm{ice}} \simeq 1\,\mathrm{g\,cm}^{-3} .$$

The available mass of condensible material is determined by the gas phase abundances of C, N and O and their hydrides. Standard gas abundances imply a ratio of volatile to refractory mass of

$$\frac{M_{\mathrm{ice}}}{M_{\mathrm{ref}}} = 1 \ldots 2 .$$

Consequently, the ice volume is 2.5 to 5 times bigger than the volume of the refractories.

2.7.2 The influence of grain size, ice and porosity

The above described composite grains differ in three fundamental aspects from their compact silicate and carbon subparticles:

- they are bigger because of coagulation

- they contain ices because of frosting

- they are porous, again because of coagulation.

Coagulation, deposition of ice, and porosity each affect the absorption coefficient and we now illustrate how. In the examples below, we fix λ to 1 mm, which is a wavelength where protostellar cores are frequently observed.

2.7.2.1 Grain size

We first explore the influence of grain size. The mass absorption coefficient K^{abs} (defined in section 2.1.3) is for a fixed wavelength a function of grain radius a, so $K^{\mathrm{abs}} = K^{\mathrm{abs}}(a)$. To demonstrate the size effect, we normalize K^{abs} to its value when the grains are small ($a \ll \lambda$); it is then insensitive to a. We adopt $a = 0.1\,\mu\mathrm{m}$, which is a typical radius of an interstellar grain and

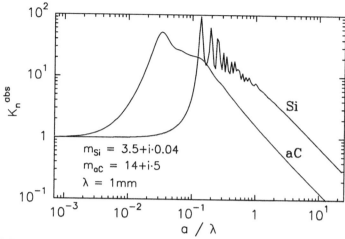

FIGURE 2.6 The influence of grain growth on the mass absorption coefficient K^{abs} for silicate (Si) and amorphous carbon spheres (aC) at a wavelength of 1 mm. The curves show the normalized coefficient K_n^{abs} defined in the text. For example, the mass absorption coefficient K^{abs} of carbon grains increases by a factor $K_n^{abs} \sim 50$ when the radius of the particles grows from 0.1μm $(a/\lambda = 10^{-4})$ to 30μm $(a/\lambda = 0.03)$.

certainly smaller than the wavelength (1 mm). The normalized coefficient is denoted K_n^{abs} and defined as

$$K_n^{abs}(a) = \frac{K^{abs}(a)}{K^{abs}(0.1\mu m)} .$$

It gives the relative change with respect to ordinary interstellar dust particles and it is plotted in figure 2.6 for silicate and carbon spheres. The spikes in figure 2.6 for silicates are resonances that disappear when the grains are not all of the same size, but have a size distribution. When $a \ll \lambda$, the normalized coefficient K_n^{abs} equals one and is constant because this is the Rayleigh limit. When $a \gg \lambda$, the normalized coefficient $K_n^{abs} \propto a^{-1}$ because $Q^{abs} \approx 1$; big lumps are not efficacious in blocking light. For sizes in between, one has to do proper calculations. They reveal an enhancement in the mass absorption coefficient K^{abs} which can be very significant (> 10) and which would strongly boost millimeter dust emission because the latter is proportional to K^{abs} (see section 6.1.1).

One can create plots like those in figure 2.6 for other wavelengths and optical constants. The *qualitative* features stay the same, but some details are quite interesting. For example, if all particles in the diffuse interstellar medium had a radius of 1μm (without changing the total dust mass), the extinction optical depth at 2.2μm (K band) would be almost ten times larger than it really is.

2.7.2.2   Ice mantles

Next we illustrate the influence of ice in the grain material, again for a wavelength $\lambda=1$mm. We take a silicate sphere of arbitrary radius $a \ll \lambda$ and deposit on it an ice mantle of the same mass; such a mass ratio of ice to refractory core is suggested by the cosmic abundances in the case of strong freeze-out. The total grain has then a three and a half times bigger volume and twice the mass of the bare silicate core. The relevant parameter for the absorption coefficient $C^{\rm abs}$ is $k_{\rm ice}$, the imaginary part of the optical constant of ice. It depends on how much the ice is polluted by impurities. Estimates for $k_{\rm ice}$ at this wavelength are around 0.01 (figure 5.14), but uncertain. Defining the normalized absorption coefficient of a single grain by

$$C_{\rm n}^{\rm abs} = \frac{C^{\rm abs}({\rm core} + {\rm ice})}{C^{\rm abs}({\rm core})} \, ,$$

we learn from figure 2.7 that an ice mantle enhances $C^{\rm abs}$ by a factor $C_{\rm n}^{\rm abs} \sim 3$ if $k_{\rm ice} = 0.01$. Note that an ice mantle increases $C^{\rm abs}$ even when $k_{\rm ice} = 0$ because the mantle grain is bigger than the refractory core and collects more light. We also show in figure 2.7 the value of $C_{\rm n}^{\rm abs}$ when the ice is not in a mantle, but mixed throughout the grain.

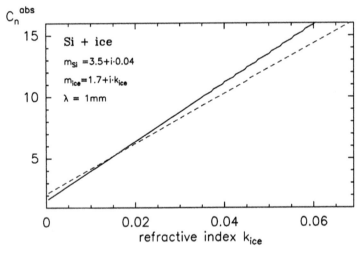

FIGURE 2.7   The influence of $k_{\rm ice}$, the imaginary part of the optical constant of ice, on the absorption coefficient $C^{\rm abs}$ of a grain that consists of silicate and ice with a mass ratio 1:1. The grain radius a is much smaller than the wavelength; here $\lambda = 1$ mm. The ordinate $C_{\rm n}^{\rm abs}$ gives the increase in the cross section with respect to the bare silicate core. The solid curve refers to a *coated sphere* with a silicate core and an ice mantle, the dashed line to a *homogeneous sphere* where ice and silicate are mixed and the optical constant $m_{\rm av}$ of the mixture is computed from the Bruggeman theory.

2.7.2.3 Fluffiness

A porous grain also has a greater absorption cross section C^{abs} than a compact one of the same mass. We consider in figure 2.8 silicates with vacuum inclusions; the normalized cross section C_n^{abs} is defined by

$$C_n^{abs} = \frac{C^{abs}(\text{fluffy grain})}{C^{abs}(\text{compact grain of same mass})} .$$

When the volume fraction of vacuum, f^{vac}, equals zero, the grain is compact. Fluffy grains are obviously better absorbers because they are bigger. A porosity parameter f^{vac} between 0.4 and 0.8, which may or may not be a reasonable estimate for interstellar conditions, suggests an increase in the absorption coefficient by a factor of two.

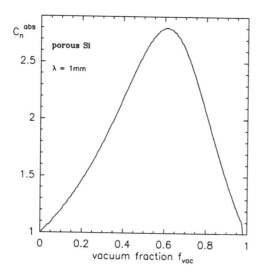

FIGURE 2.8 The influence of porosity on the grain cross section. The normalized cross section C_n^{abs} is defined in the text. The mass of the grain is kept constant and does not vary with the vacuum fraction f_{vac}, while the optical constant of the grain, which is calculated here after Bruggeman, changes with fluffiness. The compact particle has $m_{Si} = 3.5 + i\,0.04$.

3

Very small and very big particles

In sections 2.3 and 2.4, we presented general solutions of the field equations in the case when a plane wave interacts with a particle of simple geometry; we there stressed the mathematical aspects. Usually, when we want to extract numbers from the theory for a specific astronomical application, we have to deliver ourselves to the mercy of a computer. No matter how efficient such a machine, it is wise to retain some mental independence and bring to one's mind the physical aspects: computing must not be confused with understanding. In a few simple configurations, analytical solutions are possible and we turn to them in this chapter.

3.1 Tiny spheres

We derive the efficiencies for small spheres of dielectric material. This is the basic section for understanding how and why interstellar dust absorbs and scatters light.

3.1.1 Approximating the efficiencies

3.1.1.1 When is a particle in the Rayleigh limit?

When a sphere has a radius, a, small compared to the wavelength λ, i.e., when the size parameter

$$x = \frac{2\pi a}{\lambda} \ll 1 \ , \tag{3.1}$$

the calculation of cross sections becomes easy. The particle itself is not required to be small, only the ratio a/λ. In fact, a may even be big. So the heading of this section is suggestive, but not precise. With regard to thermal emission of interstellar grains, which occurs at wavelengths where the condition $\lambda \gg a$ is usually very well fulfilled, one may use for the computation of cross sections the approximations given below. If we stipulate additionally that the product of size parameter times optical constant, $m = n + ik$, be small,

$$|m| x \ll 1 \ , \tag{3.2}$$

we ensure two things:

- Because $kx \ll 1$, the field is only weakly attenuated in the particle.

- Because $nx \ll 1$, the wave traverses the particle with the phase velocity $v_{ph} = c/n$ in a time $\tau \simeq nx\omega^{-1}$ which is much shorter than the inverse circular frequency ω^{-1}.

Grains for which conditions (3.1) and (3.2) hold are said to be in the Rayleigh limit. The concept can be applied to any particle, not just spheres, if one understands by the size parameter the ratio of typical dimension over wavelength.

3.1.1.2 Efficiencies of small spheres from Mie theory

One finds simple expressions for the efficiencies of small spheres if one uses only the first coefficients a_1, b_1 in the expressions (2.37) and (2.38) for Q^{ext} and Q^{sca} and develops them into powers of x neglecting terms higher than x^6 [Pen62],

$$a_1 = -i\frac{2x^3}{3}\frac{m^2 - 1}{m^2 + 2} - i\frac{2x^5}{5}\frac{(m^2 - 1)(m^2 - 2)}{(m^2 + 2)^2} + \frac{4x^6}{9}\left(\frac{m^2 - 1}{m^2 + 2}\right)^2 \quad (3.3)$$

$$b_1 = -i\frac{x^5}{45}(m^2 - 1) . \quad (3.4)$$

- Usually one considers nonmagnetic materials ($\mu = 1$). Then the term with x^3 in the coefficient a_1 yields electric dipole absorption, the one with x^6 electric dipole scattering,

$$Q^{\text{ext}} \simeq \frac{6}{x^2} \operatorname{Re}\{a_1\} \simeq 4x \operatorname{Im}\left\{\frac{m^2 - 1}{m^2 + 2}\right\} = 4x \operatorname{Im}\left\{\frac{\varepsilon - 1}{\varepsilon + 2}\right\}$$

$$= \frac{8\pi a}{\lambda} \cdot \frac{6nk}{(n^2 - k^2 + 2)^2 + 4n^2 k^2} = \frac{8\pi a}{\lambda} \cdot \frac{3\varepsilon_2}{|\varepsilon + 2|^2} \quad (3.5)$$

$$Q^{\text{sca}} \simeq \frac{6}{x^2}|a_1|^2 \simeq \frac{8}{3}x^4 \left|\frac{m^2 - 1}{m^2 + 2}\right|^2 . \quad (3.6)$$

If x is small, both Q^{ext} and Q^{sca} approach zero. Because

$$Q^{\text{abs}} \propto x \qquad \text{and} \qquad Q^{\text{sca}} \propto x^4 ,$$

scattering is negligible at long wavelengths and extinction reduces to absorption,

$$Q^{\text{abs}} \simeq Q^{\text{ext}} .$$

It is generally not correct to assume $Q^{\text{sca}}_\lambda \propto \lambda^{-4}$ or $Q^{\text{abs}}_\lambda \propto \lambda^{-1}$ because $m(\lambda)$ is not constant.

- Because the optical constant m is after (1.46) symmetric in ε and μ,

$$m^2 = (n + ik)^2 = \varepsilon\mu, \qquad (k > 0),$$

equation (3.5), which contains only the coefficient a_1, is for purely magnetic material ($\mu \neq 1, \varepsilon = 1$) replaced by

$$Q_\lambda^{\text{abs}} = \frac{8\pi a}{\lambda} \cdot \frac{3\mu_2}{|\mu + 2|^2}. \qquad (3.7)$$

The dissipation process refers to magnetic dipole oscillations and is relevant only at frequencies $\omega < 10^{12}\,\text{s}^{-1}$.

- The coefficient b_1 presents magnetic dipole absorption of a nonmagnetic conductor (metal). Although an electric field does not penetrate into a metallic body, the magnetic field is also present in the wave and it does penetrate. Its time variability inside the sphere induces an electric field $\mathbf{E}$ and the presence of $\mathbf{E}$ in the conductor implies a current and therefore ohmic losses, i.e., absorption. b_1 translates into an absorption efficiency

$$Q^{\text{abs}} = \frac{8\pi\sigma}{15c^3}\,\omega^2 a^3. \qquad (3.8)$$

3.1.1.3 A dielectric sphere in a constant electric field

Cutting off the series expansion in Mie theory after the first term is a mathematical approach to obtain Q^{sca} and Q^{ext} for small particles. A physical one is the following. When $x \ll 1$ and $|mx| \ll 1$, the electric field in the grain changes in a quasi-stationary fashion. Therefore, to calculate the field in such a configuration, one is reduced to a simple exercise in electrostatics (figure 3.1). The basic equations valid everywhere are

$$\text{rot}\,\mathbf{E} = 0, \qquad \text{div}\,\mathbf{D} = 0.$$

They impose the boundary conditions that the tangential component of the electric field $\mathbf{E}$ and the normal component of the displacement $\mathbf{D}$ are continuous on the grain surface. For a homogeneous medium ($\varepsilon = \text{const}$), this leads to the Laplace equation

$$\Delta\varphi = 0,$$

where φ is the potential related to the field through $\mathbf{E} = -\nabla\varphi$.

When we place a sphere of radius a, volume V and dielectric constant ε into a constant field $\mathbf{E}$, the field becomes deformed. We label the field inside the sphere by $\mathbf{E}^{\text{i}}$ and outside it (external) by $\mathbf{E}^{\text{e}}$. For the potential φ of the deformed field we make the ansatz

$$\varphi = \begin{cases} -\mathbf{E}\cdot\mathbf{r} + c_1\,\dfrac{\mathbf{E}\cdot\mathbf{r}}{r^3} & : \quad \text{outside sphere} \\[2mm] -c_2\,\mathbf{E}\cdot\mathbf{r} & : \quad \text{inside sphere}, \end{cases} \qquad (3.9)$$

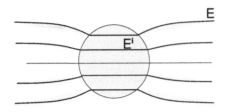

FIGURE 3.1 A dielectric sphere in a constant electric field **E**.

where **r** is the position coordinate of length $r = |\mathbf{r}|$, and c_1, c_2 are constants. Equation (3.9) reflects the expected behavior of the field. Inside the sphere, $\mathbf{E}^i$ is constant and parallel to **E**; outside, $\mathbf{E}^e$ is the sum of a dipole, which is induced by **E** and goes to zero at large distances, plus the constant field **E**. One finds for the internal field

$$\mathbf{E}^i = \frac{3}{\varepsilon + 2} \mathbf{E} \quad \text{for} \quad r \leq a . \tag{3.10}$$

Note that the field is smaller inside the body than outside, $E^i < E$. Because the polarization $\mathbf{P} = (\varepsilon - 1)\mathbf{E}^i/4\pi$ (see 1.6), the induced dipole moment of the grain is

$$\mathbf{p} = \mathbf{P} V = a^3 \frac{\varepsilon - 1}{\varepsilon + 2} \mathbf{E} = \alpha_e V \mathbf{E} , \tag{3.11}$$

and the electric polarizability α_e for a sphere becomes (see 1.8)

$$\alpha_e = \frac{3}{4\pi} \frac{\varepsilon - 1}{\varepsilon + 2} . \tag{3.12}$$

For other grain geometries there would be other dependences of α_e on ε. Equation (3.10) gives the internal field. If **e** denotes the unit vector in the direction **r**, the outer field $\mathbf{E}^e$ is the sum of **E** plus a dipole field,

$$\mathbf{E}_{\text{dip}} = \frac{3\,\mathbf{e}\,(\mathbf{e} \cdot \mathbf{p}) - \mathbf{p}}{r^3} , \tag{3.13}$$

so

$$\mathbf{E}^e = \mathbf{E} - \alpha_e V \cdot \text{grad} \left(\frac{\mathbf{E} \cdot \mathbf{r}}{r^3} \right) = \mathbf{E} + \mathbf{E}_{\text{dip}} . \tag{3.14}$$

3.1.1.4 Absorption in the electrostatic approximation

We now immediately get the absorption coefficient of the small dielectric grain from (1.55). Setting the absorbed power equal to Poynting vector S times C^{abs} gives

$$C^{\text{abs}} = \frac{4\pi}{c} \omega V \, \text{Im} \{\alpha_e\} . \tag{3.15}$$

But can we use the electric polarizability α_e of (3.12) which was obtained in the electrostatic approximation in the case when $\omega \neq 0$? Yes, we can because the variations of the field are assumed to be slow, so the electron configuration is always relaxed. This means the electrons have always sufficient time to adjust to the momentary field, just as in the static case. Of course, the polarization of the medium is not that of a static field. Instead, the dielectric permeability must be taken at the actual frequency ω of the outer field.

Another physical way to derive C^{abs} is to imagine the grain as an ensemble of NV harmonic oscillators excited in phase, V being the volume and N the oscillator density. The power W absorbed by the oscillators can be expressed in two ways, either by the dissipation losses of the harmonic oscillators given in (1.69), or via the Poynting vector S times absorption cross section C^{abs}, so

$$W = VN \frac{\gamma E_0^2 e^2}{2m_e} \cdot \frac{\omega^2}{(\omega_0^2 - \omega^2)^2 + \omega^2\gamma^2} = S C^{\mathrm{abs}} .$$

When we substitute the dielectric permeability of the harmonic oscillator after (1.77), we can transform the above equation into

$$\frac{Vc}{4\lambda} \varepsilon_2 E_0^2 = S C^{\mathrm{abs}} . \tag{3.16}$$

For the electric field E_0 which drives the oscillators, we must insert the field $\mathbf{E}^i$ inside the grain according to (3.10), and not the outer field $\mathbf{E}$. When we do this, we get, in agreement with (3.5), the cross section for electric dipole absorption

$$C^{\mathrm{abs}} = \frac{6\pi V}{\lambda} \cdot \mathrm{Im}\left\{\frac{\varepsilon - 1}{\varepsilon + 2}\right\} = \frac{V\omega}{c} \frac{9\varepsilon_2}{|\varepsilon + 2|^2} . \tag{3.17}$$

3.1.1.5 Scattering in the electrostatic approximation

As the field $\mathbf{E}$ oscillates proportionally to $E_0 e^{-i\omega t}$, so does synchronously the dipole moment $\mathbf{p}$. The oscillating dipole, which is now the grain as a whole, emits radiation. Its average power integrated over all directions is $W = |\ddot{\mathbf{p}}|^2/3c^3$ and follows from (1.92). On the other hand, W must also equal the total power scattered from the incident wave by the particle, therefore

$$W = \frac{1}{3c^3} |\ddot{\mathbf{p}}|^2 = S C^{\mathrm{sca}} . \tag{3.18}$$

For the dipole moment, we have to insert $\mathbf{p} = \alpha_e V \mathbf{E}$ and for the time-averaged Poynting vector $S = (c/8\pi) E_0^2$ after (1.44), therefore

$$C^{\mathrm{sca}} = \frac{8\pi}{3} \left(\frac{\omega}{c}\right)^4 V^2 |\alpha_e|^2 . \tag{3.19}$$

We thus find

$$C^{\mathrm{sca}} = \pi a^2 \cdot Q^{\mathrm{sca}} = \frac{24\pi^3 V^2}{\lambda^4} \cdot \left|\frac{\varepsilon - 1}{\varepsilon + 2}\right|^2 , \tag{3.20}$$

which agrees with equation (3.6) for the scattering efficiency Q^{sca}.

3.1.2 Polarization and angle-dependent scattering

For small spheres, the scattering matrix given in (2.54) reduces further (we drop the factor in front of the matrix):

$$
\begin{pmatrix} I_s \\ Q_s \\ U_s \\ V_s \end{pmatrix} = \begin{pmatrix} \frac{1}{2}(1+\cos^2\theta) & -\frac{1}{2}\sin^2\theta & 0 & 0 \\ -\frac{1}{2}\sin^2\theta & \frac{1}{2}(1+\cos^2\theta) & 0 & 0 \\ 0 & 0 & \cos\theta & 0 \\ 0 & 0 & 0 & \cos\theta \end{pmatrix} \begin{pmatrix} I_i \\ Q_i \\ U_i \\ V_i \end{pmatrix}
\tag{3.21}
$$

$\theta = 0$ gives the forward direction. There are several noteworthy points:

- The scattering pattern depends no longer on wavelength, as the matrix elements contain only the angle θ.

- Scattering is symmetrical in θ about $\pi/2$ and has two peaks, one in the forward, the other in the backward ($\theta = \pi$) direction.

- The matrix element $S_{12} < 0$ (see 2.54), so polarization is perpendicular to the scattering plane.

- If the incident radiation has unit intensity and is unpolarized ($I_i = 1$, $Q_i = U_i = V_i = 0$), we have

$$
I_s = \frac{1}{2}(1+\cos^2\theta), \quad Q_s = -\frac{1}{2}\sin^2\theta, \quad U_s = V_s = 0 .
$$

The degree of linear polarization becomes

$$
p = \frac{\sin^2\theta}{1+\cos^2\theta} ,
\tag{3.22}
$$

so the light is completely polarized at a scattering angle of 90°.

- The intensity of the scattered light, and thus the phase function $f(\theta, \phi)$, is proportional to $1 + \cos^2\theta$. As a result, the integral in (2.7) vanishes and the asymmetry factor becomes zero, $g = 0$, although scattering by a small sphere is not isotropic.

3.1.3 Small-size effects beyond Mie theory

A real grain is not a homogeneous continuum, but a crystal built up of atoms, rather regularly spaced and separated by a distance r_0, the lattice constant. Mie theory, which is based on classical electrodynamics of a continuous medium, fails when the structure of matter or quantum effects become important. A more general theory is then needed to describe the optical behavior of particles. Here we only remark on the influence of the surface in case of small grains.

Whereas atoms inside the particle are surrounded from all sides, the situation is different for those on the surface which have bonds only towards the particle's interior. This has consequences, for example, for their ability to bind to gas atoms or for the specific heat of the grain. If the particle has a diameter a, the ratio of the number of surface atoms N_{surf} to all atoms N in the grain is roughly

$$\frac{N_{surf}}{N} \simeq 6 \frac{r_0}{a} .$$

As the lattice constant is of order 2Å, a substantial fraction of atoms is on the surface only when the particle is small. One way in which the surface affects the optical grain properties can be understood when we interpret the damping constant γ in the motion of an electron (see 1.59) as a collision frequency with atoms of the crystal. If the particle is metallic and its size smaller than the mean free path of the electrons, an additional term must be added to the damping constant that comes from collisions (reflections) at the surface. Free electrons in a metal move with the Fermi velocity

$$v_F = \frac{\hbar}{m_e} \sqrt[3]{3\pi^2 n_e} ,$$

which is the threshold speed of fully degenerate (non-relativistic) electrons and follows from the Fermi energy of (A.100). The collision rate γ therefore increases roughly by v_F/a. For example, bulk graphite has an electron density $n_e \simeq 10^{20} \, \mathrm{cm}^{-3}$ and $\gamma = 5 \times 10^{12} \, \mathrm{s}^{-1}$. One estimates that noticeable changes in γ, and thus in the dielectric permeability, occur for sizes $a \lesssim 50\mathrm{Å}$.

3.2 Tiny ellipsoids

The treatment of a sphere in a constant electric field may be extended to ellipsoids. In analogy to the previous discussion, we can determine their scattering and absorption cross section once we have worked out the dipole moment that they acquire and the strength of their internal field. An ellipsoid has three principal axes a, b, c. We use the convention

$$a \geq b \geq c .$$

- When two principal axes in the ellipsoid are of equal length, it is called a spheroid.

- If then the two equally long axes are larger than the third one, $a = b > c$, the body has the shape of a *pancake*.

- Otherwise, if $a > b = c$, it resembles a *cigar*.

More educated terms are *oblate* and *prolate* spheroids. Pancakes and cigars, besides being nourishing or fragrant, are needed to explain why stellar light is polarized by dust clouds. Optically isotropic grains would not do.

3.2.1 Cross section and shape factor of pancakes and cigars

When the electric field is parallel to axis a of an ellipsoid, the scattering and absorption cross section, C^{sca} and C^{abs}, and the polarizability, α, are

$$C^{\text{abs}} = \frac{2\pi V}{\lambda} \cdot \text{Im}\left\{\frac{\varepsilon - 1}{1 + L_a(\varepsilon - 1)}\right\} \tag{3.23}$$

$$C^{\text{sca}} = \frac{8\pi^3 V^2}{3\lambda^4} \cdot \left|\frac{\varepsilon - 1}{1 + L_a(\varepsilon - 1)}\right|^2 \tag{3.24}$$

$$\alpha = \frac{\varepsilon - 1}{4\pi\left[1 + L_a(\varepsilon - 1)\right]}. \tag{3.25}$$

When the electric field is parallel to axis b or c, one has to replace L_a by L_b or L_c. In the above equations, V is the grain volume and L_a the shape factor. The latter is plotted in figure 3.2 as a function of the eccentricity e defined by

$$e^2 = 1 - \frac{c^2}{a^2}. \tag{3.26}$$

The sum over all shape factors equals one,

$$L_a + L_b + L_c = 1,$$

so only two of the three L values are independent. Obviously, when $a = b = c$, the eccentricity $e = 0$, all L's are equal to $\frac{1}{3}$, and we recover the formulae for spheres. For very long cigars (needles) and very flat pancakes $e = 1$ and $L_a = 0$, so $L_b = L_c = \frac{1}{2}$.

Only in the Rayleigh limit does the cross section C depend solely on the direction of the electric field and not on the direction of wave propagation. Consider, for example, the cigar in figure 3.3 and let the electric vector **E** swing parallel to axis c. The cross section is then the same for light that falls in parallel to axis a or parallel to axis b (which is perpendicular to a and c). In the first case, the projected surface is a small circle, in the second it is a broad ellipse and much bigger.

When the small ellipsoid is very transparent ($|\varepsilon| \simeq 1$), the shape factors loose their importance, the electric field inside and outside are more or less equal, $\mathbf{E}^i = \mathbf{E}$, and the polarizability $\alpha = (\varepsilon - 1)/4\pi = \chi$. The absorption and

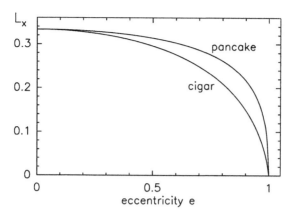

FIGURE 3.2 The shape factor for oblate (pancakes) and prolate (cigars) spheroids.

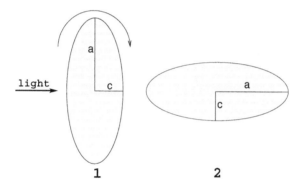

FIGURE 3.3 A cigar rotating about axis b, which is perpendicular to the page of the book. Light is traveling in the indicated direction. After a quarter of the rotation cycle, the cigar has changed from position 1 to position 2.

scattering coefficient are then independent of the axial ratios and of orientation,

$$C^{\text{abs}}(\lambda) = \frac{2\pi V}{\lambda}\, \varepsilon_2(\lambda) \qquad (3.27)$$

$$C^{\text{sca}}(\lambda) = \frac{8\pi^3 V^2}{3\lambda^4}\, |\varepsilon - 1|^2 \;. \qquad (3.28)$$

3.2.2 Randomly oriented ellipsoids

When an ellipsoid of fixed orientation in space is illuminated by an electromagnetic wave, the grand principle of superposition allows to split the electric field vector **E** of the wave into components along the orthogonal ellipsoidal axes a, b, c:

$$\mathbf{E} = (E \cos \alpha, \, E \cos \beta, \, E \cos \gamma) \, .$$

Here $E = |\mathbf{E}|$ and $\cos^2 \alpha + \cos^2 \beta + \cos^2 \gamma = 1$. Interestingly, for arbitrary grain orientation the internal field $\mathbf{E}^i$ is not parallel to the outer field $\mathbf{E}$, even if the grain material is isotropic; $\mathbf{E}^i$ and $\mathbf{E}$ are parallel only when $\mathbf{E}$ is directed along one of the principal axes.

If C_a, C_b, C_c denote the cross sections of the ellipsoid when the principal axes a, b, c are parallel to the electric vector $\mathbf{E}$ of the incoming wave, the total cross section of the grain can be written as

$$C = C_a \cos^2 \alpha \, + \, C_b \cos^2 \beta \, + \, C_c \cos^2 \gamma \, . \tag{3.29}$$

In random rotation, all directions are equally likely and the mean of $\cos^2 x$ over 4π is

$$\langle \cos^2 x \rangle = \frac{1}{4\pi} \int_0^{2\pi} dy \int_0^{\pi} dx \, \cos^2 x \sin x = \tfrac{1}{3} \, .$$

As the terms on the right of (3.29) are independent of each other, the average cross section for identical ellipsoids under random orientation is equal to the arithmetic mean,

$$\langle C \rangle = \tfrac{1}{3} C_a \, + \, \tfrac{1}{3} C_b \, + \, \tfrac{1}{3} C_c \, . \tag{3.30}$$

Equations (3.29) and (3.30) are true only in the electrostatic approximation.

Figure 3.4 displays what happens to the absorption cross section, or the optical depth along a line of sight, when small spheres are replaced by randomly oriented small spheroids. The spheroids are all of identical shape, but not necessarily size, and their total volume is the same as that of the spheres. The plot covers a broad wavelength range and is valid as long as the particles are in the Rayleigh limit. Because the spheroids are randomly oriented, their cross section, C_{spheroid}, is an average over directions. We see that ellipsoids have usually a higher cross section than spheres and the effect increases with eccentricity.

The difference between cigars and pancakes is mild. The influence of particle elongation on the optical depth is more pronounced in the far than in the near infrared. For example, grains with an axial ratio of two, which implies an eccentricity of 0.866, are at $\lambda \geq 100 \mu m$ by some 30% better emitters or absorbers than spheres.

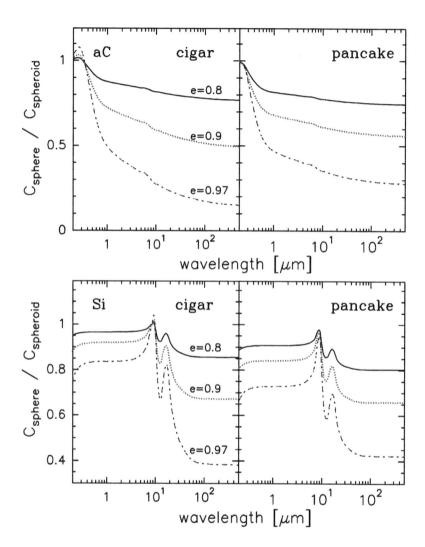

FIGURE 3.4 The cross section of spheres over that of randomly oriented spheroids of the same volume. Calculations are done in the electrostatic approximation implying grains much smaller than the wavelength. The eccentricity of the spheroids defined in (3.26) is indicated. The grains consist of amorphous carbon (aC, ***top***) or silicate (Si, ***bottom***). Note how the resonances at 10 and 18μm in the optical constants, m, of the silicate material (figure 5.10) are reflected in the particle cross section; m for aC is from figure 5.11.

3.3 The fields inside a dielectric particle

3.3.1 Internal field and depolarization field

We determined in equation (3.10) the field $\mathbf{E}^i$ inside a dielectric sphere that sits in a time-constant homogeneous outer field $\mathbf{E}$. In section 3.2 we generalized to ellipsoids. In both cases, the polarization $\mathbf{P}$ of the medium is constant and the field inside smaller than outside. Writing the *internal field* $\mathbf{E}^i$ in the form

$$\mathbf{E}^i \ = \ \mathbf{E} \ + \ \mathbf{E_1}$$

defines a new field $\mathbf{E_1}$. It arises from all atomic dipoles and is directed opposite to the polarization $\mathbf{P}$. Because $E^i < E$, one calls $\mathbf{E_1}$ the *depolarization field* (see figure 3.5). For example, from (3.10) we find for a sphere

$$\mathbf{E_1} \ = \ -\frac{4\pi}{3}\,\mathbf{P} \ . \tag{3.31}$$

It is important to make a distinction between the local field at exactly one point and macroscopic averages. $\mathbf{E_1}$ and $\mathbf{E}^i$ are such averages over many atoms, a hundred or so, but at least over one unit cell in a crystalline structure. On a microscopic level, the field has tremendous gradients. Atoms are not at random positions, but at privileged sites (lattice grid points are loci of minimum potential energy) and the local field $\mathbf{E}^{loc}$ acting on an atom is usually different from the average field $\mathbf{E}^i$, so generally

$$\mathbf{E}^{loc} \ \neq \ \mathbf{E}^i \ .$$

To find $\mathbf{E}^{loc}$ at a specific locus in the grain, say at $\mathbf{r_0}$, we have to add to $\mathbf{E}$ the fields from all dipoles. As the dipoles are at *discrete* lattice points, their distribution is not smooth. At least, it does not appear to be so close to $\mathbf{r_0}$. We therefore imagine a spherical cavity around $\mathbf{r_0}$ of such a size that beyond the cavity the dipole distribution may be regarded as smooth, whereas inside it, it is discontinuous, so

$$\mathbf{E}^{loc} \ = \ \mathbf{E} \ + \ \sum_{\text{outside}} \mathbf{E}_{\text{dip}} \ + \ \sum_{\text{cavity}} \mathbf{E}_{\text{dip}} \ . \tag{3.32}$$

The field arising from the smooth distribution outside the cavity can be expressed as a volume integral. The field from the dipoles within the cavity has to be explicitly written as a sum.

3.3.2 Depolarization field and surface charges

A body of constant polarization $\mathbf{P}$ has on its outside a surface charge σ of strength

$$\sigma \ = \ \mathbf{e} \cdot \mathbf{P} \ , \tag{3.33}$$

where $\mathbf{e}$ is the outward surface normal. This expression for σ follows when we recall that according to (1.20) a non-uniform polarization creates a charge $\rho_{pol} = -\mathrm{div}\,\mathbf{P}$. Inside the body, the divergence of the polarization vector is zero and $\rho_{pol} = 0$, but on its surface, $\mathbf{P}$ is discontinuous and a charge appears.

There is a theorem which says that for any body of *constant polarization* $\mathbf{P}$, the depolarization field $\mathbf{E}_1$ is identical to the field that arises in vacuum from the distribution of surface charges as given by (3.33). Let $\varphi(\mathbf{r})$ be the electrostatic potential from all dipoles in the body. We prove the theorem by writing $\varphi(\mathbf{r})$ as a volume integral,

$$\varphi(\mathbf{r}) = -\int \left(\mathbf{P} \cdot \nabla r^{-1}\right) dV .$$

This is correct because a single dipole $\mathbf{p}$ has a potential

$$\varphi_{dip} = -\mathbf{p} \cdot \nabla \left(\frac{1}{r}\right) .$$

Now we transform the volume integral into a surface integral employing the relation $\mathrm{div}\,(f\mathbf{P}) = f\,\mathrm{div}\,\mathbf{P} + \mathbf{P} \cdot \nabla f$. With $f(\mathbf{r}) = 1/r$ and $\mathrm{div}\,\mathbf{P} = 0$, we get

$$\varphi(\mathbf{r}) = -\oint \frac{1}{r}\mathbf{P} \cdot d\mathbf{S} = \oint \frac{\sigma}{r} dS .$$

The second integral sums up the potentials from all surface charges and their total field is thus equivalent to the field of all atomic dipoles.

3.3.3 The local field at an atom

The "outside-sum" in (3.32) may be evaluated by an integral. According to our theorem, it is equivalent to the field from the surface charges. However, there are now two boundaries. The outer, S_1, gives the depolarization field $\mathbf{E}_1$. The inner, S_2, has the same surface charge as a sphere in vacuum of constant polarization $\mathbf{P}$, only of inverted sign. Comparing with (3.31), we find that the surface charge on S_2 produces the field $(4\pi/3)\,\mathbf{P}$.

It remains to include the atoms in the spherical cavity S_2. Suppose we have a *cubic lattice*, the dipoles are at positions (x_j, y_j, z_j), their moments are all of strength p and aligned in z-direction, and the point $\mathbf{r}_0$ is at the origin. Then the z-component of the total field from all dipoles at $\mathbf{r}_0$ is (see 3.13)

$$\sum_{\text{cavity}} (\mathbf{E}_{dip})_z = p\sum_j \frac{3z_j^2 - r_j^2}{r_j^5} = p\sum_j \frac{2z_j^2 - x_j^2 - y_j^2}{r_j^5} = 0 ,$$

where $r_j = (x_j^2 + y_j^2 + z_j^2)^{1/2}$ is the distance of dipole i to the origin. The sum is zero because of the symmetry of the grid. Likewise the x- and y-components

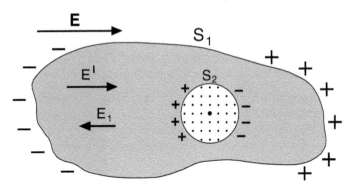

FIGURE 3.5 A field **E** produces on a dielectric grain a surface charge σ. According to a theorem of electrostatics, σ gives rise to a field $\mathbf{E_1}$ that combines with **E** to form the average internal field $\mathbf{E^i}$. The latter is not necessarily equal to the local field $\mathbf{E}^{loc}$ at a particular atom or unit cell (central larger dot in inner circle) in a crystal. To find $\mathbf{E}^{loc}$, one has to take into account the regularly, but discontinuously arranged dipoles (small dots) in the vicinity as well as those farther away which can be considered as being distributed smoothly (shaded area). S_1 and S_2 denote surfaces.

of the total field vanish. So we can neglect the influence of the nearest atoms altogether. We expect that we may also neglect it if there is no grid order at all, i.e., in an amorphous substance. For such a situation therefore

$$\mathbf{E}^{loc} = \mathbf{E}^i + \frac{4\pi}{3}\mathbf{P} . \qquad (3.34)$$

Given the outer field **E**, the local field $\mathbf{E}^{loc}$ at an atom depends on the shape of the particle. For example, for a spherical grain we derive from (3.34) with the help of (3.10) and (3.11)

$$\mathbf{E}^{loc} = \mathbf{E} .$$

For a body in the shape of a thin slab with parallel surfaces perpendicular to **E**, which is the configuarion of a parallel-plate condenser, one finds

$$\mathbf{E}^{loc} = \mathbf{E} - \frac{8\pi}{3}\mathbf{P} .$$

3.3.4   The relation of Clausius-Mossotti

The local field $\mathbf{E}^{loc}$ produces in each atom of volume V a dipole moment

$$p = \alpha E^{loc} .$$

Hence α is called atomic polarizability. The formula is analogous to (1.8) which we applied to a grain as a whole. If there are N atoms per unit volume, the polarization of the matter is in view of (3.34)

$$P = N\alpha E^{\text{loc}} = N\alpha \left(E^{\text{i}} + \frac{4\pi}{3} P \right) .$$

Because $P = \chi E^{\text{i}}$ (see 1.6), one can relate the dielectric susceptibility χ of the medium to the polarizability α of the atoms. This is done in the Clausius-Mossotti formula

$$\chi = \frac{N\alpha}{1 - \frac{4\pi}{3} N\alpha} . \qquad (3.35)$$

So the proportionality factor χ between the polarization P and the field E^{i} is not just equal to the total number of atoms, N, times α, but it is greater. The denominator $0 < 1 - 4\pi N\alpha/3 < 1$ implies an amplification due to the mutual interaction of the atoms.

3.4 Very large particles

A particle is by definition very large when its size is much bigger than the wavelength. As in the case of tiny grains, the definition is relative and the same particle may be both, small or large, depending on the wavelength. To quantify the diffraction phenomena that occur around very big grains, we have to study basic optical principles.

3.4.1 Babinet's theorem

For any very large particle the extinction efficiency,

$$Q^{\text{ext}} = Q^{\text{abs}} + Q^{\text{sca}} ,$$

approaches 2, independently of the chemical composition or shape of the particle. This important result is called Babinet's theorem or extinction paradox. For spheres,

$$Q^{\text{ext}} \rightarrow 2 \qquad \text{when} \qquad x = \frac{2\pi a}{\lambda} \rightarrow \infty . \qquad (3.36)$$

We illustrate Babinet's theorem with an experiment carried out in three steps as sketched in figure 3.6:

(1) When a parallel wave front falls on an orifice, which is much bigger than the wavelength, it produces on a far-away screen a bright spot with a blurred rim. Outside the bright spot and the rim, the screen is dark. We restrict the discussion to this dark area.

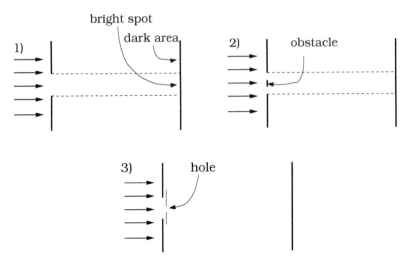

FIGURE 3.6 A diffraction experiment to explain Babinet's theorem.

(2) If one places a small, but still much bigger than λ, obstacle into the orifice, a diffraction pattern appears and light is scattered beyond the blurred rim into the former dark area.

(3) If we cover the orifice with black paper leaving just a hole of the same size and shape as the obstacle, the diffraction patterns of the hole and the obstacle are identical. According to Huygens' principle for wave propagation (this result will be shown below), the diffraction pattern of the obstacle arises because each point in the plane of the orifice, except for the obstacle itself, is the origin of a spherical wave; the diffraction pattern of the hole arises because each point in the plane of the hole is the origin of a spherical wave.

In the case of the completely uncovered orifice *(1)*, both diffraction patterns are present simultaneously. Because the region beyond the blurred rim is then dark, the patterns from the obstacle and the hole must cancel each other exactly, i.e., they must have the same intensity, but be phase-shifted by 180°. As the hole scatters all the light falling onto it, the obstacle must scatter exactly the same amount. Altogether the obstacle thus removes twice as much light as that which corresponds to its projected geometrical surface: half of it through scattering, the other half by absorption and reflection.

Scattering at the edge of a large obstacle is, however, predominantly forward. Therefore $Q^{\text{ext}} = 2$ can only be verified at far distances; it is always valid for interstellar grains. At short distances, we know from everyday experience that a brick removes only as much sunlight as falls onto its projected surface, and not twice as much.

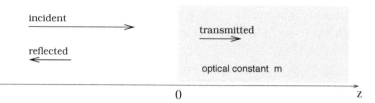

FIGURE 3.7 Light falls perpendicular on a plane surface, some is reflected and some transmitted. The medium to the right $(z > 0)$ has an optical constant $m = n + ik$.

3.4.2   Reflection and transmission at a plane surface

3.4.2.1   Normal incidence

A ray of light travels in positive z-direction and hits under normal incidence a large particle as shown in figure 3.7. We wish to evaluate which fraction of the incident flux is reflected. This quantity is denoted by r and called reflectance. If $k_1 = \sqrt{\varepsilon_1 \mu_1}\, \omega/c$ is the wavenumber after (1.38) in the medium on the left $(z < 0)$, and $k_2 = \sqrt{\varepsilon_2 \mu_2}\, \omega/c$ in the medium on the right $(z > 0)$, we obtain for the electric field

$$E = E_i\, e^{i(k_1 \cdot z - \omega t)} + E_r\, e^{i(-k_1 \cdot z - \omega t)} \qquad \text{for} \quad z < 0 \qquad (3.37)$$
$$E = E_t\, e^{i(k_2 \cdot z - \omega t)} \qquad \text{for} \quad z > 0 . \qquad (3.38)$$

On the left, we have the incident and the reflected wave (indices i and r), to the right only the transmitted wave (index t). At the surface of the particle $(z = 0)$, the tangential components of the electric and magnetic field are continuous. When we express the magnetic field via the rotation of the electric field (see 1.32), we find at $z = 0$

$$E_i + E_r = E_t$$
$$E_i - E_r = E_t\, \sqrt{\varepsilon_2 \mu_1 / \varepsilon_1 \mu_2} .$$

This immediately allows one to calculate the reflectance $r = |E_r/E_i|^2$. The most common case is the one in which there is vacuum on the left and the grain is nonmagnetic $(\mu_2 = 1)$. The optical constant of the particle then equals $m = \sqrt{\varepsilon_2}$ and the reflectance is given by

$$r = \left| \frac{1 - m}{1 + m} \right|^2 = \frac{(n - 1)^2 + k^2}{(n + 1)^2 + k^2} . \qquad (3.39)$$

Should n or k be large, the reflectivity is high and absorption low. Metal surfaces make good mirrors because they have a large optical constant m and $n \simeq k$ (see 1.107). On the other hand, if $k \simeq 0$, the reflectivity grows with n. A pure diamond sparkles at visual wavelengths because k is small and $n = 2.4$, so the reflectivity is high ($r = 0.17$). If the stone in a ring is a fake, of standard glass which has $n = 1.5$, it will catch less attention because it reflects only 4% of the light, not 17%.

We see from (3.38) how the light that enters into a large and absorbing particle is attenuated. The amplitude of its electric field is weakened proportionally to $\exp(-2\pi k z/\lambda)$, so the intensity I_t of the transmitted light diminishes like $\exp(-4\pi k/\lambda z)$. Per wavelength of penetration, the intensity decreases by a factor $\exp(-4\pi k)$. Unless k is very small, the transmitted light is removed very quickly and the penetration depth is only a few wavelengths.

One may wonder about the implications of formula (3.39) for a blackbody. By definition, it has zero reflectance and would therefore require the optical constant of vacuum ($m = 1$). But vacuum is translucent. So no real substance is a perfect absorber. A blackbody can be approximated by a particle with $n \to 1$ and $k \to 0$; the particle must also have a very large size d such that kd/λ is much greater than one, despite the smallness of k.

3.4.2.2 Oblique incidence

We generalize the reflectance r of (3.39) to the case of oblique incidence. If the incident beam is inclined to the normal of the surface element of the particle by some angle $\theta_i > 0$, the reflected beam forms an angle $\theta_r = \theta_i$ with the normal. The angle of the transmitted beam is given by *Snell's law*

$$\sin \theta_t = \frac{\sin \theta_i}{m}. \tag{3.40}$$

Whereas for a non-absorbing medium, m and θ_t are real, in the generalized form of (3.40) m and also θ_t are complex. One now has to specify in which direction the incident light is polarized which is not necessary under normal incidence. For incident *unpolarized light* the reflectance is

$$r = \frac{1}{2} \left| \frac{\cos \theta_t - m \cos \theta_i}{\cos \theta_t + m \cos \theta_i} \right|^2 + \frac{1}{2} \left| \frac{\cos \theta_i - m \cos \theta_t}{\cos \theta_i + m \cos \theta_t} \right|^2. \tag{3.41}$$

Using this formula, we can determine the limiting value of Q^{abs} and Q^{sca} for very large spheres. If the particle is translucent ($k = 0$), it only scatters the light; then $Q^{abs} = 0$ and $Q^{sca} = Q^{ext} = 2$ according to Babinet's theorem. If $k > 0$, any light that enters the grain will under the assumption of ray optics eventually peter out within it. One can define a reflection efficiency Q^{ref} for spheres by

$$\pi a^2 Q^{ref} = \int_0^a 2\pi r(x) \, dx \tag{3.42}$$

where $r(x)$ is the reflectance after (3.41) for an incident angle $\theta_i = \arcsin(x/a)$ and a is the grain radius. Q^{ref} lies between 0 and 1 and is related to Q^{abs} through $Q^{abs} = 1 - Q^{ref}$. So the absorption efficiency for large particles can never be greater than one, only smaller if some fraction of light is reflected off the surface.

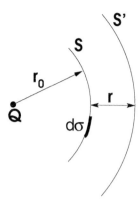

FIGURE 3.8 A spherical wave front traveling from a point source Q at velocity c.

3.4.3 Huygens' principle

In figure 3.8, a point source Q emits isotropically radiation. On any spherical surface S around Q, the oscillations of the electromagnetic field are in phase forming a wavefront. According to Huygens' principle, each surface element $d\sigma$ is the source of an elementary (or secondary) spherical wave. The strength of the elementary waves is proportional to the amplitude of the primary field and to the area $d\sigma$. It also depends on direction being greatest radially away from Q and zero towards the rear; a more detailed description is not given. The superposition (interference) of all elementary waves of equal radius originating from the surface S fixes the position of the wavefront in the future and thus describes the propagation of the primary wave. For instance, if the wavefront is at time t on the surface S of figure 3.8, a time $\Delta t = r/c$ later it will be again on a spherical surface, but of radius $r_0 + r$. The new surface S' is the envelope of all elementary waves of radius r. Outside S', the field of elementary waves is extinguished by interference.

Applied to plane waves, which are spherical waves of very large radius, Huygens' principle makes understandable why light propagates along straight lines. It also explains the laws of refraction and reflection in geometrical optics. To derive them from Huygens' principle, one just has to assume that all surface elements on the plane separating two media of different optical constants, n_1 and n_2, oscillate in phase and emit elementary waves which propagate with the phase velocity $v = c/n$ of the respective medium.

If there is an obstacle in the way of the primary wave, some region on the surface of its wavefront, corresponding in size and shape to the projected area of the obstacle, does not create secondary waves. Then the superposition of the remaining elementary waves leads to diffraction patterns. Interesting examples are the intensity distribution of light behind a straight wire or a slit,

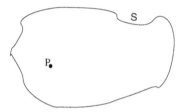

FIGURE 3.9 Huygen's principle allows one to derive the field u at any point P inside a region, if one knows u on its surface S.

or the fuzzy edges of shadows. Light does not always travel on straight lines, but can bend around corners.

3.4.3.1 Kirchhoff's strict formulation of Huygens' principle

Kirchhoff put Huygens' principle in a mathematically rigorous form. He was able to determine the electromagnetic field u at any point P inside a region when u is known on its surface S (figure 3.9). For its value at P, u_p, he found

$$4\pi\, u_\mathrm{p} \;=\; -\oint_S \left(u\, \frac{\partial}{\partial n}\, \frac{e^{ikr}}{r} - \frac{e^{ikr}}{r}\, \frac{\partial u}{\partial n} \right) d\sigma \;. \tag{3.43}$$

$\partial/\partial n$ denotes the normal derivative taken on the surface S and directed outwards from the region. The terms e^{ikr}/r appear because an outward-going spherical wave (emanating from a point on the surface S) has the form

$$u \;=\; \frac{e^{i(kr-\omega t)}}{r} \;. \tag{3.44}$$

It satisfies the general wave equation

$$\Delta u \,+\, k^2\, u \;=\; 0 \;,$$

like plane waves, $u = e^{i(\mathbf{k}\cdot\mathbf{r}-\omega t)}$, which we discussed in section 1.1.6.

3.4.3.2 Diffraction by a circular hole or a sphere

We apply equation (3.43) to the case where the surface S in figure 3.9 is opaque, so that everywhere on it $u = 0$, except for a circular hole of radius a. There light from a source Q can pass through as shown in figure 3.10). The intensity u_p at point P is then determined as a superposition of secondary waves from the hole. To first order, geometrical optics is valid and the hole deflects light only by small angles. With $r, r_0 \gg k^{-1}$, one finds from (3.43)

$$u_\mathrm{p} \;=\; -\frac{i}{\lambda}\, \frac{\cos\theta}{r r_0} \oint_S e^{ik(r+r_0)}\, d\sigma \;, \tag{3.45}$$

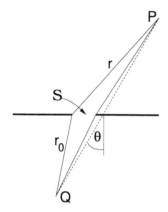

FIGURE 3.10 Light falls from
Q though an orifice S and is observed at P.

where θ is the angle between $\overline{QP}$ and the normal of the plane. The distances r and r_0 are almost constant and may stand before the integral, however, $e^{ik(r+r_0)}$ varies over the hole because it is many wavelengths across.

The symmetry of formula (3.45) with respect to r and r_0 implies that when a source at Q produces a field u_p at P, the same source brought to the point P produces the same field at Q. This is the reciprocity theorem of diffraction theory.

The evaluation of the integral in (3.45) gives

$$|u_\text{p}|^2 = \frac{4\pi^2 a^4}{\lambda^2 r^2 r_0^2} \cdot \left(\frac{J_1(ak\alpha)}{ak\alpha}\right)^2 . \tag{3.46}$$

The Bessel function $J_1(x)$ is tabulated in mathematical encyclopediae. For small x, $J_1(x) \to \frac{1}{2}x$; at $x = 1$, this approximation has an error of roughly 10%. The intensity $|u_\text{p}|^2$ drops to half from its maximum value at $\alpha = 0$ at a deflection angle

$$\alpha_\text{half} \simeq 1.617\, \frac{\lambda}{2\pi a} . \tag{3.47}$$

As a spherical obstacle produces according to Babinet's theorem the same diffraction pattern, we find, for example, that a grain with 1 mm diameter has in the visual band an α_half of about 1 arcmin.

It may seem puzzling that the intensity of the light which one observes at P and which is equal to $|u_\text{p}|^2$ rises with the fourth power of the hole radius a, although the flux passing through the hole increases only with a^2. But as the hole becomes bigger, the diffraction pattern gets smaller, and in the end, as much energy as falls onto the hole from the source, comes out on the other side; energy is conserved.

3.5 Grains of small refractive index

3.5.1 Rayleigh Gans particles

A grain that fulfills the conditions that

- its optical constant is close to unity,

$$|m - 1| \ll 1 \quad \text{or} \quad n \simeq 1 , \; k \simeq 0 , \quad (3.48)$$

- and that it is more or less transparent,

$$\frac{d}{\lambda} |m - 1| \ll 1 , \quad (3.49)$$

is called a *Rayleigh-Gans* particle. Otherwise it may be of *arbitrary shape and size*; it may therefore be big compared to the wavelength. Inside such a particle, the electromagnetic field is only weakly deformed and practically the same as in the incoming wave. If we imagine the total grain volume V to be divided into many tiny subvolumes v_i with $\sum v_i = V$, each of them absorbs and scatters independently of the others. For example, if the subvolumes are ellipsoids, their absorption and scattering cross sections are given by (3.27) and (3.28), which we derived in the Rayleigh limit. Consequently, the absorption cross section C^{abs} of the total grain is also given by formula (3.27). Therefore, the absorption efficiency of a Rayleigh-Gans sphere for any x is

$$Q^{\text{abs}} = \frac{8x}{3} \text{Im}\{m - 1\} = \frac{16\pi}{3} \frac{ak}{\lambda} . \quad (3.50)$$

To find the *scattering* cross section or the scattered intensity in a certain direction, one cannot simply sum up the contributions of the subvolumes because there is a phase difference between them which leads to interference. It is this effect which, for large bodies, tends to reduce $C^{\text{sca}}(\lambda)$ with respect to (3.28). The formalism of how waves originating from different points (subvolumes) are added up is basically the same as discussed in section 2.2 on the optical theorem. The analytical computation of the scattering cross section or the phase function is, however, complicated, even for simple grain geometries (see examples in [Hul57]). Of course, everything can be extracted from the full Mie theory, but not in analytical form.

Here are some results for spheres: If x is the size parameter and γ denotes Euler's constant ($\gamma \simeq 0.57721566$),

$$Q^{\text{sca}} = |m - 1|^2 \, \varphi(x)$$

with

$$\varphi(x) = \frac{5}{2} + 2x^2 - \frac{\sin 4x}{4x} - \frac{7(1 - \cos 4x)}{16x^2} + \frac{1 - 4x^2}{2x^2} \left[\gamma + \log 4x + \int\limits_{4x}^{\infty} \frac{\cos u}{u} \, du \right] .$$

If $x \ll 1$, we obtain the Rayleigh limit

$$Q^{\text{sca}} = \frac{32}{27} |m - 1|^2 \, x^4 \tag{3.51}$$

and if $x \gg 1$, we get

$$Q^{\text{sca}} = 2 \, |m - 1|^2 \, x^2 . \tag{3.52}$$

The latter is much smaller than for large, non-transparent spheres which always have $Q^{\text{sca}} \sim 1$. For the scattered intensity, $I(\theta)$, we have the proportionality

$$I(\theta) \; \propto \; G^2(2x \sin \tfrac{1}{2}\theta) \cdot (1 + \cos^2 \theta)$$

with

$$G(u) \; = \; \left(\frac{9\pi}{2u^3} \right)^{1/2} J_{3/2}(u) ,$$

where $J_{3/2}(u)$ is a half-integer Bessel function.

3.5.2 X-ray scattering

We apply now the results obtained for particles of small refractive index to X-rays (see optical constants in figure 5.10 to 5.14). At a wavelength of, say 10Å, interstellar dust particles are always big because $x = 2\pi a/\lambda \gg 1$. The value of the dielectric permeability in this range may be estimated from the high-frequency approximation (1.79), which gives

$$\varepsilon_1 = 1 - \frac{\omega_p^2}{\omega^2} \simeq 1, \qquad \varepsilon_2 = \frac{\omega_p^2 \gamma}{\omega^3} \ll 1 ,$$

where ω_p is the plasma frequency and γ a damping (not Euler's) constant. As the frequency ω is large, we expect ε_1 to be close to one and ε_2 to be small, so $|\varepsilon - 1| \ll 1$; likewise for the optical constant, $|m - 1| \ll 1$. The grains are therefore very transparent and even satisfy the condition (3.49),

$$\frac{2\pi a}{\lambda} |m - 1| \ll 1 .$$

For big spheres, scattering is very much in the forward direction defined by $\theta = 0$. Therefore, the angular dependence of the scattered intensity is given by

$$I(\theta) \; \propto \; G^2(u) \qquad \text{with} \qquad u = 2x \sin \tfrac{1}{2}\theta .$$

When one numerically evaluates the function $G^2(u)$, one finds that it has its maximum at $u = 0$, drops monotonously, like a bell-curve, reaches zero at about 4.5 and then stays very small. Consequently, there is forward scattering over an angle such that $\theta x \sim 2$. For grains of 1000Å radius, this corresponds to $\theta \sim 10$ arcmin. One can therefore observe towards strong X-ray sources located behind a dust cloud an X-ray halo of this size. The intensity towards it follows from the fact that the total scattered X-ray flux is smeared out over the halo.

3.5.2.1   Compton scattering

When a photon of energy $h\nu$ hits an electron at rest, the photon energy decreases to $h\nu'$, depending on the scattering angle θ,

$$\frac{\nu - \nu'}{\nu'} = \nu \, \frac{2h}{m_e c^2} \, \sin^2 \frac{\theta}{2} \, . \tag{3.53}$$

This follows readily from the conservation of the total (photon plus electron) relativistic momentum and energy. The energy transferred to the electron, $h(\nu - \nu')$, is zero in the forward direction and increases with the angle θ. The change in wavelength, $\Delta\lambda = \lambda' - \lambda$, is independent of the initial photon energy. It has its maximum for backward scattering ($\theta = 2\pi$) where it is equal to twice the Compton wavelength of the electron $h/m_e c$.

When $h\nu$ is large compared to its binding energy, an atomic electron can be considered to be essentially free and at rest. At photon energies $h\nu/m_e c^2 \ll 1$, the scattering cross section is equal to the Thomson σ_T of (1.97); this holds in all of our applications, even for X-rays. However, at high energies when $\epsilon = h\nu/m_e c^2 \gg 1$, the scattering cross section approaches after *Klein* and *Nishina* the value $\frac{3}{8}\sigma_T \, \epsilon^{-1}(\frac{1}{2} + \ln 2\epsilon)$.

3.5.3   X-ray absorption

We discuss the physical processes that lead to absorption of X-rays by dust. An energetic photon interacting with an interstellar grain is either Compton-scattered by an electron or it ejects an electron from a deep atomic (K or L) shell.

3.5.3.1   X-ray ionization

To estimate the cross section, K_ν, for photo-ejection of an electron from an inner atomic shell, we assume a crude picture (see [Voi91] for accurate formulae). The electron is a point mass vibrating as an harmonic oscillator at frequency ν_0 with an elongation x_0; within a factor of unity, $h\nu_0$ is also the ionization energy. Classically, the electron emits the power $W = e^2 x_0^2 \omega_0^4/3c^3$ (see 1.92).

The photon of energy $h\nu$ has a cross section $\sigma_{phot} = \lambda^2$, where $\lambda = c/\nu$ is the wavelength of the incident radiation; for a 1 keV photon, $\lambda \simeq 10\,\text{Å}$. It couples to the electron with a probability $P \ll 1$ given by the ratio of the crossing time $\tau_{cross} = \lambda/c$ over the time τ_{react} which the electron needs to react with the photon, therefore we write

$$K_\nu = \sigma_{phot} \frac{\tau_{cross}}{\tau_{react}}.$$

τ_{react} should be about equal to the time the oscillator needs to emit a photon which, in view of the reversibility of elementary processes, is also the time it takes to absorb one, so τ_{react} follows from

$$W \tau_{react} = e^2 x_0^2 \omega_0^4 / 3c^3 = h\nu_0.$$

Substituting the fine structure constant $\alpha = e^2/\hbar c$, we find

$$K_\nu \sim \frac{8\pi^3}{3} \alpha x_0^2 \left(\frac{\nu_0}{\nu}\right)^3. \tag{3.54}$$

Because $x_0 = a_0/Z$ and $h\nu_0 \propto Z^2$, where a_0 is the Bohr radius and Z the atomic number, K_ν increases rapidly with the nuclear charge, proportionally to Z^4, and falls off with frequency like ν^{-3}. For hydrogen atoms, the above simplistic formula yields $K_\nu \sim 2 \times 10^{-17}\,\text{cm}^2$ at the ionization threshold $h\nu = 13.6\,\text{eV}$; the correct value is $6.1 \times 10^{-18}\,\text{cm}^2$. For lead ($Z = 82$), the binding energy $h\nu_0$ is already around 100 keV and the X-ray cross section is very high. That is the reason why lead is used for protection in medical X-ray diagnosis. When the photon energy $h\nu$ increases and electrons from a deeper atomic shell can be ionized, ν_0 in (3.54) makes an upward jump. This explains the sawtooth profile of the optical constant k at X-rays (figure 5.10). A description and references of how to more correctly calculate the photo-ionization cross section is given in [Mih78].

After an X-ray photon has created a gap in the K shell, the gap will be filled by a downward transition of an electron from an upper (say, L) shell. The energy may escape either as a photon or non-radiatively, through ejection of an *Auger* electron from another shell, for example, L or M. Initially, there was only one electronic vacancy (in the K shell); afterwards there are two. The non-radiative process dominates in light atoms ($Z < 30$), the radiative in heavy ones ($Z > 30$). An Auger electron is characterized by three subshells, for example, K L1 M2. The initial vacancy is in K; the downward transition to K occurs from L1, the Auger electron is ejected from M2. The numbers after the shell letter specify the orbital angular momentum of the electron in the subshell (1 for s, 2 for p, 3 for d).

3.5.3.2   Energy deposition by X-ray photons

Whereas usually an absorbed photon deposits its full energy $h\nu$ in the grain, this need not be so at X-rays where the ejected electron is very energetic and may leave the grain. Momentum conservation requires that it travels initially in the direction of the photon. It is then scattered and in these scattering events loses energy. The stopping column density d is approximated between 20 eV and 1 MeV by the polynomial [Dwe96]

$$\log(\ell) \;=\; \sum_{i=0}^{4} a_i \, \log^i(E) \tag{3.55}$$

with

$$(a_0, a_1, a_2, a_3, a_4) = (-8.1070,\ 1.0596,\ -0.27838,\ 0.11741,\ -0.010731)\,.$$

The electron energy E in formula (3.55) is measured in eV and ℓ in g cm^{-2}. It follows that an electron of 0.1 keV will deposit its total energy in a dust grain ($\rho \simeq 2.5$ g cm^{-3}) if its radius is greater than 0.002 μm. For an energy of 1 keV and 10 keV, the minimum radius is 0.03 μm and 1.1 μm, repectively.

4

Case studies of Mie calculus

We have collected the tools to calculate how an electromagnetic wave interacts with a spherical grain. Let us apply them. With the help of a small computer, the formulae presented so far allow us to derive numbers for the cross sections, scattering matrix, phase function, and so forth. In this chapter, we present examples to deepen our understanding. To the student who wishes to perform similar calculations, the figures also offer the possibility to check his computer program against the one used in this text.

4.1 Efficiencies of bare spheres

4.1.1 Scattering and absorption

4.1.1.1 Pure scattering

As a first illustration, we consider the scattering efficiency of a non-dissipative sphere, i.e., one that does not absorb the incident light. The imaginary part k of the optical constant $m = n + ik$ is therefore zero, the absorption coefficient vanishes and

$$Q^{\mathrm{sca}} = Q^{\mathrm{ext}} .$$

The efficiency depends only on m and on the size parameter $x = 2\pi a/\lambda$. To better interpret the dependence of Q^{sca} on x, we may envisage the wavelength λ to be fixed so that x is proportional to grain radius a. Figure 4.1 demonstrates two features valid for any particle:

- When a particle is small, we naturally expect the cross section C to be small, too. But even the extinction *efficiency*, which is the cross section over projected area, goes to zero,

$$Q^{\mathrm{ext}} \to 0 \quad \text{for} \quad x \to 0 .$$

- In accordance with Babinet's theorem, when the grain is very large, the extinction efficiency approaches two (see right frame of figure 4.1),

$$Q^{\mathrm{ext}} \to 2 \quad \text{for} \quad x^{-1} \to 0 ,$$

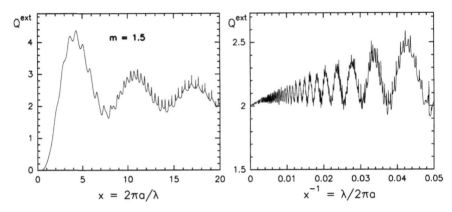

FIGURE 4.1 The extinction efficiency Q^{ext} as a function of size parameter x and and of its inverse x^{-1} for an optical constant $m = 1.5$; this value is appropriate for glass in the visible part of the spectrum. The left box shows the range $0 \le x \le 20$, the right the range $0.002 \le x^{-1} \le 0.05$ corresponding to $20 \le x \le 500$. Because in any real material, m depends on wavelength, whereas we keep it fixed in the figure, it is best to envisage λ as constant. Then the abscissa of the left frame is proportional to the grain radius.

There are two interesting aspects pertaining to a pure scatterer or weak absorber:

- Q^{ext} displays an overall undulating character with broad waves on which semi-regular ripples are superimposed. The first maximum of Q^{ext}, in our particular example, occurs at $x \sim 4$ and it is followed by gradually declining maxima around $x = 11, 17, \ldots$. The minima in between stay more or less all around two.

- Q^{ext} can be quite large. In figure 4.1, its maximum value is 4.4. The extinction cross section is then 4.4 times (!) bigger than the projected surface area of the particle.

4.1.1.2   A weak absorber

Figure 4.2 exemplifies the effects of adding impurities to the material. The top frame is identical to the left box of figure 4.1 and displays a medium with $m = 1.5$. In the middle, the optical constant is complex, $m = 1.5 + i\,0.02$. Now the particle not only scatters the light, but also absorbs it. Impurities effectively suppress the ripples and they also lead to a decline in the amplitudes of the waves. The damping is sensitive to k, as can be seen by comparing the two bottom panels where k changes from 0.02 to 0.05.

For interstellar grains, the ripples and the waves are irrelevant. For example, when we measure the intensity of a star behind a dust cloud, the observation is not monochromatic, but comprises a certain bandwidth $\Delta\lambda$. The optical constant $m(\lambda)$ will change somewhat over this bandwidth and, more important, interstellar dust displays a range of sizes, so the peaks produced by particles of a certain diameter are compensated by minima of slightly smaller or bigger ones.

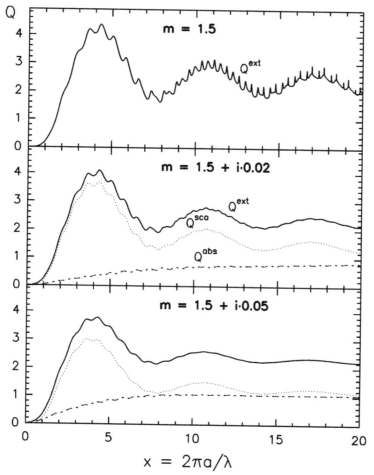

FIGURE 4.2 The efficiencies for absorption (dashed), scattering (dots) and extinction (solid) for three substances: one is not dissipative; the others are weakly absorbing. An optical constant $m = 1.5 + 0.05i$ may be appropriate for astronomical silicate around $\lambda = 6\mu m$.

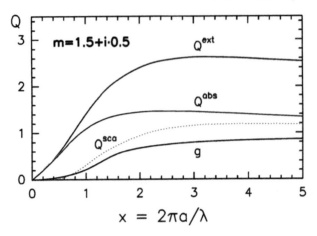

FIGURE 4.3 The efficiency for absorption, scattering and extinction, and the asymmetry factor g for a strongly absorbing material with $m = 1.5 + 0.5i$.

4.1.1.3   A strong absorber

In figure 4.3 we retain the value of $n = 1.5$ of the previous two plots, but choose for the imaginary part of the optical constant $k = 0.5$. This implies strong absorption. The efficiencies change now very smoothly and there are no signs of ripples or waves left. Because of the strong absorption, Q^{abs} increases quickly with x at $x < 1$. We discern in this particular example general features of small-sized grains. When $x \to 0$:

- Q^{abs} and x are proportional, as predicted by (3.5).
- The scattering efficiency changes after (3.6) as $Q^{sca} \propto x^4$, so Q^{sca} falls off more steeply than Q^{abs} (see logarithmic plot of figure 4.5).
- Absorption dominates over scattering, $Q^{abs} \gg Q^{sca}$.
- The asymmetry factor g of (2.7), which describes the mean direction of the scattered light, vanishes. Although isotropic scattering implies $g = 0$, the reverse is not true.

Figure 4.3 plots only the range up to a size parameter of five, but the behavior for larger x is smooth and can be qualitatively extrapolated. In this particular example of $m = 1.5 + i\,0.5$, a very large particle ($x \gg 1$)

- scatters mostly in the forward direction, so g is not far from one;
- has a large absorption efficiency and therfore a small reflectance $Q^{ref} = 1 - Q^{abs}$, as defined in equation (3.42). The asymptotic values for $x \to \infty$ are: $g = 0.918$, $Q^{abs} = 0.863$ and $Q^{ref} = 0.137$.

Surprisingly, the absorption efficiency can exceed unity, which happens here when $x > 0.9$. Unity is the value of Q^{abs} of a blackbody and may thus appear to be an upper limit for any object. Nevertheless, the calculations are correct; only the concept of a blackbody is reserved to sizes much larger than the wavelength.

4.1.1.4 A metal sphere

- Ideally, a metal has (at low frequencies) a purely imaginary dielectric permeability so that $n \simeq k$ goes to infinity (see 1.107).

- A metallic sphere is therefore very reflective and Q^{abs} tends to zero.

- When the sphere is big, of all the light removed, half is scattered isotropically and it alone would yield an asymmetry factor $g = 0$. The other half is scattered at the particle's rim entirely in the forward direction, which alone would produce $g = 1$. The combination of both effects gives $g = \frac{1}{2}$.

- When the sphere has a size comparable to or smaller than the wavelength, backward scattering dominates ($g < 0$).

- For small values of $|mx|$, the metallic sphere obeys of course also the relation $Q^{abs} \gg Q^{sca}$. But as x grows, scattering increases and soon takes over, whereas absorption stays at some low level.

Figure 4.4 illustrates these items in case of large, but not extreme metallicity $(m = 20 + i\,20)$.

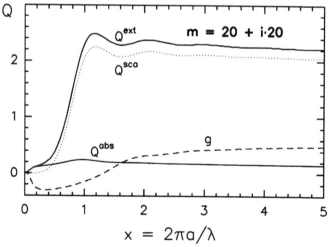

$$x = 2\pi a/\lambda$$

FIGURE 4.4 Same as figure 4.3, but now for a metal for which n and k are large and equal. The situation applies approximately to amorphous carbon at millimeter wavelengths. Note that g becomes negative.

TABLE 4.1 The asymptotic dependence of efficiency, cross section and mass or volume coefficient on the radius, a, of a spherical grain.

	symbol	large radii $a \gg \lambda$	small radii $a \ll \lambda$
efficiency	Q	$Q^{abs} = \text{const}$ $Q^{sca} = \text{const}$	$Q^{abs} \propto a$ $Q^{sca} \propto a^4$
cross section per grain	C	$C^{abs} \propto a^2$ $C^{sca} \propto a^2$	$C^{abs} \propto a^3$ $C^{sca} \propto a^6$
mass or volume coefficient	K	$K^{abs} \propto a^{-1}$ $K^{sca} \propto a^{-1}$	$K^{abs} = \text{const}$ $K^{sca} \propto a^3$

4.1.2 Efficiency vs. cross section and volume coefficient

Figure 4.5 compares efficiencies, cross sections and volume coefficients. The optical constant $m = 2.7 + i\,1.0$ applies to amorphous carbon at a wavelength $\lambda = 2.2\mu$m. The presentation is different from the preceding figures as we choose for the abscissa the grain radius a, and not the size parameter x; the wavelength is kept constant at 2.2μm. The logarithmic plots reveal the limiting behavior of small and large particles. These features, compiled in table 4.1, are independent of the choice of m or λ.

- Spheres of small radii:

  - The efficiencies depend on wavelength as given by (3.5) and (3.6). If m is constant, $Q^{abs} \propto \lambda^{-1}$ and $Q^{sca} \propto \lambda^{-4}$. Scattering is therefore negligible relative to absorption.

  - As interstellar dust has typical sizes of 0.1μm, one may assume the Rayleigh limit to be valid at all wavelengths greater than 10μm.

  - The mass absorption coefficient K_λ^{abs} is proportional to the grain volume V, irrespective of the size distribution of the particles.

  - The mass scattering coefficient K_λ^{sca} goes with V^2. Although the scattering process is identical in all subvolumes of a grain, the net effect is not linear because of interference of the scattered waves.

  - The emission coefficient $\epsilon_\lambda = K_\lambda^{abs} B_\lambda(T_{dust})$ depends also linearly on the grain volume; $B_\lambda(T_{dust})$ is the Planck function at the dust temperature (see 6.1). Provided the infrared photons are not trapped in the cloud, a condition that is usually fulfilled, the observed flux is directly proportional to the total dust volume.

• Spheres of large radii: The efficiencies, Q, reach a finite value with $Q^{\text{ext}} = Q^{\text{abs}} + Q^{\text{sca}} = 2$. The cross sections per grain, C, increase without bounds proportional to the projected area of the grain. The mass or volume coefficients, K, decrease inversely proportional to the radius.

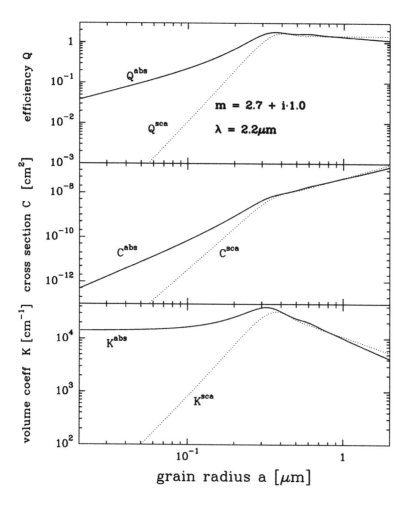

FIGURE 4.5 *Top:* absorption and scattering efficiency Q. *Middle:* cross section per grain $C = \pi a^2 Q$. *Bottom:* volume coefficient K referring to $1\,\text{cm}^3$ of grain volume. Optical constant $m = 2.7 + i\,1.0$, wavelength $\lambda = 2.2\mu m$.

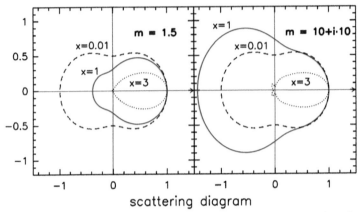

scattering diagram

FIGURE 4.6 The intensity distribution of scattered light for three size parameters x and two optical constants m. Light enters from the left and the grain is at the position (0,0). The diagrams are normalized such that the intensity equals one in the forward direction ($\theta=0$); its value in any other direction is given by the distance between the frame center and the contour line. All contours are mirror-symmetric with respect to the horizontal arrows; see also table 4.2.

4.2 Scattering by bare spheres

4.2.1 The intensity pattern of scattered light

The change of the intensity of scattered light with angle is described by the phase function of (2.45) and visualized in figure 4.6. In these examples, the left frame refers to a non-dissipative sphere with $m = 1.5$. The curve labeled $x = 0.01$ represents a dipole pattern which is characteristic for a particle small compared to the wavelength irrespective of the choice of the optical constant m. It is symmetric relative to the scattering angle of 90° (vertical line in the figure) and has therefore equal maxima in the forward ($\theta = 0$) and backward ($\theta = 180°$) direction. Note that scattering is not isotropic, although $g = 0$. The pattern is sensitive to the size parameter $x = 2\pi a/\lambda$. When $x \geq 1$, scattering becomes very forward, and for a grain diameter just equal to the wavelength ($x \simeq 3$), almost all light goes into a forward cone.

In the right frame where $m = 10 + i\,10$, we are dealing with a reflective material. The change in m relative to the non-dissipative sphere shown left does not affect the curve $x = 0.01$. There is also not much difference for $x = 3$, although curious tiny rear lobes appear. However, at the intermediate value $x = 1$, we now have preferentially backscatter and the asymmetry factor is negative ($g = -0.11$), whereas in the left panel, at $x = 1$, the scattering is mainly forward.

TABLE 4.2 Scattering efficiency and asymmetry factor for spheres whose scattering diagram is displayed in figure 4.6.

$x = 2\pi a/\lambda$	m	Q^{sca}	asymmetry factor g
0.01	1.5	2.31×10^{-9}	1.98×10^{-5}
1	1.5	2.15	0.20
3	1.5	3.42	0.73
0.01	$10 + i\,10$	2.67×10^{-8}	1.04×10^{-5}
1	$10 + i\,10$	2.05	-0.111
3	$10 + i\,10$	2.08	0.451

4.2.2 The polarization of scattered light

Figure 4.7 depicts the degree of linear polarization p of light scattered by a sphere (see 2.56) using the same optical constants and size parameters as in figure 4.6.

- The polarization curve $p(\theta)$ of tiny and weakly absorbing spheres (x=0.01, $k \ll 1$; dashes in top box of figure 4.7) is symmetric around 90° and has the shape of a bell. The polarization is 100% at a scattering angle of 90°. For weak absorbers, $p(\theta)$ hardly changes as x grows from 0.01 up to 1.

- Tiny metallic particles ($x = 0,01$, $m = 10 + i\,10$) have the same pattern as tiny dielectrics, but at $x = 1$, the polarization $p(\theta)$ is no longer symmetric and attains a maximum value of only 78%.

- The behavior of $p(\theta)$ at intermediate sizes ($x = 3$) is more complicated in both materials. The polarization changes sign and there are several maxima and minima of varying height. In our examples, p vanishes at three intermediate angles. Here the matrix element S_{12} switches sign (see 2.54) and the direction of polarization with respect to the scattering plane changes from perpendicular to parallel or back. The scattering plane is defined by the unit vectors in the direction from where the light is coming and where it is going.

  When one measures the linear polarization of scattered light at various locations around the exciting star of a reflection nebula, the observed values are always an average along the line of sight. If the particles are small, $S_{12} < 0$ and the polarization vector is perpendicular to the radial vector pointing towards the star. However, as the figure suggests, it can happen, although it is an unfamiliar scenario to astronomers, that the radial and polarization vectors are parallel.

- Very big dielectric spheres ($x \simeq 1000$) show again a simple pattern with no sign reversals. There is 100% polarization in a slightly forward direction ($\theta \simeq 65°$). For metals of this size, $p(\theta)$ has more structure.

- Polarization is always zero in the forward and backward direction ($\theta = 0$ or 180°).

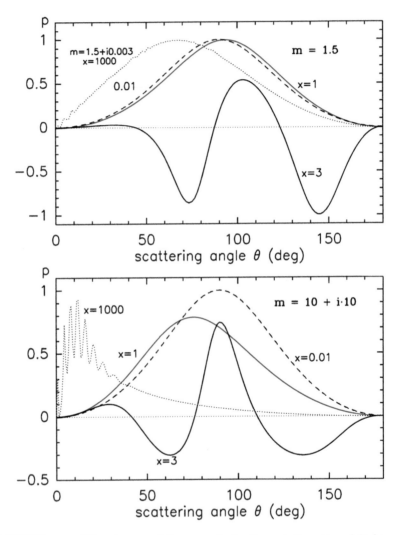

FIGURE 4.7 The degree of linear polarization p of scattered light as a function of scattering angle for various size parameters x. **Top:** Optical constant $m = 1.5$. For intermediate sizes ($x = 3$), the polarization goes at certain scattering angles through zero and becomes negative. The polarization vector flips at points where $p = 0$. The dotted line depicts a very large sphere ($x = 1000$) with slight contaminations ($m = 1.5 + i\,0.003$). **Bottom:** Optical constant $m = 10 + i\,10$.

An impressive example of a very regular polarization pattern is presented by the bipolar nebula S106 (figure 4.8). This is a famous star forming region. At optical wavelengths, one observes light from two nebulous lobes separated by a dark lane that is interpreted as a disk inside which sits the central and exciting source named IRS4. We are viewing the disk from the side. There is plenty of visual obscuration towards IRS4 (~20 mag), but in the near infrared one begins to penetrate most of the foreground and outer disk material. IRS4 is an infrared source and its radiation is reflected by the dust particles that surround it. When looking directly towards IRS4, the angle of deflection is close to 0° or 180° and the ensuing degree of linear polarization is small. But at positions a few arc seconds off, scattering occurs at right angles and the light becomes strongly polarized. Note that S106 must be optically thin to scattering in the K band as otherwise multiple scattering would lead to depolarization.

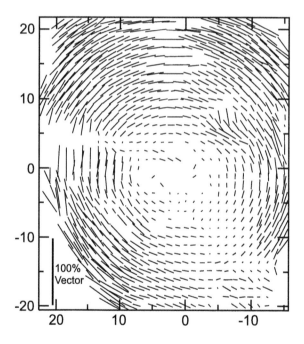

FIGURE 4.8 The infrared source IRS4 (at figure center) in the bipolar nebula S106 illuminates its vicinity. Some light is scattered by dust particles producing in the K band (2.2μm) a polarization pattern of wonderful circular symmetry (adapted from [Asp90]).

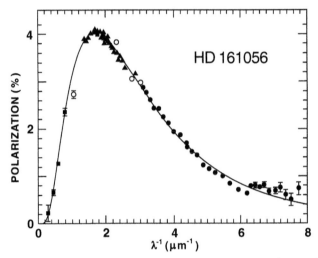

FIGURE 4.9 Linear polarization of the star HD 161056 [Som94].

4.3   Linear polarization through extinction

When a grain is optically anisotropic, for instance, when it is elongated like a cigar, its extinction efficiency, Q^{ext}, depends on the orientation of the electric field vector of the incident light. All interstellar grains are to some degree optically anisotropic because of their irregular structure. If in a cloud such grains are aligned, unpolarized light from a background star will become linearly polarized as it passes through it (recommended reading [Whi03]).

The change of the amount of polarization with wavelength, $p(\lambda)$, can for most stars be nicely fit by the empirical expression

$$\frac{p(\lambda)}{p_{max}} = \exp\left[-k \ln^2\left(\frac{\lambda}{\lambda_{max}}\right)\right] , \qquad (4.1)$$

called Serkowski curve. An example is shown in figure 4.9. There are three parameters: λ_{max}, p_{max} and k. The maximum percentage of polarization, p_{max}, is never large ($< 10\%$). A typical value for the parameter k, which determines the width of the curve, is 1.15, the wavelength of maximum polarization, λ_{max}, usually falls into the visual band.

The left box in figure 4.10 shows the Q's of infinite cylinders when the electric vector is parallel ($Q_{\parallel}$) and perpendicular ($Q_{\perp}$) to the cylinder axis as well as their difference ΔQ. The latter attains its maximum when the size parameter is around one, and it is quite well fit by the Serkowski curve (right box). If the cylinders have a certain size distribution, the spikes in ΔQ disappear and the fit to (4.1) further improves.

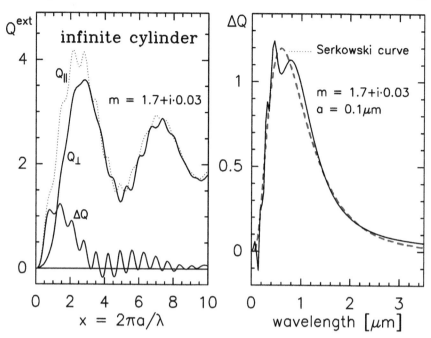

FIGURE 4.10 *Left:* The extinction efficiencies $Q_\parallel$ and $Q_\perp$ and their difference, $\Delta Q = Q_\parallel - Q_\perp$, as a function of the size parameter $x = 2\pi a/\lambda$ for infinite cylinders of radius a and optical constant $m \simeq 1.7 + i\,0.03$. *Right:* A rearrangement of ΔQ in the left box assuming that the cylinder radius, a, is constant and the wavelength λ varies.

Infinite cylinders are, of course, only constructs of the mind, covenient for computing the Q's, but otherwise unrealistic. However, we may conclude that cigar-shaped grains with a minor axis of $\sim 0.1\mu$m could produce the observed polarization. The question of how the elongated grains become aligned has not been fully settled, but it is believed that it results from magnetic dissipation as the grains, which are weakly magnetic because they contain iron atoms, are spinning in the interstellar magnetic field.

4.4 Coated spheres

The mathematical formalism of Mie theory presented in section 2.3 can be extended to coated spheres. The only thing new is the appearance of two more boundary conditions at the interface between the two dust materials.

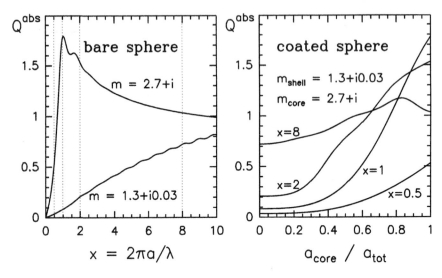

FIGURE 4.11 *Left:* Q^{abs} as a function of size parameter x for homogeneous spheres. $m = 2.7+i$ is representative of amorphous carbon at $\lambda = 2.2\mu$m and $m = 1.3+0.03i$ of dirty ice at the same wavelength. Dotted vertical lines for easy comparison with right box. *Right:* Q^{abs} of a coated sphere, x denotes the size parameter of the total grain.

The relevance of calculating coated particles stems from the observational fact that grains in cold clouds acquire ice mantles (section 8.4). Such particles may be approximated by two concentric spheres. There is a refractory core of radius a_{core} and optical constant m_{core} surrounded by a volatile shell of thickness d and optical constant m_{shell}. So the total grain radius is

$$a_{\mathrm{tot}} = a_{\mathrm{core}} + d .$$

The left box in figure 4.11 shows the absorption efficiency Q^{abs} as a function of size parameter x for homogeneous particles, the right one for coated spheres as a function of

$$r = a_{\mathrm{core}}/a_{\mathrm{tot}} .$$

For $r = 0$, the grain consists only of mantle material with $m = 1.3 + 0.03i$; for $r = 1$, only of core matter ($m = 2.7 + i$). The four curves represent size parameters $x = 2\pi a_{\mathrm{tot}}/\lambda$ of the total sphere between 0.5 and 8. The right figure demonstrates how the absorption efficiency of a coated grain changes when the core grows at the expense of the mantle while the outer radius a_{tot} is fixed.

4.5 Surface modes in small grains

4.5.1 Small graphite spheres

When grains display at some wavelength an extinction peak, the reason is generally a resonance in the grain material. However, in the case of grains which are small compared to the wavelength, the extinction cross section may be large without an accompanying feature in the bulk matter. As the extinction efficiency is in the Rayleigh limit acording to (3.5) proportional to $\varepsilon_2/|\varepsilon + 2|^2$, we expect a big value of Q^{ext} at a wavelength where

$$\varepsilon + 2 \simeq 0 .$$

Because $\varepsilon = m^2$, this happens when the real part n of the optical constant is close to zero and $k^2 \approx 2$. The phenomenon bears the name *surface mode.*

For the major components of interstellar dust, amorphous carbon and silicate, there is no wavelength where the condition of small $\varepsilon + 2$ is approximately fulfilled (figures 5.10 and 5.11), but graphite around 2200Å is a candidate (figure 5.12). Graphite is optically anisotropic and the effect appears when the electric vector of the incoming wave lies in the basal plane of the stacked carbon sheets. The dielectric constant is then denoted by $\varepsilon_\perp$ (because the electric vector is perpendicular to the normal of the basal plane). The left side of figure 4.12 shows $\varepsilon_\perp = \varepsilon_1 + i\varepsilon_2$ and the resulting extinction efficiency for spheres of different sizes. The wavelength interval stretches from 1670 to 3330Å or from 6 to $3\mu m^{-1}$. To underline the effect of the surface mode, Q^{ext} is normalized to one at $\lambda^{-1} = 3\mu m^{-1}$. Small particles ($a \leq 300$Å) clearly display a resonance around 2200Å, which is absent in the bulk material because particles with a radius greater than 1000Å do not show it. The graphite surface mode is invoked to explain the strong feature around 2200Å in the interstellar extinction curve (figure 10.1).

For the Rayleigh approximation to be valid, the particles must have diameters of 100Å or less. But the position of the resonance contains further information on grain size. The left upper box of figure 4.12 illustrates how the peak in Q^{ext} drifts towards smaller wavelengths (greater λ^{-1}) with decreasing radius a; it moves from 2280Å to 2060Å as the radius shrinks from 300Å to 30Å. There is no further shift for still smaller particles.

In astronomical observations of the extinction curve, the relevant quantities are, however, not the extinction efficiencies Q^{ext}, but the extinction coefficients per cm^3 of dust volume, K^{ext} (upper right of figure 4.12). The dependence of K^{ext} on λ^{-1} is qualitatively similar to that of Q^{ext}, but the discrimination against size is less pronounced and not possible below 100Å.

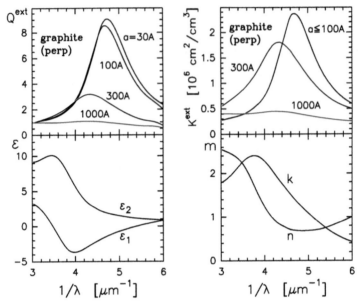

FIGURE 4.12 *Upper left:* Extinction efficiency of graphite spheres around $\lambda = 2200\text{Å}$ when the electric vector lies in the basal plane. The Qs are normalized to one at $3\mu\text{m}^{-1}$. The increase for small radii, a, is due to a surface mode and not to an intrinsic quality of the bulk material. *Lower left:* dielectric constant $\varepsilon = \varepsilon_1 + \varepsilon_2$. *Upper right:* Volume coefficients, not normalized. *Lower right:* optical constant $m = n + ik$.

4.5.2   Ellipsoids and metals

The condition $\varepsilon + 2 \simeq 0$, necessary for the appearance of a surface mode in spheres, must for ellipsoidal particles be replaced by (see 3.12 and 3.25)

$$1 + L_a(\varepsilon - 1) \simeq 0 ,$$

or similar expressions for the b- or c-axis. The form of the particle therefore influences via the shape factor L_a the strength of the surface mode as well as its center wavelength. Silicon carbide, SiC, a likely minor dust component with a resonance around $11\mu\text{m}$ is a candidate for this phenomenon [Boh83].

A surface mode is inevitable for metallic grains when damping is weak (graphite is metallic in the basal plane). With the dielectric permeability for metals given in equation (1.113), the denominator in (3.5) vanishes when

$$\left[3\left(\omega^2 + \gamma^2\right) - 1\right]^2 \omega^2 + \gamma^2 = 0 .$$

Here the frequency ω and the damping constant γ are in units of the plasma frequency ω_p. As γ is usually much smaller than ω_p, a surface mode, i.e., a strong enhancement of Q^abs, appears at $\omega = \omega_\text{p}/\sqrt{3}$.

5

Structure and composition of dust

We summarize in section 5.1 how atoms are arranged in a solid and in section 5.2 what holds them together. In section 5.3, we study the atomic structure and bonding in the two major dust constituents: silicate and carbon grains, and in section 5.4, we present their optical constants as well as absorption and extinction efficiencies for typical grain sizes. Finally, we discuss the size distribution of grains which is able to explain the observed interstellar extinction curve and sketch how grains can acquire such a distribution in grain-grain collisions.

5.1   Crystal structure

5.1.1   Translational symmetry

5.1.1.1   The lattice and the base

Interstellar grains probably contain crystalline domains of sizes 10Å to 100Å. In a crystal, the atoms are regularly arranged in a lattice which means that the crystal can be built up by periodic repetition of identical cells. A cell is a parallelepiped defined by three vectors $\mathbf{a}$, $\mathbf{b}$ and $\mathbf{c}$ such that for any two points $\mathbf{r}, \mathbf{r}'$ in the crystal whose difference can be written as

$$\mathbf{r}' - \mathbf{r} = u\mathbf{a} + v\mathbf{b} + w\mathbf{c}, \tag{5.1}$$

with integral u, v, w, the environment is exactly the same. If *any* two points $\mathbf{r}$ and $\mathbf{r}'$ with an identical environment are connected through (5.1), the vectors $\mathbf{a}, \mathbf{b}, \mathbf{c}$ are said to be primitive and generate the *primitive cell*. It is the one with the smallest possible volume. There is an infinite number of ways how to construct cells or vector triplets $(\mathbf{a}, \mathbf{b}, \mathbf{c})$, even the choice of the primitive cell is not unique; figure 5.3 gives a two-dimensional illustration. Among the many possibilities, one prefers those cells whose sides are small, whose angles are the least oblique and which best express the symmetry of the lattice. They are called conventional or elementary or unit cells.

For a full specification of a crystal, one needs besides the lattice points $u\mathbf{a} + v\mathbf{b} + w\mathbf{c}$, a description of the three-dimensional cell structure, i.e., of the distribution of charge and matter in the cell. The cell structure is called the base. The spatial relation between the lattice points and the base is irrelevant. So lattice points may or may not coincide with atomic centers. A primitive cell may contain many atoms and have a complicated base, but it always contains only one lattice point.

5.1.1.2　Physical consequences of crystalline symmetry

Geometrically, the existence of a grid always implies spatial anisotropy because in various directions structures are periodic with different spacings. Furthermore, the periodic arrangement of atoms can become evident by the morphological shape of the body, in non-scientific language: crystals are beautiful. The physical relevance of crystalline order reveals itself, among others, by

- the appearance of long range, i.e., intensified forces which lead to sharp vibrational resonances observable at infrared wavelengths;

- bonding strengths that are variable with direction: in one or two directions, a crystal cleaves well, in others it does not;

- a generally anisotropic response to fields and forces. For example, the way crystals are polarized in an electromagnetic field or stretched under mechanical stress depends on direction.

5.1.2　Lattice types

5.1.2.1　Bravais cells and crystal systems

In the general case, the lengths of the vectors **a**, **b**, **c** in the conventional cell and the angles between them are arbitrary. When lengths or angles are equal, or when angles have special values, like $90°$, one obtains grids of higher symmetry. There are 14 basic kinds of translation lattices (*Bravais* lattices) which correspond to 14 elementary cells. These 14 lattice types can be grouped into seven crystal systems which are described in table 5.1 and shown in figure 5.1.

The cubic system has the highest symmetry: **a**, **b**, **c** are orthogonal and their lengths are equal, $a = b = c$. The polarizability is then isotropic, whereas for other crystal systems, it depends on direction. Figure 5.2 shows the three lattice types of the cubic family: simple (sc), **body** centered (bcc) or **face** centered (fcc) of which only the sc cell is primitive. The bcc cell is two times bigger than the primitive cell and the fcc cell four times; furthermore, the primitive cells do not have the shape of a cube (see two-dimensional analog in figure 5.5).

TABLE 5.1 Lengths and angles in the unit cells of the seven crystal systems. See figure 5.1 for definition of sides and angles. The monoclinic, orthorhombic, tetragonal and cubic system are further split into lattice types that are either primitive or centered. For the cubic crystal system, this is shown in figure 5.2. The cells of the triclinic and rhombohedral system are always primitive, without centering. Altogether there are 14 lattice types or Bravais cells, only seven of them are primitive.

1. triclinic:	$a \neq b \neq c$	$\alpha \neq \beta \neq \gamma$	lowest symmetry
2. monoclinic:	$a \neq b \neq c$	$\alpha = \gamma = 90°, \beta \neq 90°$	not in figure 5.1
3. orthorhombic:	$a \neq b \neq c$	$\alpha = \beta = \gamma = 90°$	
4. tetragonal:	$a = b \neq c$	$\alpha = \beta = \gamma = 90°$	
5. hexagonal:	$a = b \neq c$	$\alpha = \beta = 90°, \ \gamma = 120°$	
6. rhombohedral:	$a = b = c$	$\alpha = \beta = \gamma \neq 90°$	not in figure 5.1
7. cubic:	$a = b = c$	$\alpha = \beta = \gamma = 90°$	highest symmetry

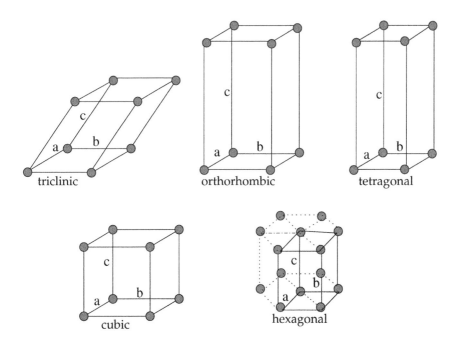

FIGURE 5.1 There are seven crystal systems of which five are shown here. The angle between the vector pair $(\mathbf{b}, \mathbf{c})$ is denoted α. If we write symbolically $\alpha = (\mathbf{b}, \mathbf{c})$, we define likewise $\beta = (\mathbf{c}, \mathbf{a})$ and $\gamma = (\mathbf{a}, \mathbf{b})$. See table 5.1.

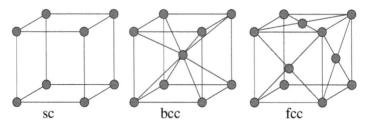

FIGURE 5.2 The three lattices types of the cubic crystal system: simple cubic (sc), body-centered cubic (bcc) and face-centered cubic (fcc). To avoid confusion, for the fcc lattice the face centering atoms are shown only on the three faces directed towards the observer.

5.1.2.2 Microscopic and macroscopic symmetry

Under a symmetry operation, a crystal lattice is mapped onto itself. A linear translation by one of the vectors $\mathbf{a}, \mathbf{b}, \mathbf{c}$ is one such operation, but it becomes apparent only on an atomic scale because the shifts are of order 1Å and thus much smaller than the dimensions of a real (even an interstellar) crystal. The translational symmetry is revealed from X-ray images and is described by the Bravais lattices. However, there are symmetry operations that are evident also on a macroscopic scale, such as

- Inversion at a center. If the center is in the origin of the coordinate system, the operation may be symbolized by $\mathbf{r} = -\mathbf{r}'$.

- Rotation about an axis. One can readily work out that the only possible angles compatible with translational symmetry, besides the trivial 360°, are 60°, 90°, 120° and 180°. However, for a single molecule (not a crystal), other angles are possible, too.

- Reflection at a plane.

In each of these symmetry operations, there is a symmetry element that stays fixed in space (the inversion center, the rotation axis or the mirror plane, respectively).

Macroscopically, a crystal has the shape of a polyhedron, like a gem in an ornament or the grains in a salt shaker. It can differ from another of the same type in size and shape (it is usually "distorted"; i.e., not all faces have the same distance to the crystal center). However, the angles between corresponding faces are the same, and when one subjects a crystal to a symmetry operation, the normals of the faces of the polyhedron do not change their directions.

All crystals can be divided into 32 classes where each class represents a mathematical group whose elements are the symmetry operations above.

These 32 classes completely define the morphological (macroscopic) appearance of crystals. Macroscopic symmetry is, of course, the result of the microscopic structure, although the latter is not used explicitly in the derivation of the 32 crystal classes. Without proof, when the 14 Bravais lattices of the atomic world are combined with the above symmetry elements, including the two further symmetry elements screw axis and glide plane, one arrives at 230 space groups.

5.1.2.3 Two-dimensional lattices

The classification scheme is much simpler and easier to apprehend in two dimensions. The basic unit of a two-dimensional crystal is a parallelogram. Now there are only five (translational) Bravais lattices and 10 crystal classes. When the Bravais lattices are combined with the symmetry elements permitted for a plane (rotation axis, mirror plane and glide plane), one obtains 17 planar space groups. The parallelograms of Bravais lattices may be

- oblique ($a \neq b$, $\gamma \neq 90°$, figure 5.3);

- hexagonal ($a = b$, $\gamma = 120°$);

- quadratic ($a = b$, $\gamma = 90°$);

- rectangular ($a \neq b$, $\gamma = 90°$, figure 5.4);

- centered rectangular ($a \neq b$, $\gamma = 90°$, figure 5.5).

Any two vectors $\mathbf{a}'$, $\mathbf{b}'$ with lengths $a' = b'$ and angle γ between them describe a centered cell of a rectangular lattice, but the cell with vectors $\mathbf{a}, \mathbf{b}$ as in figure 5.5 seems to better convey to us the symmetry. Likewise in three dimensions, the body-centered (bcc) and face-centered (fcc) cubic type can be generated from primitive rhombohedra (where all three lattice constants are equal, $a = b = c$), but then again, one loses the advantage of a rectangular coordinate system because the angles are not $90°$ any more.

5.1.3 The reciprocal lattice

Because the structure in a crystal is periodic, it is natural to expand the density $\rho(\mathbf{x})$ in a Fourier series. If $\mathbf{a}, \mathbf{b}, \mathbf{c}$ form the primitive vectors of the lattice according to (5.1), then

$$\rho(\mathbf{x}) = \sum_{\mathbf{G}} \rho_{\mathbf{G}} \, e^{i\mathbf{G} \cdot \mathbf{x}} . \tag{5.2}$$

The $\rho_{\mathbf{G}}$ are the Fourier coefficients and the sum extends over all vectors

$$\mathbf{G} = u \, \mathbf{a}^* + v \, \mathbf{b}^* + w \, \mathbf{c}^* , \tag{5.3}$$

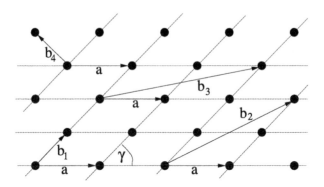

FIGURE 5.3 A two-dimensional lattice. There is an infinite number of ways to generate cells. Here it is done by the vector pairs $(\mathbf{a}, \mathbf{b}_i)$ with $i = 1, 2, 3, 4$. All define primitive cells, except $\mathbf{a}, \mathbf{b}_2$ which has an area twice as large.

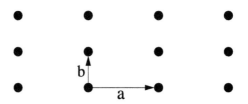

FIGURE 5.4 A simple rectangular lattice is generated by the primitive vectors $\mathbf{a}$ and $\mathbf{b}$.

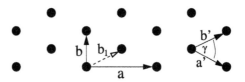

FIGURE 5.5 In a centered rectangular lattice, a primitive cell is generated by $\mathbf{a}, \mathbf{b}_1$ and has half the area of a conventional cell generated by $\mathbf{a}, \mathbf{b}$. Note that $\mathbf{a}', \mathbf{b}'$ with $a' = b'$ are also primitive vectors.

where u, v, w are integers and $\mathbf{a}^*, \mathbf{b}^*, \mathbf{c}^*$ are the primitive vectors of the reciprocal lattice. The first, $\mathbf{a}^*$, is defined by

$$\mathbf{a}^* = 2\pi \frac{\mathbf{b} \times \mathbf{c}}{\mathbf{a} \cdot (\mathbf{b} \times \mathbf{c})} \tag{5.4}$$

so that $\mathbf{a}^*$ is perpendicular to $\mathbf{b}$ and $\mathbf{c}$ and $\mathbf{a}^* \cdot \mathbf{a} = 2\pi$. The unit of $\mathbf{a}^*$ is one over length. There are corresponding definitions and relations for $\mathbf{b}^*$ and $\mathbf{c}^*$.

Suppose a plane wave of wave vector $\mathbf{k}$ falls on a crystalline grain. Any small subvolume in the grain, let it be at locus **x**, scatters the incoming light. The amplitude of the scattered wave is proportional to the electron density $\rho(\mathbf{x})$. Two outgoing beams with the same wavenumber $\mathbf{k}'$ which are scattered by subvolumes that are a distance **r** apart have a phase difference $e^{i\Delta\mathbf{k}\cdot\mathbf{r}}$ with

$$\Delta\mathbf{k} \;=\; \mathbf{k}' \;-\; \mathbf{k} \;.$$

The vectors **k** and $\mathbf{k}'$ have the same length, but point into different directions. To find the total scattered amplitude E_s, one must integrate over the whole grain volume V, so modulo some constant factor

$$E_\mathrm{s} \;=\; \int_V \rho(\mathbf{x}) \, e^{-i\Delta\mathbf{k}\cdot\mathbf{x}} \, dV \;.$$

With respect to an arbitrary direction $\mathbf{k}'$, the integral usually vanishes as a result of destructive interference. It does, however, not vanish if $\Delta\mathbf{k}$ is equal to a reciprocal lattice vector **G** of (5.3), i.e.,

$$\Delta\mathbf{k} \;=\; \mathbf{k}' - \mathbf{k} \;=\; \mathbf{G} \;. \tag{5.5}$$

Equation (5.5) is another way of formulating the Bragg law for reflection in crystals.

5.2 Binding in crystals

Solids are held together through electrostatic forces. Five basic types of bonds exist and they are discussed below. All five types are realized in cosmic dust.

5.2.1 Covalent and ionic bonding

In a *covalent or homopolar bond* between atoms, a pair of electrons of anti-parallel spin, one from each atom, is shared. The more the orbitals of the two valence electrons in the pair overlap, the stronger the atoms are tied together. The bonding is directional and follows the distribution of the electron density which has a significant high concentration *between* the atoms. The wave functions in covalent bonding are not extended and only nearest neighbors interact. One finds covalent binding in solids and in molecules.

In *ionic or heteropolar bonds*, the adhesion is due to long-range Coulomb forces. In a crystal, the ions of opposite sign attract and those of the same sign repel each other; but in the end, attraction wins. At short distances, there is an additional kind of repulsion, but it acts only between adjacent atoms. It results from Pauli's exclusion principle which restricts the overlap of electron clouds.

Let us take a sodium chloride crystal (NaCl) as an example of ionic bonding. One partner, the alkali metal, is relatively easy to ionize (5.14 eV) and the other, the halogen Cl, has a high electron affinity (3.61 eV). Although the transfer of the electron from Na to Cl requires $5.14 - 3.61 = 1.53$ eV, the Na^+ and Cl^- ions can supply this deficit. They can provide even more than that in the form of potential energy by coming close together. Indeed, at the equilibrium distance of 2.8Å, the binding energy per molecule equals 6.4 eV.

As the ions of heteropolar bonds have acquired full electron shells, like inert gases, their electron clouds are fairly spherical and the charge distribution between the ions is low. Ionic bonding is therefore non-directional and usually found only in solids.

Pure ionic bonding is, however, an idealization; one finds it neither in molecules nor in crystals. Even when compounding alkali metals with halogens, there is some homopolar component or overlap of orbitals. In the extreme case of CsF, homopolar binding still amounts to 5%. There are also no pure covalent bonds, even in H_2 or diamond, because of the fluctuations of the electron clouds. In the real world, there are only blends between the two bonding types.

5.2.2 Metals

Another type of strong binding exists in metals. One may imagine a metal as a lattice of positive ions bathed in a sea of free electrons; metallic binding is therefore non-directional. For instance, in the alkali metal sodium (Na), the eleventh element of the Periodic Table, the two inner shells with $2 + 8 = 10$ electrons are complete, while the $3s$ valence electron is responsible for binding. All $3s$ electrons form the conduction band.

In an *isolated* sodium atom, the $3s$ electron is bound to the Na^+ ion by an ionization energy $\chi_{Na} = 5.14$ eV. The electron is localized near the atomic nucleus and its wave function ψ goes to zero at large distances from the nucleus. When, on the other hand, the $3s$ electron is in the potential of a lattice of Na^+-ions, as in a metal, its wave function ψ is very extended and subject to a different boundary condition: symmetry requires that the gradient of the wave function vanishes midway between atoms. The quantum mechanical calculations now yield a binding energy of the electron to the ionic lattice of 8.2 eV.

Because sodium has a density of 1.01 g/cm^3, the concentration of $3s$ electrons is 2.65×10^{22} cm^{-3} and the Fermi energy after (A.100) $E_F = 3.23$ eV. From (A.100) also follows the density of states,

$$\rho(E) = \frac{dN}{dE} = \frac{V}{2\pi^2} \left(\frac{2m_e}{\hbar^2} \right)^{3/2} E^{1/2} .$$

The *average* energy $\langle E \rangle$ of a conduction electron is therefore

$$\langle E \rangle = N^{-1} \int \rho(E) \, dE = \tfrac{3}{5} E_F \simeq 1.9 \, \text{eV} \ .$$

Consequently, in a metal each sodium atom is bound by $8.2 - \chi_{\text{Na}} - 1.9 \simeq 1.16 \, \text{eV}$.

5.2.2.1 Energy bands and Bloch functions

It is a general property of crystals that the electronic states are clustered in bands, with energy gaps between them. The basics of their formation is explained in section A.8.3 on one-dimensional periodic potentials. To derive the energy states and eigenfunctions of an electron in the 3-dimensional potential $U(\mathbf{x})$, we first note that as $U(\mathbf{x})$ comes from the lattice ions and is strictly periodic, it may after (5.2) be written as

$$U(\mathbf{x}) = \sum_{\mathbf{G}} U_{\mathbf{G}} \, e^{i\mathbf{G} \cdot \mathbf{x}} \ , \tag{5.6}$$

where $\mathbf{G}$ are the reciprocal lattice vectors of (5.3). The general solution for the wavefunction $\psi(\mathbf{x})$ is sought in a Fourier expansion,

$$\psi(\mathbf{x}) = \sum_{\mathbf{k}} C_{\mathbf{k}} \, e^{i\mathbf{k} \cdot \mathbf{x}} \tag{5.7}$$

where the wavenumber $\mathbf{k}$ fulfills the periodic boundary conditions and is connected to the momentum $\mathbf{p}$ by

$$\mathbf{p} = \hbar \mathbf{k} \ .$$

The coefficients $C_{\mathbf{k}}$ are found by inserting (5.7) into the Schrödinger equation

$$\hat{H}\psi(\mathbf{x}) = \left[\frac{\hat{\mathbf{p}}^2}{2m_e} + \hat{U}(\mathbf{x}) \right] \psi(\mathbf{x}) = E \, \psi(\mathbf{x}) \tag{5.8}$$

and solving the resulting set of algebraic equations. In (5.8), the interaction among electrons is neglected. After some algebra, one is led to *Bloch's theorem*, namely that the eigenfunction $\psi_{\mathbf{k}}(\mathbf{x})$ for a given wavenumber $\mathbf{k}$ and energy $E_{\mathbf{k}}$ must have the form

$$\psi_{\mathbf{k}}(\mathbf{x}) = u_{\mathbf{k}}(\mathbf{x}) \, e^{i\mathbf{k} \cdot \mathbf{x}}$$

where $u_{\mathbf{k}}(\mathbf{x})$ is a function periodic with the crystal lattice. So if $\mathbf{T}$ is a translation vector as given in (5.1), then

$$u_{\mathbf{k}}(\mathbf{x}) = u_{\mathbf{k}}(\mathbf{x} + \mathbf{T}) \ .$$

The wave function $\psi_{\mathbf{k}}(\mathbf{x})$ itself does generally not display the periodicity. It is straightforward to prove Bloch's theorem in a restricted form, for a one-dimensional ring of N atoms, where after N steps, each of the length of the grid constant, one is back to the starting position.

One can understand the formation of bands, for example, from the point of view of the tightly bound inner electrons of an atom. When we imagine compounding a solid by bringing N free atoms close together, the inner electrons are afterwards still localized and bound to their atomic nuclei, only their orbits have become disturbed. Because of the disturbance, a particular energy state E_i of an isolated atom will be split into a band of N substates due to the interaction with the other N–1 atoms. There will be a $1s, 2s, 2p, \ldots$ band.

To approximately calculate the width and average energy of a band, one considers just one crystal electron, bound to one atom but perturbed by all others, and assumes that its wavefunction is a linear combination of the unperturbed eigenfunctions of all crystal atoms. The expansion coefficients follow from the condition that it is a Bloch function. One then finds that the average band energy is lower than the corresponding energy in a free atom implying binding in the crystal, and that the shorter the internuclear distance, the stronger the overlap of the wave functions and the broader the band; the energetically deeper the electrons, the narrower the band.

5.2.3   van der Waals forces and hydrogen bridges

The van der Waals interaction is much weaker than the bonding types discussed above and typically only $0.1\,\text{eV}$ per atom. It arises between neutral atoms or molecules and is due to their dipole moments. The dipoles have a potential U that rapidly falls with distance ($U \propto r^{-6}$, see section 8.1).

If the molecules do not have a permanent dipole moment, like CO_2, the momentary quantum mechanical fluctuations of the electron cloud in the neighboring atoms induces one; its strength depends of course on the polarizability of the molecules. This *induced* dipole moment is usually more important even for polar molecules. Its attraction holds CO_2 ice together and binds the sheets in graphite.

Hydrogen can form bridges between the strongly electron-negative atoms N, O or F; the binding energy is again of order $\sim 0.1\,\text{eV}$ per atom. For example, when a hydrogen atom is covalently bound to oxygen, it carries a positive charge because its electron has been mostly transferred to the oxygen atom. The remaining proton can thus attract another negatively charged O atom.

The double helix of the DNA is bound this way, but so also is water ice. In solid H_2O, every oxygen atom is symmetrically surrounded by four others with hydrogen bonds between them. Because of these bonds, the melting and evaporation point of water is unusually high compared, for instance, to H_2S, which otherwise should be similar as sulphur stands in the same Group VI of the Periodic Table right below oxygen. This thermodynamic peculiarity of water is crucial for the existence of life.

5.3 Carbonaceous grains and silicate grains

5.3.1 Origin of the two major dust constituents

Interstellar grains, at least their seeds, cannot be made in the interstellar medium; they can only be modified there or destroyed. Observational evidence points towards the wind of red giants on the asymptotic giant branch (AGB) of the Hertzsprung-Russell diagram as the place of their origin. At first, tiny refractory nuclei are created in the inner part of the circumstellar envelope which then grow through condensation as they traverse the cooler outer parts (section 7.4 and 12.5). The sort of dust that is produced depends on the abundance ratio of carbon to oxygen in the envelope.

It can vary and be different from the interstellar one for the following reason. AGB stars have an inert, electron-degenerate core of C and O surrounded by a He layer and farther out by a H shell that is largely convective. The bottom of the He layer periodically ignites in a thermal runaway (a pulse with $L \gtrsim 10^6 \, L_\odot$). This enormous luminosity cannot be carried away radiatively. A convection zone arises that extends to the outer convective envelope and thus brings up C and O, which have fused in the He burning shell, to the surface.

- If there is less carbon than oxygen, $[C]/[O] < 1$, as in M-type and in OH/IR stars, no carbon is available for the solid phase because all C atoms are locked up in CO which is a diatomic molecule with an exceptionally strong bond ($E_{\text{bind}} = 11.1$ eV).

  In such an environment, mainly silicates form, besides minor solid constituents like MgS, MgO and FeO. The observational characteristic of silicates is the presence of an Si–O stretching mode at 9.8μm and an O–Si–O bending mode at 18μm (section 9.3). Silicon has a cosmic abundance $[Si]/[H] = 3 \times 10^{-5}$; it is heavily depleted in the gas phase (depletion factor $f_{\text{depl}}(Si) \sim 30$), so almost all of it resides in grains.

- If $[C]/[O] > 1$, as in C-type stars (these are carbon-rich objects like C,R,N stars), carbonaceous grains form.

  Carbon has a cosmic abundance $[C]/[H] \simeq 3.6 \times 10^{-4}$ and a depletion factor $f_{\text{depl}}(C) \sim 3$. Two out of three atoms are built into solids; one is in the gas phase. There are several forms of carbonaceous grains: mainly amorphous carbon, graphite and PAHs.

Carbon and silicates are the major constituents of interstellar dust. Other types of grains, for whose existence there are either observational or theoretical indications, include metal oxides (MgO, FeO, Fe_3O_4, Al_2O_3), MgS and SiC. The last substance, silicon carbide, is only seen in C stars through its emission signature at $11.3\mu m$, and is not detected in the interstellar medium.

5.3.2 The bonding in carbon

Carbon, the sixth element of the periodic system, has four electrons in its second ($n = 2$) uncompleted shell: two s electrons denoted as $2s^2$ with angular momentum quantum number $l = 0$ and two unpaired p electrons ($2p^2$) with $l = 1$, in units of $\hbar$.

The ground state of carbon is designated $^3P_0$. It is a triplet (prefix 3) with total spin $S = 1$, total orbital angular momentum $L = 1$ (letter P) and total (including spin) angular momentum $J = 0$ (suffix 0). In the ground state, there are only two unpaired p electrons. However, one $2s$ electron may be promoted to a $2p$ orbital and then all four electrons in the second shell become unpaired; this is the chemically relevant case where carbon has four covalent bonds. The promotion requires some energy ($4.2\,eV$), but it will be more than returned in the formation of a molecule.

The bonding of carbon to other atoms is of the covalent type and there are two kinds:

- In a σ-bond, the electron distribution has rotational symmetry about the internuclear axis.

- In a π-bond, the wave function of the two electrons has a lobe on each side of the internuclear axis.

The s and p wave functions can combine linearly resulting in hybrid orbitals. The combination of one s and three p electrons is denoted as an sp^3 hybrid, likewise there are sp^2 and sp hybrids. Here are examples (see figure 5.6):

- In methane, CH_4, the coupling of carbon is completely symmetric to all four H atoms and the difference between s and p valence electrons has disappeared. In the identical hybridized sp^3 orbitals, the C–H binding is through σ-bonds. They point from the C atom at the center of a tetrahedron towards the H atoms at its corners. The angle α between the internuclear axes follows from elementary geometry; one gets $\cos\alpha = -\frac{1}{3}$, so $\alpha = 109.5°$.

- The situation is similar in ethane, H_3C–CH_3, although the symmetry is no longer perfect. Each of the two C atoms has again four sp^3 hybrid orbitals connecting to the other carbon atom and to the neighboring three hydrogen atoms. Because of the σ-bond between the two C's, the CH_3 groups can rotate relative to each other (full rotation is, however, prevented by potential barriers).

- An sp^2 hybrid is present in ethylene, H_2C-CH_2, which is a planar molecule, where the angle from C over C to H equals 120°. There are σ-bonds between C–H and C–C. But the two remaining $2p$ orbitals in the carbon atoms, besides the sp^2 hybrids, overlap to an additional π-bond, so altogether we have a C=C double bond (one σ and one π), which makes the CH_2 groups stiff with respect to rotation.

- In the linear molecule acetylene, HC–CH, the C atoms are sp hybridized yielding C–C and C–H σ-bonds. There are two more π-bonds, whose orbitals are perpendicular to each other and to the internuclear axis; in total, the carbon atoms are held together by a triple bond, C≡C. Acetylene is believed to be the basic molecule for the nucleation of carbonaceous grains in the wind of giant stars.

- Benzene, C_6H_6, is an especially simple and symmetric hydrocarbon. In this planar ring (aromatic) molecule, each carbon atom has three sp^2 hybrid orbitals connecting through σ-bonds to the adjacent two C atoms and one H atom. There is still one unhybridized $2p$ electron per C atom left. The carbon atoms are sufficiently close together so that neighboring $2p$ orbitals can pair to form an additional π-bond; the p-orbitals are perpendicular to the plane. Because of the complete symmetry of the ring, the unhybridized electrons in the $2p$-orbitals are not localized, but equally shared (resonant) in the ring. So on average, two adjacent C atoms are coupled by a σ and half a π-bond. The binding is therefore intermediate in strength between a single and a double bond. The more bonds between two C atoms, the stronger they are bound and the shorter their internuclear distance.

5.3.3   Carbon compounds

5.3.3.1   Diamond

In diamond, a crystalline form of carbon, the bonding is similar to methane. Each C atom is linked tetrahedrally to its four nearest neighbors through sp^3 hybrid orbitals in σ-bonds. The distance between atoms is 1.54Å, the angle between the internuclear axes again 109.5°. As a C atom has only four nearest (and 12 next nearest) neighbors, the available space is filled to only 34% *assuming* that the atoms are hard spheres. This is to be compared with a close-packed structure, either face-centered cubic or hexagonal close-packed (see figure 5.8), where in both cases each atom has 12 nearest neighbors and the volume filling factor amounts to 74%.

The bonding in diamond is strong (7.3 eV per atom) and due to its overall bonding isotropy, diamond is extremely hard with hardness number 10; diamond is used for cutting and drilling. The stiffness of the bonds also explains

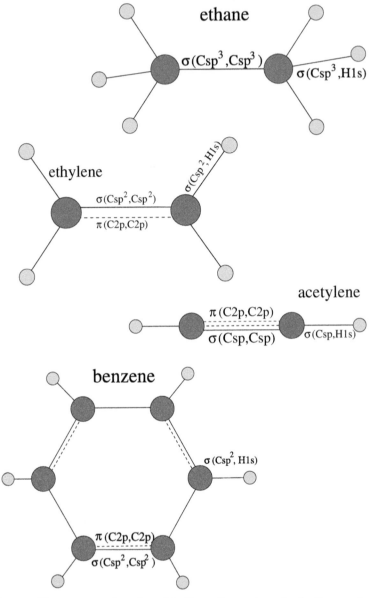

FIGURE 5.6 Four types of carbon binding, *from top to bottom:* single bond in ethane (C_2H_6); double bond in ethylene (C_2H_4); triple bond in acetylene (C_2H_2); resonance structure in benzene (C_6H_6) with three π-bonds shared among six C atoms. *Symbols:* big circles: C atoms; small ones: H atoms. σ-bonds are drawn with full lines, π-bonds broken. In the description of the bond, the type (σ or π) is followed by the atomic symbol and the participating orbitals, for instance, $\sigma(Csp^2, H1s)$.

the exceptionally high thermal conductivity (five times greater than copper). As there are no free electrons, diamond is electrically insulating.

On Earth, diamonds are formed under high pressure and temperature at a depth of about 200 km. They are mined in Kimberlite pipes which are vents associated with volcanoes where the deep material is transported upwards so that it can be reached by man. The upward transport during the volcano eruption occurs sufficiently rapidly without a phase transition. In fact, diamond is not in equilibrium at the conditions at the surface of the Earth; it is metastable in our environment, but well separated by a large activation barrier from graphite, which is the energetically lower phase. So diamonds are long lasting presents that will outlive the affluent donors as well as the pretty recipients of the gems. Big raw diamonds are cleft parallel to octahedral crystal faces. In the subsequent cutting, however, into brilliants, which is the standard gem with a circular girdle and 33 facets above and 25 facets below the girdle, the angles between the facets are not related to the crystalline structure, but to the refractive index n. The angles are chosen to maximize the fire (brilliance) of the stone.

5.3.3.2 Graphite

Carbon atoms combine also to planar structures consisting of many C_6 hexagonals. The distance between adjacent C atoms is 1.42Å, the angle between them 120°. The bonding is similar to benzene as again the π-bonds are not localized in the rings and the electrons of the unhybridized $2p$ orbitals are free to move in the plane. This mobility endows graphite with an electric conductivity within the sheets. Graphite consists of such C_6 sheets (basal planes) put on top of each another. The stacking is arranged in such a way that above the center of an hexagon in the lower sheet, there is always a C atom in the upper one. The distance from one sheet to the next is 3.35Å (figure 5.7). The sheets are only weakly held together by van der Waals forces. Therefore, graphite cleaves well along these planes and owing to this property, it got its name from the Greek word for writing: When gently pressing a pen over a piece of paper, tiny chunks peel off to form sentences of rubbish or wisdom, depending on who is pressing.

Graphite has a density of 2.23 g cm^{-3} and a very high sublimation temperature ($T_{ev} > 2000$ K at interstellar pressures). We reckon that in interstellar space about 10% of the carbon which is in solid form is graphitic.

5.3.3.3 PAHs

A polycyclic aromatic hydrocarbon, briefly PAH, is a planar compound made up of a not too large number of aromatic (C_6) rings (chapter 9). At the edge, hydrogen atoms are attached, but only sporadically; the coverage need not be complete, as in benzene. Possibly one also finds at the periphery other species, besides hydrogen, like OH or more complex radicals. PAHs are unambiguously

graphite sheets

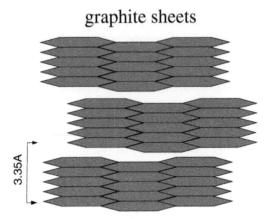

FIGURE 5.7 Graphite is formed by sheets of carbon atoms, each with an hexagonal honeycomb structure. The side length of the hexagons equals 1.42Å; the distance between neighboring sheets is 3.35Å. In analogy to close-packed spheres (figure 5.8), besides the sheet sequence $ABABAB\dots$ also $ABCABC\dots$ is possible and combinations thereof.

next higher layer B

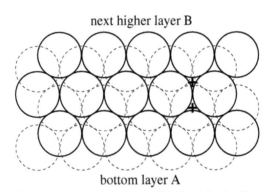

bottom layer A

FIGURE 5.8 A ground layer A of equal balls (***dashed***), each touching its six nearest neighbors, is covered by an identical but horizontally shifted layer B (***solid***). There are two ways to put a third layer on top of B; one only needs to specify the position of one ball in the new layer; all other locations are then fixed. ***a)*** When a ball is centered at the lower cross, directly over a sphere in layer A, one obtains by repetition the sequence $ABABAB\dots$. This gives an hexagonal close-packed structure (hcp). ***b)*** When a ball is over the upper cross, one gets by repetition a sequence $ABCABC\dots$ and a face-centered cubic lattice.

identified in the interstellar medium and account for a few percent by mass of the interstellar dust.

A fascinating kind of PAHs are *fullerenes*. The most famous, C_{60}, has the shape of a football. It is made of 12 pentagons (C_5), which produce the curvature, and 20 hexagons (C_6). The diameter of the spheres is ~ 7.1Å. Fullerines do not contain hydrogen. Their existence in interstellar space is hypothetical.

5.3.3.4 Amorphous carbon

Diamond and graphite are very regular (crystalline). If there are many defects and if regularity extends only over a few atoms, the material is said to be amorphous. Most interstellar carbon grains are probably amorphous (aC). These amorphous particles should also contain hydrogen because it is present in the environment where they are created (wind of mass loss giants); pure carbon grains could only come from the hydrogen deficient atmospheres of WC stars or R CrB stars. We may tentatively think of such hydrogenated amorphous carbon, abbreviated HAC, as a potpourri of carbon compounds with all types of bonds discussed above, or as a disordered agglomerate of PAHs, some of them stacked, with many unclosed rings and an endless variety of carbon-hydrogen connections.

5.3.4 Silicates

5.3.4.1 Silicon bonding and the origin of astronomical silicates

Silicon stands right below carbon in Group IV of the Periodic Table. Its inner two shells are filled; in the third ($n = 3$) it has, like carbon, two s and two p electrons ($3s^2\, 3p^2$ configuration). The ground state is also a $^3P_0$. There are two basic reasons why silicon is chemically different from carbon. First, it takes more energy to promote an s electron in order that all four electrons become available for covalent bonding. Second, silicon atoms are bigger, so their p orbitals cannot come together closely enough to overlap and form a π-bond. Therefore, there are no silicon ring molecules and double bonding (Si=Si) is rare (it involves d electrons).

The dust particles formed in oxygen rich atmospheres are predominantly silicates. Silicates consist of negatively charged SiO_4-groups in the form of tetrahedra with a side length of 2.62Å. They are the building blocks which form an ionic grid together with positively charged Mg and Fe cations. We may idealize an SiO_4 tetrahedron as an Si^{4+} ion symmetrically surrounded by four O^{2-} ions; however, there is also substantial covalent bonding in the tetrahedron.

TABLE 5.2 The limiting ratio of the radius of the central ion to that of its nearest neighbors determines the coordination number and the kind of polyhedron around it.

ratio r of ionic radii	coordination number	polyhedron around central ion
0.15	3	flat triangle
0.23	4	tetrahedron
0.41	6	octahedron
0.73	8	cube
1	12	hexagonal or cubic close-packed

5.3.4.2 Coordination number

It is intuitively clear that the relative size of adjacent atoms in a crystal is important for the crystal structure. Let us define

$$r \ = \ \text{ratio of the radii of neighboring atoms.}$$

In silicates, the radius of the central Si^{4+} ion is 0.41Å and much smaller than the radius of the O^{2-} ion which is 1.40Å, therefore $r = 0.41/1.4 \simeq 0.3$. This implies, first, that oxygen fills basically all the volume; and, second, that the coordination number N of a Si^{4+} ion, which is the number of its nearest equal neighbors, is 4.

In other crystals, r is different. The coordination number and the coordination polyhedron for the limiting values of r, assuming that the ions are hard spheres touching each other, are summarized in table 5.2. For crystals with $0.23 < r \leq 0.41$, the coordination number N equals 4 and the nearest neighbors form a tetrahedron; for $0.41 < r \leq 0.73$, one has $N = 6$ and an octahedron as coordination polyhedron. The case $r = 1$ is depicted in figure 5.8.

5.3.4.3 Olivine, bronzite and other silicate minerals

The (negative) tetrahedral $[SiO_4]$ anions and the (positive) metal cations can be regularly arranged in various ways. The basic silicate types are determined by whether the tetrahedrons share oxygen atoms or not.

- No oxygen atom is common to two tetrahedra.

  They are then isolated and one speaks of islands silicates. The prototype is olivine, $(Mg,Fe)_2SiO_4$ which belongs to the orthorhombic crystal family. It is probably the most common interstellar silicate; in space, it is, however, rather amorphous. By writing (Mg,Fe), one indicates that magnesium may be replaced by iron. As Si, Mg and Fe have very similar cosmic abundances (table 5.3), the forms with pure Mg^{2+} or Fe^{2+} cations should also be present in space, which are

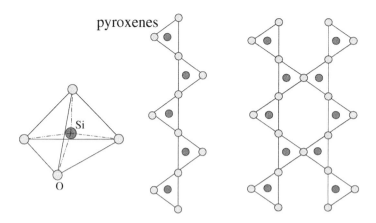

FIGURE 5.9 *Left:* The building blocks of silicates are SiO_4 tetrahedra. *Middle:* In pyroxenes, which have a chain structure, two adjacent tetrahedra share one oxygen atom; the chain repeats after two tetrahedra. The triangles represent the SiO_4 units. In interstellar grains, the chains are usually not very regular and linear, as in crystals, but may more resemble worms because of widespread disorder. *Right:* Other chain types are realized in silicates, too; here is a double chain where more than one O atom per tetrahedron is shared (amphiboles).

- forsterite: Mg_2SiO_4 and
- fayalite: Fe_2SiO_4.

To visualize the three-dimensional structure of olivine, imagine that each Mg^{2+} or Fe^{2+} cation lies between six O atoms of two independent tetrahedra. These six O atoms are at corners of an octahedron around the cation. The ratio r of the ionic radius of the metal cations to that of O^{2-} is about 0.5. Because the O atoms are so big, they form approximately an hexagonal close packed structure (figure 5.8).

- One, two, three or even all four oxygen atoms are shared among neighboring tetrahedra.

Astronomically relevant is bronzite, $(Mg,Fe)SiO_3$. It is a pyroxene (see figure 5.9), so two O atoms are common to neighboring SiO_4 units. The pure forms are

- enstatite: $MgSiO_3$ and
- ferrosilite: $FeSiO_3$.

In orthopyroxenes the unit cell is orthorhombic. They dominate at low temperatures (and when there is no big cation, like Ca^{2+}) and should therefore be favored in astronomical crystals, but a monoclinic structure of the unit cell is also possible (so-called clinopyroxenes).

TABLE 5.3 Solar abundance of elements relevant for dust, their origin, abundance in refractory grains and observed gas phase depletion in diffuse clouds. Numbers are from [Gre01] and references listed in [Whi03].

Element	Solar abundance	Origin	Abundance in dust	Gas phase depletion
H	1	big bang		
He	8.5×10^{-2}	big bang, H-burning		
C	3.6×10^{-4}	He-burning	2.4×10^{-4}	65%
N	9.3×10^{-5}	from C in CNO cycle		20%
O	6.8×10^{-4}	He-burning	1.2×10^{-4}	20%
Ne	1.2×10^{-4}	C-burning		
Mg	3.8×10^{-5}	Ne-burning	3.4×10^{-5}	90%
Al	3.0×10^{-6}	Ne-burning	?	99%
Si	3.6×10^{-5}	O-burning	3.5×10^{-5}	97%
S	2.1×10^{-5}	O-burning	?	10%
Fe	3.2×10^{-5}	supernovae type I	3.2×10^{-5}	99%
Ca	2.3×10^{-6}	Ne-burning	2.3×10^{-6}	100%
P	2.8×10^{-7}	O-burning	?	80%

5.3.5 The origin of the elements found in dust grains

Interstellar dust in diffuse clouds is made of just a handful of elements:

- C (34% by mass), Si (12%), Mg (10%), Fe (21%) and O (23%).

In dark clouds, the grains may additionally have ice mantles of H_2O, NH_3, CO, CO_2, CH_3OH, N_2, O_2 and others, which will double the dust mass.

The mass fractions are not as accurate as they look, but they total 100%. All these elements are synthesized by nuclear burning in stars. Their abundances in the solar photosphere (table 5.3) should be representative also of the interstellar medium because, when the Sun formed, the Milky Way had already two thirds of its present age and the chemical composition has not changed much since then. In the gas phase, most elements are depleted with respect to their solar abundance because they have condensed into solids. Interestingly, those with the strongest depletion have the highest condensation temperature. When one adds up the mass that is missing in the gas phase of diffuse clouds, one gets an upper limit for the mass in solids and this limit implies a maximum dust-to-gas ratio of about 0.08%.

C, O, Mg, Si, Fe are created by the fusion of α-particles (or their multiples); most other elements (except for nitrogen) and the plethora of isotopes are built by neutron capture. The heavy elements come by and large from supernovae (mainly type II; type I delivers Fe). When they blow up they disperse the products of nucleosynthesis into the interstellar medium. Much less is supplied by intermediate mass stars (< 6 M$_\odot$). Although they eject plenty of C, O, Si and others during their AGB phase, these elements have not been forged within them. The degenerate carbon-oxygen cores, where new C and O have been synthesized, are not really tapped by the wind zone.

5.4 Optical constants of dust materials

We have learnt what dust is made of and how it is glued together. Now we present a reasonable set of optical constants $m(\lambda)$ for the major interstellar solid substances: silicate, amorphous carbon, graphite and ice. Optical constants vary with time, at least, in the literature they do; but the question of whether the choice displayed in the figures is up-to-date or out-of-date degrades to secondary importance in view of the general uncertainties about interstellar grain properties.

Let us begin with silicate (figure 5.10). In the imaginary part, k, of the optical constant which causes the attenuation of a plane wave propagating in an infinite medium (see 1.36), we can discern three broad features:

- a bump centered around 1000Å arising from electronic band transitions; the energies involved are around 10 eV;

- the famous 10μm silicate band due to Si–O stretching;

- to its right, a broad hump that starts with the 18μm band from O–Si–O bending (section 9.3) and then gradually declines towards long wavelengths. It reflects the many vibrational modes in the solid.

Between the first and second bump, from 0.2 and 8μm, extinction by silicate particles is remarkably weak, and for big grains ($a \sim 0.1\mu$m) mostly due to scattering.

Of course, even when k is very small, one must not approximate it by zero. We therefore display it in the middle box of figure 5.10 on a logarithmic scale. We also show the ratio of absorption efficiency over grain radius, Q^{abs}/a, to demonstrate that this quantity does not depend on a in two cases:

– when the grains are transparent, i.e., at the shortest wavelengths (the grains may even be big so that $a/\lambda > 1$, see 3.50)

– or when one is in Rayleigh-Jeans limit ($a \ll \lambda$).

The absorption and extinction efficiencies resulting from $m(\lambda)$ for spheres are also plotted in figure 5.10 (top frame). The grains have radii of 0.01 and 0.1μm, respectively, which covers the relevant size range of interstellar dust.

The scattering efficiency, Q^{sca}, can be found as the difference between extinction and absorption. By and large, k and Q have a similar behavior with frequency. In the far infrared and beyond, for $\lambda > 40\mu$m, Q changes proportionally to $\nu^{2.0}$. The exponent is important for estimating dust temperatures (section 6.2.4). At high energies ($\lambda < 100$Å), k and Q^{abs} are directly proportional (see 3.50) and, as explained in section 3.5.3, roughly $Q \propto \nu^{-3}$, with discontinuities at wavelengths where a new inner electron shell can be ionized.

The behavior of amorphous carbon (aC, see figure 5.11) is at X-rays similar to that of silicates, but only qualitatively. From 0.01 and 0.5μm, the extinction is fairly flat, and then falls steadily towards greater λ. In the visual and UV, the albedo of big grains is of order 0.5, and quite small beyond 3μm. With respect to light from stars that are not very hot, roughly in the region from 0.2 to 8μm, carbon particles absorb much more efficiently than silicates of the same size (see also figure 10.7). In the far infrared ($\lambda > 40\mu$m), the slope in the absorptivity is more variable than for silicates: between 100 and 500μm, Q declines proportionally to $\nu^{2.1}$, and at $\lambda > 500\mu$, like $\nu^{1.6}$.

Graphite, the other major solid carbon component, is optically unisotropic. The cross section per unit mass, K, of randomly orientated spheres should be a weighted mean, $K = \frac{2}{3}K_\perp + \frac{1}{3}K_\parallel$. Optical constants and the Qs are displayed in figure 5.12 and 5.13.

An ice mantle can form around a refractory core (mainly silicate, amorphous carbon or graphite) when gases freeze out in a dense clump. The ice is always a mixture of molecules, but water dominates. The mantle will evaporate again when the grains are warmed up to about 100 K, for example, as a result of star formation in or near the clump.

During the evolution of an ice covered grain, energetic events, like absorption of UV photons or passage of cosmic rays, may transform the volatile ices in the mantle into more refractory organic compounds. These compounds are thought to withstand temperatures much above 100 K, so when the grains are returned from the molecular cloud into the diffuse medium, they would be covered by a layer of organic material. Optical constants of such organic refractories and of pure water ice are depicted in figure 5.14.

- With the help of figures 5.10 to 5.14, one can find the absorption or scattering coefficient of grains of various type and at any wavelength. The accuracy in reading off the numbers from the graphs is in all practical applications more than sufficient. If the particle radii $a < 0.01\mu$m, one may use the analytical expressions of the Rayleigh limit (section 3.1, optical constants are given in the figures). If 0.01μm $< a < 0.1\mu$m, one may interpolate between the curves. Only if $a > 0.1\mu$m *and* $a \gtrsim 0.02\,\lambda$ does one have to perform a Mie calculation.

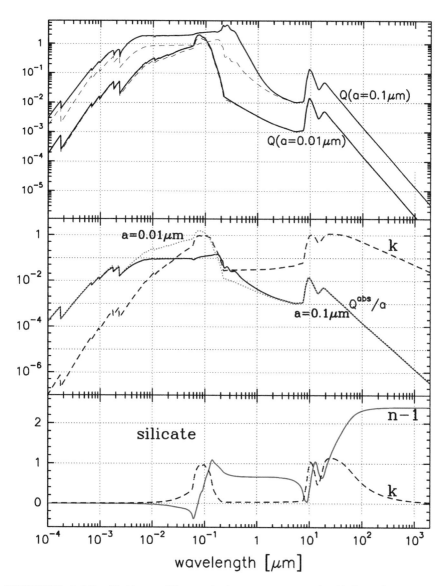

FIGURE 5.10 *Bottom:* The optical constant $m = (n, k)$ for silicate material after [Wei01]. The data fulfill the Kramers-Kronig relations of section 2.6. *Middle:* k (dashes) from bottom frame, but on a logarithmic scale, and Q^{abs}/a in the unit 10^{-6} cm^{-1} for spheres with radii $a = 0.1\mu$m (solid) and 0.01μm (dots). *Top:* Absorption (dashes) and extinction (solid) efficiency, Q^{abs} and Q^{ext}, for $a = 0.01\mu$m and 0.1μm.

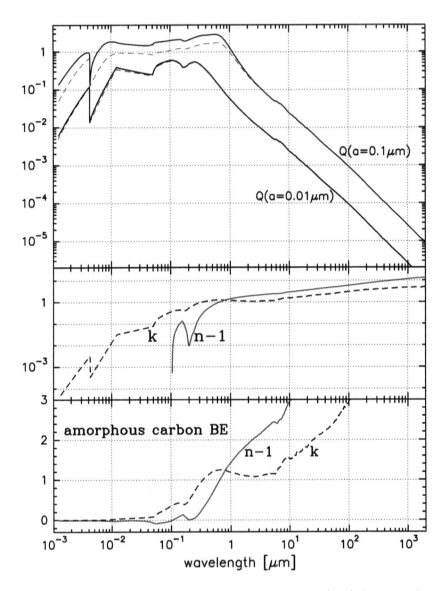

FIGURE 5.11 ***Bottom:*** The optical constant $m = (n, k)$ for amorphous carbon after [Zub96] (their type BE). ***Middle:*** k and n from bottom frame, but on a logarithmic scale. Note that $n - 1$ changes sign around 0.1μm. ***Top:*** Absorption (dashes) and extinction (solid) efficiency, Q^{abs} and Q^{ext}, for spheres with radii $a = 0.01\mu$m and 0.1μm.

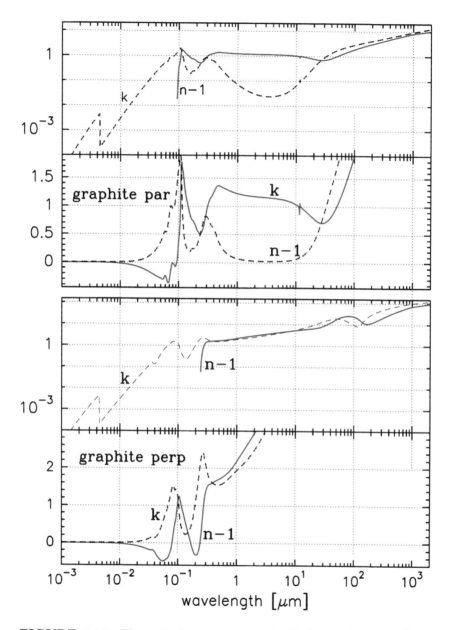

FIGURE 5.12 The optical constant $m = (n, k)$ for graphite after [Lao93] when the electric vector lies in the basal plane and is thus perpendicular to its normal ($m_\perp$, lower boxes) and when the electric vector is perpendicular to the basal plane ($m_\parallel$, upper boxes).

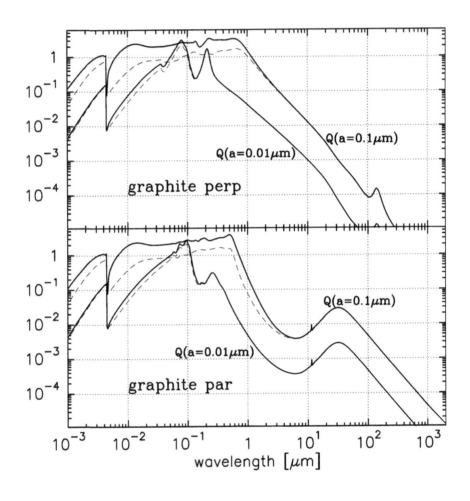

FIGURE 5.13 Absorption (dots) and extinction (solid) efficiency, Q^{abs} and Q^{ext}, of graphite spheres with radii $a = 0.01\mu$m and 0.1μm and optical constants as in figure 5.12. Note the bumps at 2200Å and 800Å for small particles.

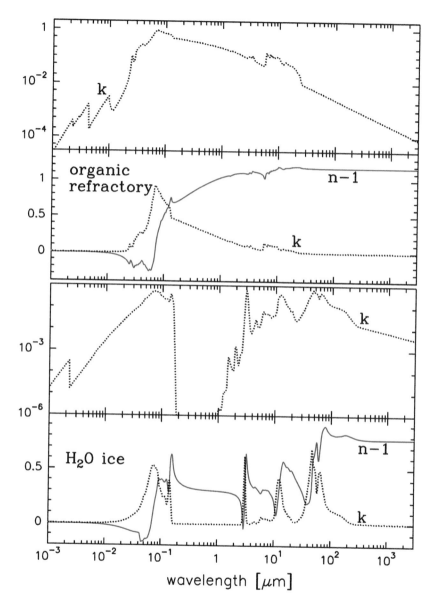

FIGURE 5.14 *Bottom panels:* The optical constant $m = (n, k)$ for pure water ice after [War84]. *Top panels:* The same for organic refractory material after [Li97].

5.5   Grain sizes

5.5.1   The MRN size distribution

Because the interstellar extinction curve (section 10.1.2) covers a large frequency range, particles of different sizes must be invoked to explain its various sections. The infrared part requires large grains, the optical region medium-sized and the UV, the 2175Å bump and the far UV small particles. Qualitatively speaking, we expect that at wavelength λ, the radii a of the relevant grains are given by $2\pi a/\lambda \sim 1$; smaller ones are inefficient as absorbers or scatterers. To quantify this supposition, one can ask how the grains have to be distributed in size in order that their wavelength dependent extinction follows the observed interstellar reddening curve. Adopting that they are spheres and defining the size distribution $n(a)$ such that

$$n(a)\,da \; = \; \text{number of grains per cm}^3 \text{ with radii in the interval } [a,\, a+da] \,,$$

one obtains a power law,

$$n(a) \; \propto \; a^{-q} \qquad \text{with} \quad q \simeq 3.5 \,. \tag{5.9}$$

This so called MRN size distribution [Mat77] is widely accepted. It follows from dividing the grain radii into intervals $[a_i, a_{i+1}]$ with $a_{i+1} > a_i$ and then determining the number of grains n_i in each bin such that one gets the best agreement with the extinction curve. For the dust composition, one assumes silicate and carbon material with optical constants similar to those in figure 5.10 and 5.11; composite grains made of these materials are also possible.

To fix the size distribution, one must additionally specify its upper and lower limit, a_+ and a_-, roughly,

$$a_- \; \simeq \; 10\text{Å} \qquad \text{and} \qquad a_+ \; \simeq \; 0.3\mu\text{m} \,, \tag{5.10}$$

but the outcome is not very sensitive to either boundary. Not to a_+, because extinction becomes gray for large grains, and even less to a_- because small grains are in the Rayleigh limit ($a_- \ll \lambda$) where only the total volume of the grains counts, and not their individual sizes. It is therefore no wonder that the smallest grains ($a < 100$Å) were detected through their emission, and not from studies of the extinction curve.

The transition from grains to molecules is probably continuous. The abrupt cutoff at the upper limit, a_+, in the MRN distribution appears artificial and to avoid it, a modification has been proposed to the more physical form $n(a) \propto a^{-q}\,e^{-a/a_0}$ with $a_0 \simeq a_+$ [Jur94].

For an MRN distribution, which has an exponent $q = 3.5$, the total mass or dust volume V is supplied by the big particles and the geometrical surface

F by the small ones:

$$V \propto \int_{a_-}^{a_+} n(a)\, a^3\, da \quad \propto \quad (\sqrt{a_+} - \sqrt{a_-}) \quad \sim \quad \sqrt{a_+} \tag{5.11}$$

$$F \propto \int_{a_-}^{a_+} n(a) a^2\, da \quad \propto \quad \left(\frac{1}{\sqrt{a_+}} - \frac{1}{\sqrt{a_-}} \right) \quad \sim \quad \frac{1}{\sqrt{a_-}}. \tag{5.12}$$

If q were greater than 4, the small grains would also contain most of the volume.

5.5.2 Collisional fragmentation

One can theoretically obtain a size distribution close to $n(a) \propto a^{-3.5}$ by studying grains undergoing destructive collisions, in which the total mass of the particles is conserved ([Hel70, Dor82]). The process is called shattering (section 7.4.2), and there is no evaporation. Consider now a *time-dependent* mass distribution $N(m,t)$ such that

- $N(m,t)\, dm$ = number of particles in the interval $m \ldots m + dm$

- $P(m,t)\, dt$ = probability that particle of mass m is collisionally destroyed during time dt

- $\xi(\mu, m)\, d\mu$ = number of fragments in the interval μ to $\mu + d\mu$ from smashing a body of mass m

If fragmentation is the only process that determines the mass distribution, $N(m,t)$ is governed by the formula

$$\frac{\partial N(m,t)}{\partial t} = -N(m,t) \cdot P(m,t) + \int_m^M N(\mu,t)\, P(\mu,t)\, \xi(m,\mu)\, d\mu . \tag{5.13}$$

M is the maximum mass of the particles. To solve (5.13), one separates the variables,

$$N(m,t) = \eta(m) \cdot \zeta(t) .$$

The probability $P(m,t)$ for destruction of a particle of mass m is obviously proportional to an integral over relative velocity v_{rel} of the colliding particles times collisional cross section. If we assume, as a first approximation, that v_{rel} is constant, we may write

$$P(m,t) = Q \int_{m'}^M N(\mu,t) \left[\mu^{\frac{1}{3}} + m^{\frac{1}{3}} \right]^2 d\mu = \zeta(t) \cdot \gamma(m) . \tag{5.14}$$

with

$$\gamma(m) = Q \int_{m'}^M \eta(\mu) \left[\mu^{\frac{1}{3}} + m^{\frac{1}{3}} \right]^2 d\mu .$$

Q is some proportionality factor. The lower boundary on the integral, m', is the minimum mass for destructive collisions; very small particles are not destroyed. (5.13) now becomes

$$\frac{\dot{\zeta}}{\zeta^2} = -\gamma(m) + \frac{1}{\eta(m)} \int_m^M \eta(\mu)\,\gamma(\mu)\,\xi(\mu, m)\,d\mu \ . \tag{5.15}$$

The equation is completely seperated as there is pure time dependence on the left, and pure mass dependence on the right. Both sides may be set to a constant. The solution for $\zeta(t)$ is

$$\zeta(t) = \zeta_0/[1 + t \cdot C\zeta_0]$$

where ζ_0 is the value of ζ at $t = 0$ and C is a positive constant. To solve (5.14) with $\dot{\zeta}/\zeta^2 = -C$ for $\eta(m)$, we need to know the mass distribution of the debris, $\xi(\mu, m)$, after each shattering event. One expects that the number of debris, $\xi(\mu, m)$, is related to the ratio of initial mass over fragment mass, m/μ. When one puts

$$\xi(\mu, m) = \theta\, m^x/\mu^{x+1} \ ,$$

empirical evidence suggests for the exponent $0.5 \le x \le 2$. The constant θ follows from the condition of mass conservation

$$\theta \int_0^m \frac{m^x}{\mu^{x+1}} \mu\, d\mu = \theta \int_0^m \left(\frac{m}{\mu}\right)^x d\mu = m \ .$$

With a power law ansatz

$$\eta(m) = m^{-y} \ ,$$

one finds with a little algebra that $y = 1.83$ over a wide range of x values $(0.5 \le x \le 3)$. Because for spheres $m \propto a^3$ and $dm \propto a^2 da$, one recovers, with respect to sizes, the MRN distribution $n(a) \propto a^{3y-2} \simeq a^{-3.5}$.

Interestingly, in the outflow of AGB stars, possibly the major source of dust, the grains, judging from their infrared spectra, already seem to follow an MRN size distribution. It has therfore been suggested that collisional fragmentation is at work there [Bie80]. However, if that is true the particles must be born uncomfortably big before they can be ground down to the final distribution $n(a) \propto a^{-3.5}$.

But even if grains are injected by AGB stars into the interstellar medium with the same size distribution that one observes there, its origin has not been fully explained. First, nothing is known about the sizes with which grains form in other, probably not negligible dust suppliers, like supernovae. Second, grains do suffer further important modifications in the interstellar medium, such as accretion and coagulation (as witnessed by ice mantles or varying values of R_V), or sputtering and evaporation, all of which affect their sizes. A satisfying explanation for the observed size ditribution can only be one in which $n(a)$ is derived from a balance between growth and destruction processes, as first postulated by [Hon78].

6

Dust radiation

We treat in this chapter the physical background of dust emission. A single grain radiates according to Kirchhoff's law and we strictly derive this law in section 6.1 under the assumption that a grain consists of an ensemble of coupled harmonic oscillators in thermal equilibrium. In section 6.2, we show how to compute the temperature to which a grain is heated in a radiation field and section 6.3 illustrate its emission. For evaluating the emission of very small grains, which are those whose temperature fluctuates, we have to know the specific heat and the internal energy of the dust material. We therefore investigate in section 6.4 the calorific properties of dust. Finally, section 6.5 presents examples of the emission of very small grains.

6.1 Kirchhoff's law

6.1.1 The emissivity of dust

A grain bathed in a radiation field acquires an equilibrium temperature T_d which is determined by the condition that it absorbs per second as much energy as it emits. According to Kirchhoff, in local thermodynamic equilibrium (LTE) the ratio of the emission coefficient ϵ_ν (do not confuse with the dielectric permeability ε_ν) to the absorption coefficient K_ν^{abs} is a function of temperature and frequency only. This discovery was already made in the nineteenth century; later it was found that $\epsilon_\nu / K_\nu^{\mathrm{abs}}$ equals the Planck function.

Of course, a grain in interstellar space is not at all in an LTE environment. But whenever it is heated, usually by a photon, the excess energy is very rapidly distributed among the very many energy levels of the grain and the resulting distribution of energy states is given by the Boltzmann distribution and depends only on the dust temperature, T_d. Therefore, grain emission is also only a function of T_d and, as in LTE, given by

$$\epsilon_\nu = K_\nu^{\mathrm{abs}} B_\nu(T_d) . \tag{6.1}$$

The emission over all directions equals $4\pi\epsilon_\nu$. The quantity ϵ_ν is called emissivity. It can refer to a single particle, or to a unit volume or a unit mass.

The dimensions change accordingly. For example, the emissivity ϵ_ν per unit volume is expressed in $\mathrm{erg\,s^{-1}\,cm^{-3}\,Hz^{-1}\,ster^{-1}}$.

6.1.2   Thermal emission of grains

To understand how and why an interstellar grain emits according to equation (6.1), we assume that it consists of N atoms and approximate it by a system of $f = 3N - 6$ (see section 6.4.2) weakly-coupled one-dimensional harmonic oscillators, each of mass m_i, coordinate q_i, momentum p_i and resonant frequency ω_i. Classically, the total energy E of the system is

$$E = E_{\mathrm{kin}} + V + H' \simeq \frac{1}{2} \sum_{i=1}^{f} \left(\frac{p_i^2}{m_i} + m_i \omega_i^2 q_i^2 \right) \qquad (6.2)$$

and consists of kinetic (E_{kin}), potential (V) and interaction energy (H'). The latter is small compared to E_{kin} and V, but it must not be neglected altogether. Otherwise the oscillators would be decoupled and not be able to exchange energy at all: starting from an arbitrary configuration, equilibrium could never be established.

In the thermal (*canonical*) description, the level population of the quantized harmonic oscillators is described by a temperature T. The i-th oscillator has levels $v_i = 0, 1, 2, \ldots$ and can make transitions of the kind $v_i \to v_i - 1$. If the energy difference is lost radiatively, a photon of frequency ν_i escapes. The probability $P_{i,v}$ to find the i-th oscillator in quantum state v is given by the Boltzmann equation (A.9),

$$P_{i,v} = \frac{e^{-\beta h \nu_i (v + \frac{1}{2})}}{Z(T)} = \left[1 - e^{-\beta h \nu_i} \right] e^{-v \beta h \nu_i} ,$$

with the partition function $Z(T)$ from (A.11) and $\beta = 1/kT$. The sum over all probabilities is of course one,

$$\sum_{v=0}^{\infty} P_{i,v} = 1 .$$

To obtain $4\pi \epsilon_{\mathrm{th}}(i, v)$, the radiation integrated over all directions from oscillator i in the transition $v \to v - 1$ at temperature T, we have to multiply $P_{i,v}$ by the Einstein coefficient $A^i_{v,v-1}$ (here the superscript i is not an exponent) and the photon energy $h\nu_i$,

$$4\pi \, \epsilon_{\mathrm{th}}(i, v) = P_{i,v} \, A^i_{v,v-1} \, h\nu_i = \left[1 - e^{-\beta h \nu_i} \right] e^{-v \beta h \nu_i} \, A^i_{v,v-1} \, h\nu_i . \qquad (6.3)$$

After (A.89), $A^i_{v,v-1} = v A^i_{1,0}$ for $v \geq 1$. Note that for an ideal harmonic oscillator, ν_i is the same for all levels v. To compute the total emission per solid angle of oscillator i, we sum up over all levels $v \geq 1$,

$$\epsilon_{\mathrm{th}}(i) = \sum_{v \geq 1} \epsilon_{\mathrm{th}}(i, v) .$$

Because of the mathematical relations (B.1), we obtain

$$\epsilon_{\text{th}}(i) \;=\; h\nu_i\, A^i_{1,0} \sum_{v\geq 1} v \left[1 - e^{-\beta h\nu_i}\right]\, e^{-v\beta h\nu_i} \;=\; \frac{h\nu_i\, A^i_{1,0}}{e^{\beta h\nu_i} - 1} \; . \tag{6.4}$$

This is the total emission of the grain at frequency ν_i. For the integrated radiation ϵ_{th} of the particle over all frequencies, one must add up all oscillators,

$$\epsilon_{\text{th}} \;=\; \sum_i \epsilon_{\text{th}}(i) \; .$$

6.1.3 Absorption and emission in thermal equilibrium

According to the theory of line radiation, the absorption coefficient, integrated over the line, in the transition $v \to v-1$ of the oscillators with frequency ν_i can be expressed through the Einstein B-coefficients and the level populations as the difference between the number of upward and downward induced transitions,

$$K^v_{\nu_i} \;=\; \left[P_{i,v-1} B^i_{v-1,v} - P_{i,v} B^i_{v,v-1}\right] \frac{h\nu_i}{c} \; .$$

As the statistical weight of the levels is one, $B^i_{v,v-1} = B^i_{v-1,v}$ and a change from B to A-coefficient according to (A.92) yields

$$K^v_{\nu_i} \;=\; \frac{c^2}{8\pi\nu_i^2}\, A^i_{v,v-1} \left[1 - e^{-\beta h\nu_i}\right] \left[e^{-(v-1)\beta h\nu_i} - e^{-v\beta h\nu_i}\right] \; .$$

Summing up over all levels $v \geq 1$ gives the absorption coefficient K_{ν_i} of the grain at frequency ν_i,

$$K_{\nu_i} \;=\; \sum_v K^v_{\nu_i} \;=\; \frac{c^2 A^i_{1,0}}{8\pi\nu^2} \; . \tag{6.5}$$

We might drop in equation (6.4) and (6.5) the subscript i, but have to be aware that the frequency ν_i refers to a certain oscillator. When we combine the two formulae, we retrieve Kirchhoff's law (6.1).

Note that the absorption coefficient $K^v_{\nu_i}$, which is taken with respect to a particular pair of levels $(v, v-1)$, still contains the temperature via $\beta = 1/kT$. However, radiation of frequency ν_i interacts with all level pairs and in the *total* absorption coefficient K_{ν_i} the temperature has disappeared. Also note that K_{ν_i} does include stimulated emission because it is incorporated in the above formula for $K^v_{\nu_i}$; however, the typical factor $(1 - e^{-h\nu_i/kT})$ has cancelled out.

6.2   The temperature of big grains

6.2.1   The energy equation

Equation (6.1) permits us in a straightforward way to calculate the dust temperature T_d from the balance between radiative heating and radiative cooling,

$$\int K_\nu^{abs} J_\nu \, d\nu \;=\; \int K_\nu^{abs} B_\nu(T_d) \, d\nu \; . \tag{6.6}$$

K_ν^{abs} is the mass (or volume) absorption coefficient of dust at frequency ν, and J_ν is the average of the radiation intensity over all directions. For spherical grains, our standard case, of radius a and absorption efficiency $Q_\nu^{abs}(a)$,

$$\int Q_\nu^{abs}(a) \, J_\nu \, d\nu \;=\; \int Q_\nu^{abs}(a) \, B_\nu(T_d) \, d\nu \; . \tag{6.7}$$

If other forms of heating or cooling are relevant, for instance, by collisions with gas particles, they have to be added in equation (6.6) and (6.7).

It is evident from (6.7) that it is not the absolute value of Q_ν^{abs} that determines the temperature of a dust grain in a radiation field J_ν. Two efficiencies, $Q_{1,\nu}^{abs}$ and $Q_{2,\nu}^{abs}$, that differ at all frequencies by an arbitrary, but constant factor, yield exactly the same T_d. Qualitatively speaking, a grain is hot when Q_ν^{abs} is high at the wavelengths where it absorbs and low where it emits.

6.2.1.1   Approximate absorption efficiency at infrared wavelengths

When one wants to determine the dust temperature from equation (6.7), one has to know the absorption efficiency $Q_\nu^{abs}(a)$ at all frequencies and then solve the integral equation for T_d. This requires a computer. However, there are approximations to Q_ν^{abs} which allow one to evaluate T_d analytically, often with only a small loss in accuracy.

For a quick estimate, one exploits the fact that dust emits effectively only in the far infrared where the Rayleigh limit is valid. According to (3.5), the absorption efficiency Q_ν^{abs} of spheres is then proportional to the size parameter $2\pi a/\lambda$ times a function that depends only on frequency. Simplifying the latter by a power law, we get

$$Q_\nu^{abs} \;=\; \frac{8\pi a}{\lambda} \cdot \mathrm{Im}\left\{ \frac{m^2(\lambda) - 1}{m^2(\lambda) + 2} \right\} \;=\; a\, Q_0 \nu^\beta \; . \tag{6.8}$$

From the discussion and the figures in section 5.4, it is clear that the exponent β is at infrared wavelengths around two, or somewhat smaller, and we therefore propose for quick estimates

$$Q_\nu^{abs} \;\simeq\; 10^{-23}\, a\nu^2 \qquad [\,a \text{ in cm, } \nu \text{ in Hz}\,] \; . \tag{6.9}$$

Values of β smaller than 1 are excluded for very long wavelengths because of the Kramers-Kronig relations (section 2.6). A blackbody would have $\beta = 0$ and $aQ_0 = 1$. When the complex optical constant $m(\lambda)$ deserves its name and is truly constant, $\beta = 1$.

Putting $Q_\nu^{abs} \propto \nu^2$, overestimates, of course, the absorption efficiency at high frequencies, but this is irrelevant for the evaluation of the integral on the right side of (6.7) describing emission, as long as the grain temperature stays below a few hundred K so that the Planck function at these frequencies is small. With the power law (6.8) for Q_ν^{abs}, the energy equation (6.7) becomes

$$\int_0^\infty Q_\nu^{abs} B_\nu(T)\, d\nu \;=\; aQ_0 \frac{2h}{c^2} \left(\frac{kT}{h}\right)^{4+\beta} \int_0^\infty \frac{x^{3+\beta}\, dx}{e^x - 1} \;.$$

The right integral is evaluated in (B.2). Approximate values are

$$\int_0^\infty \frac{x^{3+\beta}\, dx}{e^x - 1} \;\simeq\; \begin{cases} 24.89 : & \text{if} \quad \beta = 1 \\ 122.08 : & \text{if} \quad \beta = 2 \;. \end{cases} \tag{6.10}$$

6.2.2 Temperature estimates

6.2.2.1 Blackbodies

Easiest of all is to find the temperature of a blackbody, a perfect absorber with $Q_\nu^{abs} = 1$ everywhere. Equation (6.7) then reduces to

$$\int J_\nu\, d\nu \;=\; \frac{\sigma}{\pi} T^4 \;, \tag{6.11}$$

σ being the radiation constant of (A.61). The temperature is now *independent of particle size*. In a given radiation field, a blackbody is usually colder than any other object. But this is not a law; one can construct counter examples. If a blackbody is heated by a star at distance r and bolometric luminosity L, its temperature follows from

$$\frac{L}{4\pi r^2} \;=\; 4\sigma T^4 \;. \tag{6.12}$$

For example, we find that a blackbody r astronomical units from the Sun heats up to

$$T \;=\; 279 \left(\frac{r}{\text{AU}}\right)^{-1/2} \text{K} \;.$$

For $r = 1$, this value is not far from the average temperature of the surface of the Earth (285 K), although clouds reflect some 30% of insolation and of the rest, 25% are absorbed in the atmosphere of which half is radiated downwards. So the Earth, as a blackbody, should be more than 30 K colder. We owe the pleasant temperatures to the greenhouse effect.

6.2.2.2 Interstellar grains directly exposed to stars

For real grains, we may use for the absorption efficiency Q_ν^{abs} the approximation (6.9) together with the integral (6.10) for $\beta = 2$. The energy equation (6.7) then transforms into

$$\int_0^\infty Q_\nu^{\text{abs}} J_\nu \, d\nu \simeq 1.47 \times 10^{-6} \, a \, T_d^6 \qquad \text{[cgs-units]} . \qquad (6.13)$$

In cgs-units, a is expressed in cm, ν in Hz, T_d in K, and J_ν in $\text{erg cm}^{-2}\,\text{s}^{-1}$ $\text{Hz}^{-1}\,\text{ster}^{-1}$. The equation is applicable as long as the grain diameter stays below $\sim 1\mu\text{m}$, which is true in the interstellar medium.

- The heating rate of a grain and, in view of equilibrium, its total emission rate are proportional to T_d^6.

Consequently, a moderate temperature difference implies a vast change in the energy budget. For example, to warm a particle, or a whole dust cloud, from 20K to 30K requires a tenfold increase in flux.

To determine T_d in (6.13), it remains to evaluate the integral on the left. There is a further simplification if the radiation field is hard. For an early type star or even the Sun, one may put at all wavelengths *relevant for absorption*

$$Q_\nu^{\text{abs}} \simeq 1$$

so that

$$\int_0^\infty J_\nu \, d\nu \simeq 1.47 \times 10^{-6} \, a \, T_d^6 \qquad \text{[cgs-units]} . \qquad (6.14)$$

If, additionally, J_ν is diluted blackbody radiation of temperature T_*, the integral over J_ν is directly proportional to T_*^4. Consider a grain at distance r from a hot star of luminosity L, effective temperature T_* and radius R_*. From the position of the grain, the star subtends a solid angle $\pi R_*^2/r^2$ and the intensity towards it is $B_\nu(T_*)$. The average intensity over all directions is therefore

$$J_\nu = \frac{B_\nu(T_*) \, \pi R_*^2}{4\pi r^2} .$$

When we insert this expression of J_ν into (6.14) and use $L = 4\pi\sigma R_*^2 T_*^4$ (see A.63), the stellar radius disappears and we find (a, r in cm, L in erg s^{-1})

$$\frac{L}{16\pi^2 r^2} \simeq 1.47 \times 10^{-6} \, a \, T_d^6 \qquad \text{[cgs-units]} . \qquad (6.15)$$

So the grain temperature falls off with distance from the star like

$$T_d \simeq 4.0 \times \left(\frac{L}{a}\right)^{\frac{1}{6}} r^{-\frac{1}{3}} \qquad \text{[cgs-units]} . \qquad (6.16)$$

If there is extinction along the line of sight, one must replace the luminosity L by a frequency average over $L_\nu e^{-\tau_\nu}$. In view of the usual uncertainties regarding radiation field, grain properties and geometry of the configuration, estimates of the above kind are often no less trustworthy than sophisticated computations.

6.2.3 Relation between grain size and grain temperature

In the estimate (6.16), the dust temperature changes with particle radius a like

$$T_{\mathrm{d}} \propto a^{-\frac{1}{6}} .$$

However, this relation is valid only for grains of intermediate size. If the particles are very small so that the Rayleigh limit is applicable to *both emission and absorption*, i.e., to either side of (6.7), the radius a in the absorption efficiency $Q_\nu^{\mathrm{abs}}(a)$ cancels out (see 3.5) and all tiny grains of the same chemical composition have the same temperature. If the grains are big, much bigger than the particles found in interstellar space, $Q_\nu^{\mathrm{abs}}(a)$ in (6.7) becomes independent of size and the grain temperature levels off to some limit, T_{lim}. This value is similar, but not identical to that of a blackbody, T_{bb}, because Q_ν^{abs} stays under the integrals in (6.7) and is not constant.

As an example, we compute grain temperatures near a luminous star and in the interstellar radiation field (ISRF). Thereby we mean the radiation field in the Milky Way in the neighborhood of the Sun, but outside clouds and not close to any star, rather in the space between the stars. The expected average over solid angle is depicted in figure 6.1. Its main sources are stars of spectral type A and F, giants and interstellar dust. The ISRF should be uniform in the galactic disk on a scale of 1 kpc. One can discern in figure 6.1 several components. The two major ones are of comparable strength in terms of $\nu J_\nu^{\mathrm{ISRF}}$, one is from starlight peaking at $\sim 1\mu$m, the other from interstellar dust with a maximum at $\sim 200\mu$m. Integrated over frequency and solid angle,

$$4\pi \int J_\nu^{\mathrm{ISRF}} \, d\nu \sim 0.04 \quad \mathrm{erg \ s^{-1} \ cm^{-2}} .$$

Figure 6.2 demonstrates the size dependence of the dust temperature for these two environments. Generally speaking, small grains are warmer than big ones. The figure contains none of our above approximations, the absorption efficiencies are calculated from Mie theory with optical constants from figure 5.10 and 5.11. The temperature increases from tiny to huge particles by a factor of about three. For the size range supposed to prevail in the interstellar medium ($a \leq 0.3\mu$m), the maximum temperature ratio is smaller, about 1.5.

We will see in section 6.6 that really tiny grains may not attain an equilibrium temperature at all; instead their internal energy fluctuates in response to the quantum character of the absorbed photons. Then a plot of the kind in figure 6.2 becomes meaningless.

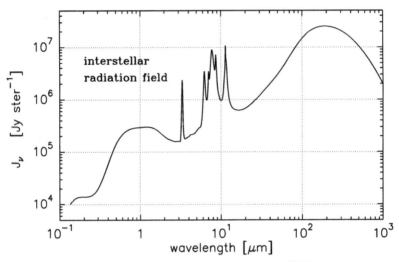

FIGURE 6.1 The approximate mean intensity J_ν^{ISRF} of the interstellar radiation field (ISRF) at the locus of the Sun. The curve is based on data from [Per87], but includes the emission of PAHs (spikes).

6.2.4 Dust temperatures from observations

Whereas gas temperatures can often be obtained rather accurately from properly chosen line ratios, the experimental determination of dust temperatures is always dubious. According to the basic equation of radiative transport (see section 11.1), the intensity I_ν at frequency ν towards a uniform dust layer of optical thickness τ_ν and temperature T_d is

$$I_\nu(\tau) = B_\nu(T_\mathrm{d}) \cdot [1 - e^{-\tau_\nu}] . \tag{6.17}$$

An observer receives from a solid angle Ω then the flux

$$F_\nu = B_\nu(T_\mathrm{d}) \cdot [1 - e^{-\tau_\nu}] \cdot \Omega . \tag{6.18}$$

The flux ratio at two frequencies, marked by the subscripts 1 and 2, becomes

$$\frac{F_1}{F_2} = \frac{B_1(T_\mathrm{d}) [1 - e^{-\tau_1}] \Omega_1}{B_2(T_\mathrm{d}) [1 - e^{-\tau_2}] \Omega_2} . \tag{6.19}$$

It is customary to extract from (6.19) the dust temperature under the following premises:

- The emission is optically thin ($\tau \ll 1$), which is usually true in the far infrared.

- The observational frequencies, ν_1 and ν_2, do not lie both in the Rayleigh-Jeans limit of the Planck function. If they did, T_d would cancel out because $B_\nu(T_\mathrm{d}) \propto T_\mathrm{d}$.

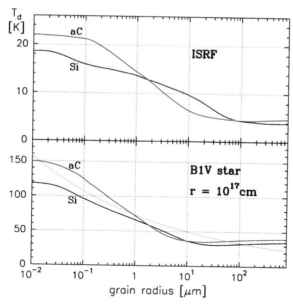

FIGURE 6.2 The temperature of spheres of amorphous carbon (aC) or silicate (Si) as a function of grain radius. *Top:* in the interstellar radiation field with mean intensity J_ν^{ISRF} after figure 6.1. *Bottom:* at a distance of 10^{17} cm from a B1V star with $L = 10^4\,L_\odot$ and $T_* = 2\times10^4$; here the integrated mean intensity of the radiation field is about 7000 times stronger. The dotted line represents T_d from the approximate formula (6.16). Optical constants are from figure 5.10 and 5.11. For comparison, a blackbody would acquire a temperature $T_\mathrm{bb} = 34.2$ K near the B1 star and $T_\mathrm{bb} = 3.8$ K in the ISRF.

- The observations at the two frequencies refer to the same astronomical object, for instance, a certain region in a galactic cloud. This apparently trivial condition is sometimes hard to fulfill when the source is extended relative to the telescope beam and the observations have to be carried out with different spatial resolution.

- The dust temperature in the source is uniform.

- The ratio of the dust absorption coefficients, K_1/K_2, over the particular frequency interval from ν_1 to ν_2 is known, or equivalently the exponent β in (6.8); K_ν itself is not needed.

If all requirements are fulfilled, (6.19) simplifies to

$$\frac{F_1}{F_2} = \frac{K_1\,B_1(T_\mathrm{d})}{K_2\,B_2(T_\mathrm{d})} \tag{6.20}$$

and yields in a straightforward manner T_d. Should the assumed value for K_1/K_2 be wrong, one gets a purely formal color temperature, without any physical correspondence. In view of the numerous restrictions and because the exponent β in (6.8) is debatable, we should always be skeptical about absolute values of T_d. However, we may trust results of the kind

$$T_d(\text{source A}) \; > \; T_d(\text{source B}) \tag{6.21}$$

and their implications for the energy budget because they depend only weakly on the ratio K_1/K_2.

The flux ratio determines also the spectral index α of the energy distribution; it is defined by

$$\frac{F_1}{F_2} \; = \; \left(\frac{\nu_1}{\nu_2}\right)^\alpha .$$

The exponent β in (6.8) and the slope α are *at very long wavelengths* related through

$$\alpha \; = \; \beta + 2 ,$$

independent of temperature. In a double-logarithmic plot of F_ν vs. ν and when λ is large, all curves are parallel for any T. The number 2 in the above equation comes from the Rayleigh-Jeans part of the Planck function. A blackbody ($\beta = 0$), like a planet, has a spectral index $\alpha = 2$.

6.3 The emission of big grains

6.3.1 Constant temperature and low optical depth

As a first illustration of the spectral energy distribution emitted by dust, figure 6.3 displays for silicate grains of 600Å radius the product $Q_\nu^{\text{abs}} B_\nu(T)$. This quantity is proportional to the emissivity ϵ_ν of (6.1). The blackbody intensity ($Q_\nu^{\text{abs}} = 1$) is shown for comparison. The temperatures chosen in figure 6.3 represent three astronomical environments:

- 20 K for the bulk of the dust in the Milky Way

- 60 K for warm dust in star forming regions

- 200 K for hot dust close to stars.

Several points deserve a comment:

- A dust particle emits at all wavelengths less than a blackbody, especially in the far infrared.

- The dust emission is very smooth. For instance, only in the hot dust do we see the 10μm resonance (figure 5.10).

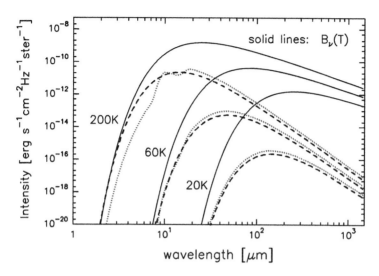

FIGURE 6.3 Emission of a blackbody (*solid*) and of a silicate grain of $600\text{Å} = 0.06\mu\text{m}$ radius at different temperatures. The efficiency Q_ν^{abs} of the silicate grain is calculated from approximation (6.9) (*dashed*) and from Mie theory with optical constants after figure 5.10 (*dots*). The Planck function $B_\nu(T_d)$ as well as $Q_\nu^{\text{abs}} B_\nu(T_d)$ have the units of an intensity.

- A seemingly unspectacular temperature change by a factor of three, say from 20 to 60 K, can boost emission by orders of magnitude.

- The dust emission is very smooth. For instance, only in the hot dust do we see the $10\mu\text{m}$ resonance (figure 5.10).

- Unless the dust is hot ($\gtrsim 100\,\text{K}$), the power law approximation for the absorption efficiency Q_ν^{abs} from (6.9) gives acceptable results.

- The spectrum of a dust grain can be crudely characterized by the wavelength of maximum emission. This is similar to Wien's displacement law for a blackbody (see discussion after A.58).

6.3.1.1 The influence of the optical depth

If the dust in a cloud is at constant temperature and the emission optically thin, the curves $Q_\nu^{\text{abs}} B_\nu(T_d)$ in figure 6.3 are proportional to the observed intensity I_ν as given by equation (6.17). To obtain the *absolute* value of I_ν, one needs the optical depth τ_ν. For small τ_ν, the spectral distribution of the intensity is

$$I_\nu = \tau_\nu B_\nu(T) \,.$$

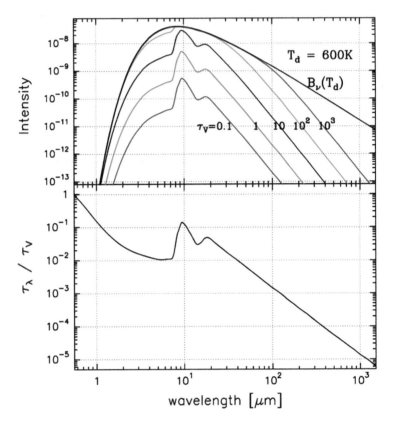

FIGURE 6.4 *Bottom:* The normalized optical depth of silicate grains with 600Å radius. The wavelength scale starts at 0.55μm where $\tau_\lambda/\tau_V = 1$. *Top:* The intensity (in units erg s^{-1} cm^{-2} Hz^{-1} ster^{-1}) towards a cloud of temperature $T = 600$ K filled with such grains for visual optical thickness τ_V from 0.1 and 1000.

In figure 6.4, the intensity, I_ν, has been calculated towards a cloud consisting of identical silicate spheres of 600 Å radius at a temperature of 600 K. The visual optical thickness of the cloud, τ_V, varies between 0.1 and 10^3. The Planck function $B_\nu(T_d)$ corresponds to $\tau_V = \infty$.

The line with $\tau_V = 0.1$ reflects the optically thin case. As τ_V increases, I_ν asymptotically approaches the Planck function. For instance, the cloud becomes optically thick at 25μm and emits like a blackbody for $\lambda \leq 25\mu$m when $\tau_V > 100$. Any information about the emitters is then lost. This is exemplified by the disappearance of the 10μm feature. At long wavelengths, the intensity curves for all τ run parallel. They are equidistant and have the same spectral slope $I_\nu \propto \nu^\alpha$ with $\alpha \simeq 4$.

6.3.2 Total emission and cooling rate of a grain

The cooling rate of a grain, $4\pi\epsilon$, is given by its total emission into all directions. In equilibrium, it is equal to the heating rate. For a sphere of radius a, absorption efficiency Q_ν, and at temperature T, equation (6.1) gives

$$4\pi\,\epsilon(T) \;=\; 4\pi\int \epsilon_\nu\,d\nu \;=\; 4\pi\cdot\pi a^2\int Q_\nu B_\nu(T)\,d\nu\;. \qquad (6.22)$$

If $U(T)$ is the internal energy of the grain, the cooling time is defined by

$$t_{\mathrm{cool}}(T) \;=\; \frac{U(T)}{\epsilon(T)}\;. \qquad (6.23)$$

The total power emitted by one grain, $4\pi\,\epsilon(T)$, its cooling time, $t_{\mathrm{cool}}(T)$, and the internal energy of 1 g of dust are plotted in figure 6.5.

- The cooling time is almost independent of particle size (also true for PAHs) because the internal energy and $\epsilon(T)$ are both proportional to the grain volume. However, silicates cool at all temperatures some three times more slowly than carbon grains.

- The emitting power is for carbon particles at $T > 1000$ K about two times higher than of silicates; below 400 K, the rates are equal.

When one divides equation (6.22) by $4\pi a^2$, the power refers to the unit surface area of the grain. The variation of this quantity (ϵ/a^2) with particle radius is shown in figure 6.6 for three representative temperatures and compared with the radiative power σT^4 of a blackbody surface (see A.62).

Grains smaller than the wavelength of peak emission $(2\pi a < \lambda_{\mathrm{max}})$ are inefficient radiators and ϵ is proportional to the grain volume (Rayleigh-Jeans limit). When they are big, their emission becomes size-independent but does not necessarily approach the blackbody value. At intermediate sizes, a grain may emit more per unit area than a blackbody of the same temperature.

6.4 Calorific properties of solids

To investigate the calorific properties of a dust particle, one treats the conglomeration of atoms in the grain as an ensemble of harmonic oscillators. In a simple-minded model, the atoms are replaced by mass points connected through springs. Their equations of motion follow from the Lagrange function (A.66). If the atomic oscillations are small, one finds for a grain of N atoms

$$f \;=\; 3N - 6$$

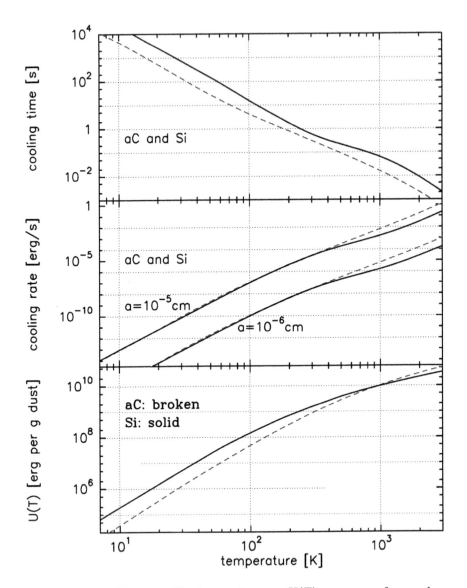

FIGURE 6.5 *Bottom:* The internal energy $U(T)$ per gram of amorphous carbon and silicate; *Middle:* the total cooling rate for one grain of such substances with 0.01 and 0.1μm radius; *Top:* the cooling time of a grain.

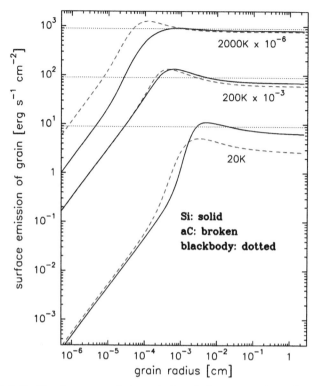

FIGURE 6.6 The frequency integrated emission per unit surface area, ϵ/a^2, (see 6.22) of silicate (Si) and amorphous carbon (aC) spheres as a function of radius for $T = 20\,\mathrm{K}$ (most of the galactic dust), $200\,\mathrm{K}$ (hot dust near star) and $2000\,\mathrm{K}$ (as hot as dust can be). The curves for $T = 200\,\mathrm{K}$ and $2000\,\mathrm{K}$ are reduced by the factors indicated. Optical constants from 5.10 and 5.11. Horizontal lines show blackbodies at the corresponding temperatures.

frequencies describing the internal oscillations of $3N - 6$ harmonic oscillators. The N atoms have, of course, altogether $f = 3N$ degrees of freedom, but one has to subtract $3 + 3 = 6$ which refer to the translatory and rotational motion of the grain as a whole. In most applications, we can put $f \simeq 3N$.

6.4.1 Traveling waves in a crystal

Simple examples are provided by crystals of the cubic sytem. In certain directions of the wave vector (labeled [100], [110] and [111] in crystallography), all atoms in a plane perpendicular to the wave vector oscillate in phase (see illustation in figure 6.7). When a plane is shifted by a wave out of its equilibrium position, the force acting on it depends only on the displacement relative to the adjacent planes. As the displacement is described by one coordinate, the dynamics of the atoms may be treated as if they were arranged in a linear chain.

6.4.1.1   Identical atoms

First, suppose all atoms have the same mass m and are connected through springs of force constant κ. At rest, they are spaced from one another by a distance d. We denote by η_s the displacement of atom s from its equilibrium position. The atoms may oscillate along the chain (longitudinal wave) or perpendicular to it (transverse wave); the force constant κ is then different. The force by which atom s is pushed or pulled depends only on the distance to its neighbor to the left and right, and its equation of motion reads

$$m\ddot{\eta}_s \;=\; \kappa(\eta_{s+1} - \eta_s) - \kappa(\eta_s - \eta_{s-1}) \;=\; \kappa(\eta_{s+1} - 2\eta_s + \eta_{s-1}) \,. \qquad (6.24)$$

Writing η_s as a traveling plane wave, $\eta_s = \eta\, e^{iskd}\, e^{-i\omega t}$, where $k = 2\pi/\lambda$ is the wavenumber and λ the wavelength, yields

$$-m\omega^2 \;=\; \eta_s \;=\; \kappa\,(e^{ikd} - 2 + e^{-ikd}) \;=\; 2\kappa\,(\cos kd - 1)$$

and hence the dispersion relation

$$\omega^2 \;=\; \frac{2\kappa}{m}\left[1 \;-\; \cos(kd)\right] \;=\; \frac{4\kappa}{m}\sin^2\frac{kd}{2}\,. \qquad (6.25)$$

As the ratio of consecutive elongations $\eta_{s+1}/\eta_s = e^{ikd}$, all independent values of the wave vector k lie in the range $-\pi < kd \le \pi$, and the smallest possible wavelength is $\lambda = 2d$. The group velocity becomes

$$v_g \;=\; \frac{d\omega}{dk} \;=\; d\sqrt{\frac{\kappa}{m}}\cos\frac{kd}{2}\,.$$

In the long wavelength limit, when the atoms are very densely packed and $kd \to 0$, as in a continuum, the group velocity is independent of frequency and all waves travel at the same speed. At the short wavelength limit ($k = \pi/d$), the group velocity is zero and one is dealing with a standing wave.

6.4.1.2   Two kinds of atoms

Next, let there be two different atoms in the chain of mass m and $M \ne m$, in alternating sequence, one of each kind per unit cell, but still only one force constant κ. The elongations of the atoms in cell s are η_s and ξ_s, respectively. Atom s of type m interacts with its neighboring atoms s and $s-1$ of type M; likewise for atom s of type M. One then has two equations similar to (6.24),

$$m\ddot{\eta}_s \;=\; \kappa(\xi_s - 2\eta_s + \xi_{s-1}) \qquad M\ddot{\xi}_s \;=\; \kappa(\eta_{s+1} - 2\xi_s + \eta_s)\,.$$

They are again solved by traveling plane waves: $\eta_s = \eta\, e^{i(skd - \omega t)}$ and $\xi_s = \xi\, e^{i(skd - \omega t)}$. The requirement of non-trivial solutions ($\eta, \xi \ne 0$) leads to the

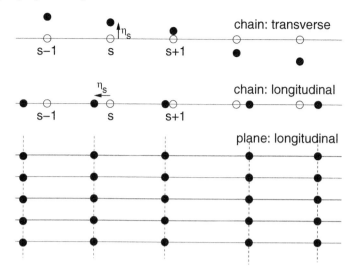

FIGURE 6.7 *Top:* transverse oscillations in a chain of atoms. The open circles show their rest positions, the filled ones their temporary elongations. *Middle:* longitudinal oscillations. *Bottom:* Atoms can move in phase in planes perpendicular to the wave vector, these planes are indicated by vertical dashed lines. The dymanics of the atoms are then the same as in a string. Here for longitudinal oscillations.

dispersion relation

$$\omega_\pm^2 = \frac{\kappa}{mM}\left[M + m \pm \sqrt{M^2 + m^2 + 2mM\cos(kd)}\right]. \qquad (6.26)$$

For each wavenumber k, there are now two eigenfrequencies: the spectrum splits into an acoustical branch (minus sign, low frequencies) and an optical branch (plus sign, high frequencies). Figure 6.8 shows the motion of the atoms, m and M, for a transverse wave.

- At long wavelengths ($kd \to 0$), one finds for the optical and acoustical branch

$$\omega_{\rm opt}^2 = \frac{2\kappa}{\mu} \qquad \omega_{\rm aco}^2 = \frac{\kappa(kd)^2}{2(m+M)},$$

  where $\mu = mM/(m+M)$ denotes the reduced mass. For the optical branch, $\eta/\xi = -M/m < 0$. This implies that the two types of atoms are always swinging towards or away from each other. When they respresent ions of opposite charge, there arises a dipole moment which leads to radiation (or absorption); hence the name *optical mode*.

- At the shortest wavelengths ($kd = \pi$), when additionally $M \gg m$,

$$\omega_{\rm opt}^2 = \frac{2\kappa}{m} \qquad \omega_{\rm aco}^2 = \frac{2\kappa}{M}.$$

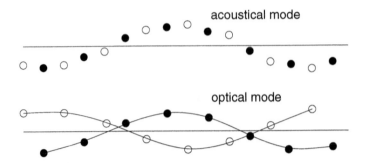

FIGURE 6.8 Optical and acoustical mode for a transverse wave in a chain with two kinds of atoms: open and filled circles.

The waves are standing whenever $v_g = d\omega/dk = 0$.

A (three-dimensional) crystal with one atom per primitive cell has one longitudinal and two transverse waves, the latter with atomic motions (polarizations) perpendicular to each another. If there are $p > 1$ atoms per primitive cell, there appear 3 acoustical and $3p - 3$ optical branches, altogether $3p$. The separation into longitudonal and transverse waves is in a crystal possible only in certain symmetry directions.

6.4.2   Internal energy of a grain

An atom in a solid can be considered as a three-dimensional harmonic oscillator. If the body has N atoms, there are $f = 3N - 6$ one-dimensional oscillators. We bin oscillators of identical frequencies. Let there be s different frequencies altogether. If n_i oscillators have the frequency ν_i, then

$$\sum_{i=1}^{s} n_i = f \, ,$$

and the possible energy levels are

$$E_{iv} = h\nu_i \left(v + \tfrac{1}{2}\right) \qquad v = 0, 1, 2, \ldots, \qquad i = 1, 2, 3, \ldots, s \, .$$

At temperature T, the mean energy of an oscillator of frequency ν_i is

$$\langle E_i \rangle = \tfrac{1}{2} h\nu_i + \frac{h\nu_i}{e^{h\nu_i/kT} - 1} \tag{6.27}$$

(see A.12). The zero-point energy $\tfrac{1}{2} h\nu_i$ is not necessarily small compared to the second term in (6.27), but drops out when calculating the specific heat and is therefore disregarded. One obtains the total energy content U by summing over all oscillators,

$$U(T) = \sum_{i=1}^{s} \frac{h\nu_i}{e^{h\nu_i/kT} - 1} \, n_i \, . \tag{6.28}$$

When N is large, the eigenfrequencies ν_i are closely packed and one may replace the sum by an integral,

$$U(T) = \int_0^{\nu_D} \frac{h\nu}{e^{h\nu/kT} - 1} \rho(\nu) \, d\nu \ , \qquad (6.29)$$

where $\rho(\nu)$ denotes the number density of oscillator frequencies and ν_D the upper integration limit.

6.4.3   The Debye temperature

The density of vibrational modes, $\rho(\nu)$, in a rectangular crystal with sides a_x, a_y, a_z equals

$$\rho(\nu) = \frac{4\pi a_x a_y a_z}{\bar{c}^3} \nu^2 \ .$$

This result follows from the condition that standing waves in the crystal must have nodes at the walls because the atoms cannot move there. The number of eigenfrequencies in the interval ν to $\nu + d\nu$ is therefore

$$dZ = \frac{4\pi a_x a_y a_z}{\bar{c}^3} \nu^2 \, d\nu \ . \qquad (6.30)$$

$\bar{c}$ denotes the average sound velocity. A disturbance can travel in the crystal

- as a longitudinal wave or
- as a transverse wave.

Because the latter has two kinds of polarization corresponding to motions with perpendicular velocity vectors, there are altogether (at least) three types of waves. If c_l and c_t are the sound velocities for the longitudinal and for both transverse waves, $\bar{c}$ is defined by

$$\frac{1}{\bar{c}^3} = \frac{1}{c_l^3} + \frac{2}{c_t^3} \ .$$

If the crystal consists of N atoms, the maximum number of possible modes equals $3N$ and there exists a maximum frequency ν_D such that

$$3N = \int_0^{\nu_D} \rho(\nu) \, d\nu \ .$$

If $n = N/(a_x a_y a_z)$ is the density of atoms, we get

$$\nu_D = \bar{c} \sqrt[3]{\frac{9n}{4\pi}} \ . \qquad (6.31)$$

This is the Debye frequency. The Debye temperature θ is related to ν_D through

$$\theta = \frac{h\nu_D}{k} \ . \qquad (6.32)$$

The frequency density may now be written as

$$\rho(\nu) = \frac{9N}{\nu_D^3} \nu^2 . \tag{6.33}$$

6.4.4 Specific heat

The specific heat at constant volume of a system of f one-dimensional oscillators follows from (6.28),

$$C_v = \left(\frac{\partial U}{\partial T}\right)_V = k \sum_{i=1}^{f} \frac{x_i^2 e^{x_i}}{[e^{x_i} - 1]^2} n_i \qquad x_i = \frac{h\nu_i}{kT} . \tag{6.34}$$

It specifies how much energy is needed to raise the temperature of a body by $1\,\mathrm{K}$. For a continuous distribution of modes, we insert $\rho(\nu)$ from (6.33) into (6.29) to find the internal energy and specific heat,

$$U(T) = \frac{9N}{\nu_D^3} \int_0^{\nu_D} \frac{h\nu^3}{e^{h\nu/kT} - 1} d\nu \tag{6.35}$$

$$C_v(T) = \left(\frac{\partial U}{\partial T}\right)_V = 9kN \left(\frac{T}{\theta}\right)^3 \int_0^{\theta/T} \frac{x^4 e^x}{(e^x - 1)^2} dx . \tag{6.36}$$

There are two important limits:

- At low temperatures, the upper bound θ/T approaches infinity and the total energy of the grain and its specific heat are (see B.2)

$$U = \frac{3\pi^4}{5} NkT \left(\frac{T}{\theta}\right)^3 \tag{6.37}$$

$$C_v = \frac{12\pi^4}{5} Nk \left(\frac{T}{\theta}\right)^3 . \tag{6.38}$$

The characteristic feature in C_v is the proportionality to T^3.

- At high T, the formulae approach the classical situation where a three-dimensional oscillator has the mean energy $\langle E \rangle = 3kT$, so

$$U = 3NkT , \tag{6.39}$$

and the specific heat is constant and given by the rule of *Dulong-Petit*,

$$C_v = 3Nk . \tag{6.40}$$

Any improvement over the Debye theory has to take the force field into account to which the atoms of the body are subjected.

6.4.5 Two-dimensional lattices

When the analysis is repeated for a two-dimensional crystal of N atoms, one finds a density of modes

$$\rho(\nu) = \frac{6N}{\nu_D^2} \nu \, , \tag{6.41}$$

which increases only linearly with frequency, not as ν^2. Therefore now $C_v \propto T^2$, and

$$U(T) = \frac{6N}{\nu_D^2} \int_0^{\nu_D} \frac{h\nu^2}{e^{h\nu/kT} - 1} d\nu$$

$$C_v = 6kN \frac{1}{y^2} \int_0^y \frac{x^3 e^x}{(e^x - 1)^2} dx \, , \qquad y = \frac{\theta}{T} \, . \tag{6.42}$$

The two-dimensional case is important for PAHs because they have a planar structure. It also applies to graphite (which consists of PAH sheets), however, only at temperatures above 20 K. Although the coupling between the sheets is relatively weak, when the temperature goes to zero, C_v eventually approaches the T^3-dependence.

For graphitic sheets, one can improve the model of the specific heat by using two Debye temperatures,

$$C_v(T) = kN \left[f(\theta_z) + 2f(\theta_{xy}) \right] \, . \tag{6.43}$$

θ_z refers to out-of-plane bending, θ_{xy} to in-plane stretching vibrations, and

$$f(y) = \frac{2}{y^2} \int_0^y \frac{x^3 e^x}{(e^x - 1)^2} dx \, .$$

Very low vibrational frequencies cannnot be excited in small grains. One can estimate the minimum frequency ν_{min} from the condition that in the interval $[0, \nu_{min}]$ the PAH has just one frequency,

$$1 = \int_0^{\nu_{min}} \rho(\nu) \, d\nu \, .$$

With $\rho(\nu)$ from (6.41), this gives

$$\nu_{min} = \frac{k\theta}{h\sqrt{3N}} \, . \tag{6.44}$$

The bigger the PAH, the smaller ν_{min}. Collisional excitation of the modes by gas atoms is possible provided $kT_{gas} \geq h\nu_{min}$. Under normal interstellar conditions, only the lowest levels are populated.

Figure 6.9 summarizes the specific heats $C_v(T)$ which we use in the calculations of the emission by small grains. At low temperatures, graphite and silicate have a T^3-dependence, for PAHs C_v changes like T^2. As the grains get warmer, the curves flatten towards the Dulong-Petit rule (6.40), $C_v = 3Nk$. Graphite has 5.0×10^{22} and silicate about 2.6×10^{22} atoms per gram.

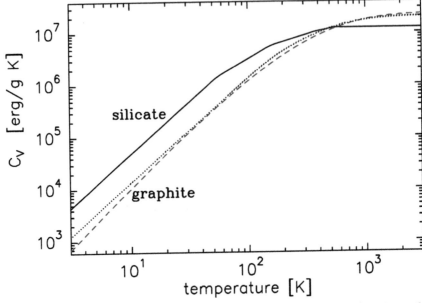

FIGURE 6.9 The specific heat per gram of dust for silicate (solid, similar to [Guh89]), graphite (broken, lab data from [Cha85], and PAHs without H atoms (dotted, using 6.43) with $\theta_z = 950\,\text{K}$ and $\theta_{xy} = 2500\,\text{K}$ [Kru53].

6.5 Temperature fluctuations of very small grains

When a dust particle is very small, its temperature will fluctuate. This happens because whenever an energetic photon is absorbed, the grain temperature jumps up by some not negligible amount and subsequently declines as a result of cooling. We will speak of **very small grains** (vsg) when we have in mind particles whose temperature is time-variable because they are tiny. To compute their emission, we need their optical *and* thermal properties. The optical behavior depends in a sophisticated way on the two dielectric functions $\varepsilon_1(\omega)$ and $\varepsilon_2(\omega)$ and on the particle shape. The thermal behavior is determined more simply from the specific heat.

6.5.1 The probability density $P(T)$

Consider a large ensemble of identical grains in some interstellar environment. Let us arbitrarily pick out one of them and denote by $P(T)\,dT$ the chance that its temperature lies in the interval from $T \ldots T + dT$ and call $P(T)$ the probability density. It is of course normalized,

$$\int_0^\infty P(T)\,dT = 1 . \tag{6.45}$$

In normal interstellar grains of average size, the temperature oscillates only a little around an equilibrium value T_{eq} and in the limit of large grains, $P(T)$ approaches the δ-function $\delta(T_{eq})$, where T_{eq} follows from the steady-state balance between emission and absorption after (6.7),

$$\int Q_\nu^{abs} J_\nu \, d\nu = \int Q_\nu^{abs} B_\nu(T_{eq}) \, d\nu .$$

Even for a very small particle we will assume that its radiation obeys at any time Kirchhoff's law (6.1), so in case of a sphere of radius a, we can express its average monochromatic emission per solid angle by

$$\epsilon_\nu = \pi a^2 \, Q_\nu^{abs} \int B_\nu(T) \, P(T) \, dT . \tag{6.46}$$

Although the emission of a single such grain is not time-constant, the whole ensemble radiates at any frequency at a steady rate. We are faced with the problem of finding $P(T)$ and describe below its solution (see [Guh89] and [Dra01]).

6.5.2 The transition matrix

When a grain absorbs or emits a photon, its internal energy $U(T)$, which is a function of temperature only, changes. We bin $U(T)$ into N states U_j of width ΔU_j with $j = 1, 2, \ldots, N$. Each state U_j corresponds to a temperature T_j or frequency $\nu_j = U_j/h$ with corresponding spreads ΔT_j and $\Delta \nu_j$. The probability P_j of finding an arbitrary grain in a large ensemble of $\mathcal{N}$ particles in state U_j is equal to the number of all grains in level j divided by $\mathcal{N}$.

An absorption or emission process implies a transition in U from an initial state i to a final state f. They occur at a rate $\mathcal{N} P_i A_{fi}$, where the matrix element A_{fi} denotes the transition probability that a single grain changes from state i to f. In equilibrium, for each level j the number of populating and depopulating events, N_{pop} and N_{depop}, must be equal,

$$N_{pop} = N_{depop} .$$

Therefore, for each level j,

$$N_{\text{pop}}/\mathcal{N} = \sum_{k>j} P_k A_{jk} + \sum_{k<j} P_k A_{jk} = \sum_{k\neq j} P_k A_{jk}$$

$$N_{\text{depop}}/\mathcal{N} = \underbrace{P_j \sum_{k<j} A_{kj}}_{\text{cooling}} + \underbrace{P_j \sum_{k>j} A_{kj}}_{\text{heating}} = P_j \sum_{k\neq j} A_{kj} \ .$$

In both formulae, the first sum after the first equal sign refers to processes that cool and the second to those which heat the grain. With the purely mathematical definition

$$A_{jj} = -\sum_{k\neq j} A_{kj} \ ,$$

we may write for all j the condition $N_{\text{pop}} = N_{\text{depop}}$ as

$$\sum_k A_{jk} P_k = 0 \ . \tag{6.47}$$

Only $N-1$ of these N equations are linearly independent. To find the probability density $P(T)$ required in equation (6.46), one may first put $P_1 = 1$, solve (6.47) for $P_2, \ldots, P_n$ and then rescale the P_j by the obvious condition that all probabilities must add up to one (see 6.45),

$$\sum_j P_j = 1 \ . \tag{6.48}$$

A matrix element A_{kj} referring to dust heating $(j < k)$ is equal to the number of photons of frequency $\nu_k - \nu_j$ which a grain absorbs per Hz and second times the width of the final bin $\Delta\nu_k$,

$$A_{kj} = \frac{4\pi C_\nu^{\text{abs}} J_\nu}{h\nu} \Delta\nu_k \ , \qquad \nu = \nu_k - \nu_j \ , \quad j < k \ . \tag{6.49}$$

J_ν stands for the mean intensity of the radiation field and C_ν^{abs} is the absorption cross section of the grain. As it should be, the number of transitions $P_j A_{kj}$ from $j \to k$ is thus proportional to the width of the initial (via P_j) and the final energy bin. Likewise we have for dust cooling from state j to a lower one k,

$$A_{kj} = \frac{4\pi C_\nu^{\text{abs}} B_\nu(T_j)}{h\nu} \Delta\nu_k \ , \qquad \nu = \nu_j - \nu_k \ , \quad k < j \ . \tag{6.50}$$

Above the main diagonal stand the cooling elements, below those for heating. The energy balance between cooling and heating for each level j reads

$$\underbrace{\sum_{k<j} A_{kj} \nu_{kj} = \int_0^\infty 4\pi C_\nu^{\text{abs}} B_\nu(T_j)\, d\nu}_{\text{cooling}} \ , \qquad \underbrace{\sum_{k>j} A_{kj} \nu_{kj} = \int_0^\infty 4\pi C_\nu^{\text{abs}} J_\nu\, d\nu}_{\text{heating}} \ ,$$

with $\nu_{kj} = |\nu_j - \nu_k|$. As cooling proceeds via infrared photons which have low energy, their emission changes the grain temperature very little. This suggests that one needs to consider in cooling from state j only the transitions to the levels immediately below. In fact, in practical applications it suffices to ignore cooling transitions with $j \to k < j - 1$. One can therefore put all matrix elements A_{fi} above the main diagonal to zero, except $A_{j-1,j}$. But the latter, in order to fulfill the energy equation, have to be written as

$$A_{j-1,j} = \int_0^\infty 4\pi C_\nu^{\text{abs}} B_\nu(T_j) \, d\nu \cdot [h(\nu_j - \nu_{j-1})]^{-1} . \tag{6.51}$$

The total matrix A_{fi} has thus acquired a new form where, above the main diagonal, only the elements A_{fi} with $f = i - 1$ are non-zero. One now immediately obtains from (6.47) the computationally rapid recursion formula (but see the simple trick described in [Guh89] to safeguard against numerical rounding errors):

$$P_{j+1} = -\frac{1}{A_{j,j+1}} \sum_{k \leq j} A_{kj} P_k \qquad j = 1, \ldots, N - 1 . \tag{6.52}$$

We mention that *heating* may not be reduced to transitions $j \to j + 1$. This would ignore the big energy jumps of the grain after UV photon absorption which are important for the probability function $P(T)$. Although heating elements of the form

$$A_{j+1,j} = \int_0^\infty 4\pi C_\nu^{\text{abs}} J_\nu \, d\nu \cdot \left[h(\nu_{j+1} - \nu_j) \right]^{-1}$$

do not violate energy conservation, they would result in an unrealistically small spread around the equilibrium temperature T_{eq} of (6.7).

6.5.3 The stochastic time evolution of grain temperature

One can also follow the temperature evolution of a grain by solving the differential equation

$$\frac{dU}{dt} = 4\pi \cdot \pi a^2 \left\{ \int Q_\nu^{\text{abs}} J_\nu(t) \, d\nu - \int Q_\nu^{\text{abs}} B_\nu(T(t)) \, d\nu \right\} . \tag{6.53}$$

Its internal energy U changes in time and the right side describes the difference between the power absorbed from the radiation field J_ν and the cooling rate. Whereas the cooling flux of low energy infrared photons may be assumed to be continuous, heating must be treated as a sequence of stochastic absorptions of single energetic photons. Having followed $U(T)$ or $T(t)$ over a long period, one can obtain the probability density $P(T)$ from the fraction of time that the grain has spent in the temperature interval $T \ldots T + dT$.

6.6   The emission spectrum of very small grains

We illustrate the stochastic temperature fluctuations and their effect on the spectrum for grains in the environment of an B1V star. It is luminous ($L = 10^4$ $L_\odot$), has a hot atmosphere ($T_{\text{eff}} = 2 \times 10^4 \, \text{K}$) and guarantees a copious amount of energetic quanta.

The emission coefficient ϵ_ν in figures 6.10 and 6.11 refers to one grain and has the unit $\text{erg s}^{-1} \, \text{Hz}^{-1} \, \text{ster}^{-1}$. The correct emission from (6.46) is displayed as a solid line. For comparison, the emission under the false supposition of constant temperature from (6.1) is shown dotted.

The probability density $P(T)$ is computed in two ways, either from the stochastic time evolution of the temperature $T(t)$ which follows from solving (6.53) or, more simply, from the formalism developed in section 6.5.2. In the first case, we depict $P(T)$ by dots, otherwise by a solid line.

6.6.1   Moderate fluctuations

The top panel of figure 6.10 displays the temperature variation $T(t)$ of a single silicate grain of 40Å radius at a distance $r = 10^{18} \, \text{cm}$ from the star. We followed $T(t)$ over $\tau = 10^5 \, \text{s}$, but show only an arbitrary section of 4000 s. The temperature excursions amount to some 50% around a mean value. As one can judge from the figure, the grain is hit by an energetic photon about once every 100 s and the temperature excursions are substantial.

The probability density $P(T)$ in the lower left panel of figure 6.10 gives the chance of finding the grain within a temperature interval of 1 K width centered on T. When using equation (6.53) to compute $P(T)$ (dotted line), the total time interval over which we integrated, was large, but finite, and temperatures far from the mean never occurred. One would have to wait very long to see the grain, say, at 170 K. Therefore the dotted line does not extend to probabilities below $\sim 10^{-5}$, is determined well only around the maximum of $P(T)$, and displays a scatter. The method of calculating $P(T)$ from the transition matrix is clearly superior; however, it does not give such a vivid idea of the temperature oscillates.

The bottom frame on the right shows the emission spectrum. When it is evaluated under the false supposition of temperature equilibrium (dotted), the spectrum is a good approximation only at far infrared wavelengths. In the mid infrared, the errors are large (two powers of ten at $10\mu\text{m}$) because the grain is occasionally at temperatures far above the average. The total frequency integrated emission is, however, the same for the dotted and the solid curve and equal to the absorbed energy.

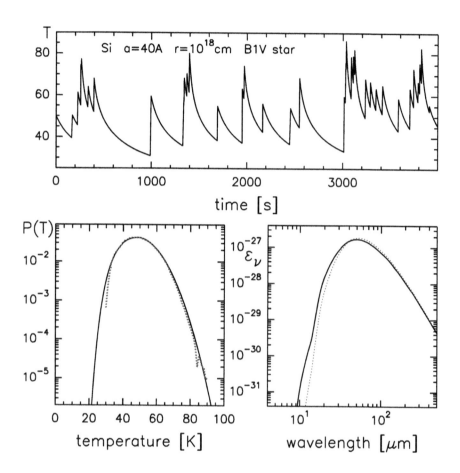

FIGURE 6.10 Moderate stochastic temperature excursions. ***Top:*** Time evolution of the temperature of a silicate grain of 40Å radius at a distance of 10^{18} cm from a B1V star. ***Lower left:*** Probability distribution $P(T)$ of the temperature shown on a linear scale. The dotted line is computed from equation (6.53), the solid from section 6.5.2. ***Lower right:*** Emissivity ϵ_ν of the grain in erg s^{-1} Hz^{-1} ster^{-1}. Dots show emission assuming a constant temperature; solid line includes temperature fluctuations.

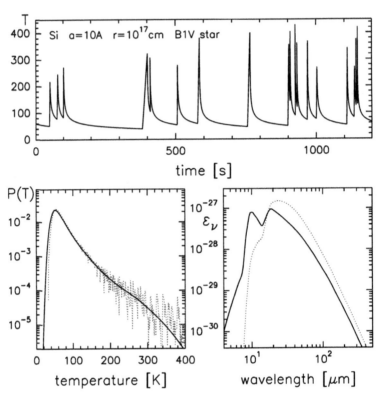

FIGURE 6.11 Strong fluctuations. As figure 6.10, but the grain is now small ($a = 10\text{Å}$) and closer to the star ($r = 10^{17}$ cm).

6.6.2 Strong fluctuations

In figure 6.11, the grain is much closer to the star ($r = 10^{17}$ cm), but has a radius of only 10Å. Compared to the preceding example where $a = 40\text{Å}$, the heat capacity of the particle is now $4^3 = 64$ times lower and photon absorption induces a much larger relative change in the internal energy. The absorption cross section is also 4^3 times smaller than before (Rayleigh limit) and photon capture is accordingly less frequent.

When looking at the temperature evolution $T(t)$, which gives a better feeling for the scatter than the probability function $P(T)$, one hesitates to assign an average temperature at all, although mathematically this can of course be done. There are now two disparate regimes: Most of the time the grain is cold and cooling is slow, but occasionally the grain is excited to a high temperature from which it rapidly cools. Absorption of lower energy photons is rare because the correpsonding Q-values are so small.

The probability density $P(T)$ has turned asymmetric; the maximum is at $T_{max} = 51.9\,K$ and far from the equilibrium temperature $T_{eq} = 116.2\,K$ after (6.7). The dotted line, representing $P(T)$ as determined from $T(t)$, is jerky and inaccurate above $120\,K$ because of the finite time over which we calculated the evolution; a longer integration time would smooth it. To evaluate the emission without taking into account the hot excursions does not make sense any more, even in the far infrared. The dotted line in the bottom right box bears no resemblance to the real emission (solid), although the total flux is, of course, in both cases the same. Note the strong $10\mu m$ silicate feature.

We can follow in figure 6.11 in detail the individual absorption events, associated with different photon energies, and the subsequent cooling. The cooling time is given by equation (6.23) as the ratio of the energy reservoir $U(T)$ of the grain over its cooling rate, and it is displayed in figure 6.5. The rapid rise of the cooling time towards lower temperatures is reflected by the broad wings of the spikes in figure 6.11. For the spikes of the present examples, t_{cool} is of order $10\,s$.

6.6.3 Temperature fluctuations and flux ratios

Very small grains display in their emission another peculiar feature. It concerns their color temperature or, equivalently, the flux ratio at two wavelengths, λ_1 and λ_2. In a reflection nebula, a grain of normal size ($> 100\text{Å}$) becomes colder when the distance to the exciting star is increased, obviously because it receives less energy. For a very small grain, the situation is more tricky. Of course, it also receives fewer photons farther from the star, but its color temperature at shorter wavelengths is only determined by the hot phases corresponding to the spikes in figure 6.11. The interesting point is: no matter whether these spikes are rare or common, as long as they do not overlap, the flux *ratio* is constant; only the emitted *power* in the wavelength band from λ_1 to λ_2 diminishes with distance.

Figure 6.12 shows the flux ratio for several wavelengths as a function of distance from the star. The grains are made of graphite and have radii between 5 and 40Å. For all grain sizes, the near infrared colors, like $K - M$ corresponding to $\epsilon_{2.2\mu m}/\epsilon_{4.8\mu m}$, first fall as one recedes from the star. But already at distances smaller than the typical dimension of a reflection nebula, they level off. Near infrared color temperatures do not change across a reflection nebula if the grain radius is $\sim 10\text{Å}$ or smaller. For mid and far IR colors, the critical particle size is pushed up a bit and the flux ratios are constant only at larger distances. They continue to stay constant far away from the star where the stellar UV radiation field resembles that of the diffuse interstellar medium.

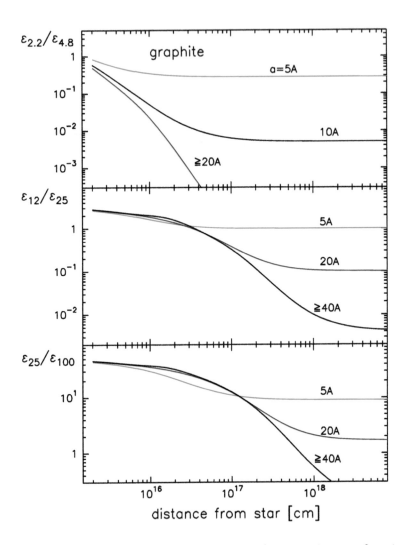

FIGURE 6.12 Flux ratios for graphite grains of various sizes as a function of distance to a B1V star. The wavelengths in microns are indicated as subscripts in the emissivity. Small particles undergo strong temperature fluctuations which lead to flux ratios that are almost independent of the distance to the star.

7

Dust and its environment

7.1 Grain charge

A grain in interstellar space is not likely to be electrically neutral. Mechanisms are at work that tend to alter its charge, notably

- electron impact

- impact of positive ions

- ejection of an electron by a UV photon

In equilibrium, the processes that make the grain positive and negative exactly balance.

7.1.1 Charge equilibrium in the absence of a UV field

First we neglect the radiation field. Consider a spherical grain of radius a and charge Z. Its effective cross section for capturing electrons of mass m_e and velocity v is

$$\sigma_e(v) = \pi a^2 \left(1 + \frac{2Ze^2}{am_e v^2}\right) . \tag{7.1}$$

The term in brackets determines the change over the pure geometrical cross section $\sigma_{\text{geo}} = \pi a^2$. The bracket has a value greater than one when the dust is positively charged ($Z > 0$). In case of negative grain charge ($Z < 0$), it is smaller than unity, and it vanishes when v equals the critical value, v_0, determined by

$$\frac{m_e v_0^2}{2} = -\frac{Ze^2}{a} \qquad (Z < 0) . \tag{7.2}$$

Slower electrons are repelled; they do not reach the surface. Equation (7.1) follows immediately from the conservation of energy and angular momentum for an electron with grazing impact,

$$\tfrac{1}{2}m_e v^2 = \tfrac{1}{2}m_e V^2 - \frac{Ze^2}{a} \tag{7.3}$$

$$a_{\text{eff}}\, v = a\, V . \tag{7.4}$$

V is the actual impact velocity and $\pi a_{\text{eff}}^2 = \sigma_e$ the effective cross section. The number of electrons striking and then staying on the grain equals

$$Y_e \, n_e \, \langle v\sigma_e(v)\rangle \;=\; Y_e \, n_e \, \pi a^2 \cdot 4\pi \left(\frac{m_e}{2\pi kT}\right)^{3/2} \int_{v_l}^{\infty} v^3 \left[1 + \frac{2Ze^2}{am_e v^2}\right] e^{-m_e v^2/2kT} \, dv$$

(7.5)

n_e is their density, Y_e their sticking probability. The bracket $\langle \ldots \rangle$ on the left hand side denotes an average over the Maxwellian velocity distribution (A.14) at the temperature T of the plasma. The lower integration limit, v_l is zero for $Z \geq 0$, and $v_l = v_0$ from (7.2) when $Z > 0$. There are corresponding formulae for ion capture where the quantities m, σ, n, Y bear the subscript i. For a plasma with equal density and sticking probability for electrons and singly charged ions, $n_e = n_i$ and $Y_e = Y_i$. In equilibrium, we can write for the brackets $\langle \ldots \rangle$ that appear in (7.5),

$$\langle v\sigma(v)\rangle_e \;=\; \langle v\sigma(v)\rangle_i \, .$$

(7.6)

On the left, the average refers to a Maxwell distribution for electrons and on the right for ions. The impact rate of electrons on a grain of charge Z is $n_e \langle v\rangle \sigma_{\text{eff}}$, where

$$\sigma_{\text{eff}} \;=\; \sigma_{\text{geo}} \cdot \begin{cases} (1 + Ze^2/akT) : \text{if } Z > 0 \\ e^{Ze^2/akT} \qquad : \text{if } Z < 0 \end{cases}$$

(7.7)

is the effective cross section and $\langle v\rangle = \sqrt{8kT/\pi m_e}$ the mean electron velocity. There is a corresponding equation for ions.

Without photoemission, the grain must be negatively charged ($Z < 0$) because the electrons are much faster than ions. When we evaluate the integrals $\langle v\sigma(v)\rangle$ using the relations (B.6), (B.8) for ions and (B.10) and (B.12) for electrons, we get

$$\exp\left(\frac{Ze^2}{akT}\right) \;=\; \left(\frac{m_e}{m_i}\right)^{\frac{1}{2}} \left[1 - \frac{Ze^2}{akT}\right] \, .$$

Solving this equation for $x = Ze^2/akT$ and assuming that the ions are protons, one finds

$$Z \;\simeq\; -2.5 \, \frac{akT}{e^2} \, .$$

(7.8)

Interestingly, the degree of ionization and the density of the plasma do not appear in this formula. For a fixed temperature, large grains bear a greater charge than small ones (figure 7.1), but the potential

$$U \;=\; \frac{Ze}{a}$$

(7.9)

does not change. To obtain the surface potential in Volt, when the electron charge e is expressed in esu (table B.3) and a in cm, one has to multiply U in (7.9) by 300.

As one would expect, Z in (7.8) adjusts itself in such a way that, by order of magnitude, the mean kinetic energy of a gas atom, kT, equals the work Ze^2/a necessary to liberate one unit charge. We notice that in a hot plasma, the charge can become very large.

7.1.1.1 The distribution of charges

Not all grains, when a and T are fixed, will carry exactly the charge given by (7.8). Instead, there will be a range of charges. To find the distribution function, we assume that no multiple ionized ions are in the gas. It then suffices to consider the balance between levels Z and $Z+1$ and to put the net "flux", J_Z, from Z to $Z+1$ to zero. This net flux is analogous to the one in grain formation in (7.49). Defining

$$x = \frac{e^2}{akT}$$

and n_Z as the number of grains carrying charge Z, we get

$$J_Z = n_{Z+1} \frac{2k^2T^2}{m_e^{1/2}} e^{(Z+1)x} - n_Z \frac{2k^2T^2}{m_i^{1/2}} (1 - Zx) = 0 .$$

The term containing n_{Z+1} is the transition rate $Z+1 \to Z$ induced by electrons, the one with n_Z the rate from $Z \to Z+1$ due to ions, altogether

$$n_Z = n_{Z+1} \left(\frac{m_i}{m_e} \right)^{1/2} \frac{e^{(Z+1)x}}{1 - Zx} .$$

The resulting charge distribution is a spread around the value of (7.8), as illustrated in figure 7.1. The inclusion of photons in the ionization balance (see below) can change the picture. Another interesting effect, not described here, comes from the polarization of grains by the colliding ions or electrons when they are at close distance [Dra87].

7.1.2 The photoelectric effect

7.1.2.1 Charge balance

In the presence of a hard radiation field of mean intensity J_ν, one has to include in the charge equilibrium equation (7.6) a term that accounts for the electrons that are chipped off the bulk material of the grain by UV photons. This certainly happens near a star of early spectral type, but also in diffuse clouds permeated by the average interstellar radiation field. If photoemission is strong, the grains will be positively charged and one can then neglect the impact of ions. Putting the sticking probability Y_e of electrons to one, we find for the charge balance with respect to impinging electrons and absorbed photons

$$n_e \langle v \rangle \left(1 + \frac{Ze^2}{akT} \right) = 4\pi \int_{\nu_t}^{\infty} \frac{J_\nu Q_\nu^{abs}}{h\nu} y_\nu \, d\nu . \qquad (7.10)$$

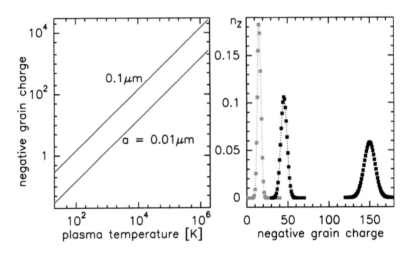

FIGURE 7.1 *Left:* The equilibrium charge of a grain (in units of the charge of an electron) in a hydrogen plasma of temperature T for two particle radii. Photoelectric processes are absent. *Right:* The charge distribution, n_Z, for $x = e^2/akT = 0.0167, 0.0557$ and 0.167. The right bell curve ($x = 0.0167$) corresponds to $a = 10^{-5}$ cm and $T = 10^4$ K. The sum of all populations n_Z is one.

We divided under the integral by $h\nu$ because we wanted the *number* of absorbed photons. To solve (7.10) for the charge Z, we still have to know the parameters ν_t and y_ν. The first, ν_t, represents a threshold frequency for photon absorption leading to electron emission; the energy $h\nu_t$ is of order $10\,\mathrm{eV}$ and includes the work to liberate an electron from the solid ($\sim 4\,\mathrm{eV}$) and to overcome the potential U in (7.9) of the positively charged grain. The second parameter, y_ν, is the yield for photoemission, to be discussed below.

When we consider a charge distribution, the flux balance between ionization stage Z and $Z+1$ is given by

$$n_{Z+1}n_e\sigma_e\langle v_e\rangle = n_Z \pi a^2 y_\nu N_{\mathrm{UV}}$$

where σ_e is from (7.1), $\langle v_e\rangle = \sqrt{8kT/\pi m_e}$ and N_{UV} is the number of photons per unit area and time capable of producing the photoelectric effect. Defining again $x = e^2/akT$ and putting $\frac{1}{2}m_e v^2 = \frac{3}{2}kT$ yields

$$\frac{n_Z}{n_{Z+1}} = \frac{n_e}{y_\nu N_{\mathrm{UV}}}\sqrt{\frac{8kT}{\pi m_e}}\left[1 + \tfrac{2}{3}x(Z+1)\right].$$

The population has its maximum where the derivative with respect to Z vanishes, or where $n_{Z+1} = n_Z + 1$, which gives

$$Z \simeq \frac{y_\nu N_{\mathrm{UV}} T^{1/2}}{n_e}\sqrt{\frac{9\pi k m_e}{32 e^4}}\, a\,. \qquad (7.11)$$

For example, the flux of photons with energy $h\nu \geq 10\,\text{eV}$ at a distance of 0.1 pc from a B1 star ($L = 10^4 \, L_\odot$, $T_{\text{eff}} = 20000\,\text{K}$) equals $N_{\text{UV}} \sim 3 \times 10^{11}\,\text{cm}^{-2}\,\text{s}^{-1}$. Assuming a photoelectric yield $y_\nu = 0.1$, a grain of radius $a = 10^{-5}$ cm in a medium where $T^{1/2}/n_{\text{e}} = 0.1\,\text{K}^{1/2}\,\text{cm}^3$ acquires a mean charge $Z \simeq 37$. The charge distribution is again bell-shaped, as in figure 7.1.

7.1.2.2 The photon yield

The yield y_ν for photoemission may be estimated from the following physical picture based on classical electrodynamics [Pep70]. The energy absorbed by a subvolume dV of a grain is given by (2.39). When one applies the Gauss theorem (B.29), one gets

$$dW_{\text{a}} = -\text{div}\,\mathbf{S}\,dV$$

where $\mathbf{S}$ is the Poynting vector. Let the subvolume be at a depth x below the surface. The likelihood P that the absorption leads to emission of a photoelectron is assumed to have the form

$$P = C \exp(-x/l_{\text{e}}) .$$

The factor C incorporates the following two probabilities: for excitation of an electron to a "free" state and, when such an electron has reached the surface, for penetrating to the outside and not being reflected. The exponential term $\exp(-x/l_{\text{e}})$ gives the probability that the electron reaches the surface at all and is not deexcited in any of the scattering processes on the way. The deeper the subvolume dV below the grain surface, the higher the chance for deexcitation. The mean free path of electrons in the bulk material, l_{e}, is of order 30Å, possibly shorter for metals and longer in dielectrics. Because of the factor $e^{-x/l_{\text{e}}}$, it is evident that small grains (radii $a < 50$Å) are much more efficient in photoemission than big ones ($a \sim 1000$Å). From Mie theory, one can compute the internal field of the grain and thus the Poynting vector.

The yield y_ν for electron emission induced by photons of frequency ν is now defined via the equation

$$y_\nu \int_V \text{div}\,\mathbf{S}\,dV = C \int_V e^{-x/l_{\text{e}}} \,\text{div}\,\mathbf{S}\,dV \tag{7.12}$$

where the integrals extend over the whole grain volume V. The material constants are uncertain, but various evidence points towards $y_\nu \sim 0.1$ [Wat72].

7.1.2.3 The photoelectric effect and gas heating

Photoemission can also be important for heating the interstellar gas. As the mean kinetic energy of a photoejected electron, E_ν, exceeds the average thermal energy $\frac{3}{2}kT$ of a gas particle, the excess energy, after subtraction of the electrostatic grain potential U, is collisionally imparted to the gas. In the end, the electron is thermalized and its average own energy will then also be $\frac{3}{2}kT$. The heating rate due to one dust grain is therefore

$$H = 4\pi \cdot \pi a^2 \int_{\nu_t}^{\infty} \frac{J_\nu Q_\nu^{abs}}{h\nu} \, y_\nu \left[E_\nu - U - \tfrac{3}{2}kT \right] d\nu \,. \qquad (7.13)$$

The heating efficiency is very much biased towards small particles, not only because the chance for an electron escaping the grain is exponentially enhanced (see above) but also because small grains dominate the far UV absorption. PAHs must therefore be important for heating the gas provided they are not so highly ionized that their ionization potential (see 7.9) exceeds the maximum energy of the photons which, outside HII regions, is 13.6 eV.

The photoeffect has to be invoked, for example, to explain the fairly high gas temperatures of 50 ... 100 K observed in HI regions (the dust there is much cooler, of order 20 K). Of course, the gas temperature does not follow from the rate H in (7.13) alone, but only from the balance with the cooling processes. Quite generally, the photoelectric effect is the prevailing heating mechnism (of the neutral gas) in an environment with a strong UV radiation field.

7.2   Grain motion

The motion of grains produces many interesting effects; we mention a few.

- The grains move relative to each other, pushed around by gas atoms. Now and then they collide; sometimes they stick and thus grow.

- Usually dust and gas are kinematically tied together when averaging over a volume containing many grains. But occcasionally the two components decouple and move relative to each other. For example, in a protostellar disk the grains sediment because they are not supported against gravity by a pressure gradient perpendicular to the plane of the disk, like the gas. This leads to dust enrichment in the mid-plane, a precondition for the formation of big particles from small ones by coagulation. For the same reason do the grains feel in their orbital (circular Keplerian) motion a head wind from the gas and therefore spiral inwards.

- There is kinematic decoupling also in the wind of old stars (section 12.5). The radiation pressure drives the dust outward and the dust pushes the gas along so that, in the end, the whole stellar envelope is expelled.

- Grain rotation is essential in the alignment of elongated dust particles in the interstellar magnetic field. Alignment leads to the polarization of light (section 4.3).

- Examples of grain motion in the solar system are given in section 7.3.

7.2.1 Random walk

Suppose one makes in three-dimensional space, starting from position $\mathbf{R}_0 = \mathbf{0}$, a sequence of steps defined by the vectors $\mathbf{L}_i$. After N steps, one has arrived at the position $\mathbf{R}_N = \mathbf{R}_{N-1} + \mathbf{L}_N$. If the vectors $\mathbf{L}_i$ are of constant length L, but in arbitrary direction, the mean square of the distance, $\langle R_N^2 \rangle$, grows like

$$\langle R_N^2 \rangle = N L^2 . \tag{7.14}$$

The mean increase per step is $\langle R_N^2 - R_{N-1}^2 \rangle = L^2$ because $\langle \mathbf{R}_{N-1} \cdot \mathbf{L}_N \rangle$ obviously vanishes. Figure 7.2 shows a two-dimensional random walk and figure 7.3 the verification of formula (7.14) in a numerical experiment.

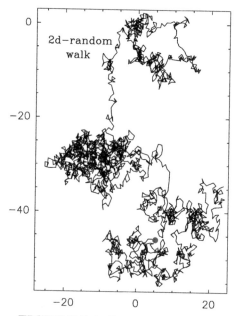

FIGURE 7.2 Random walk in two dimensions starting at the coordinate center (0,0) and ending after 2700 steps of equal length $L = 1$ at the big dot near $(5, -46)$.

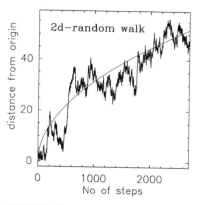

FIGURE 7.3 According to (7.14), one travels in a random walk after N steps a mean distance $\sqrt{N}L$ (smooth line). The actual distance in the numerical experiment of figure 7.2 is the jittery line.

7.2.2 The drag on a grain subjected to an outer force

Let us consider the one-dimensional motion of a heavy test particle (grain) through a fluid or gas in more detail. Suppose one applies a constant outer force F on the test particle of mass M. It then experiences a drag from the fluid or gas molecules that is proportional to its velocity $\dot{x} = V$ and the equation of motion of the test particle reads

$$F = M\ddot{x} + \mu\dot{x} . \tag{7.15}$$

On the right stands an accelerational term $M\ddot{x}$ and a dissipational term. The coefficient μ in the latter depends on the properties of both the fluid and the test particle. In a steady state, $\ddot{x} = 0$ and $F = \mu V$.

We determine the force F in (7.15) needed to move a spherical grain of geometrical cross section πa^2 at a constant velocity V through a gas. The gas has a number density N, is at temperature T, and its atoms have a mass $m \ll M$. Let the grain advance in positive x-direction. The velocity distribution $N(v_x)$ of the gas atoms along this direction *with respect to the grain* is no longer given by equation (A.15), but has an offset in the exponent,

$$N(v_x)\,dv_x = N\left(\frac{m}{2\pi kT}\right)^{1/2} e^{-m(v_x+V)^2/2kT}\,dv_x .$$

The y and z-axis are, of course, unaffected and purely Maxwellian. The maximum of $N(v_x)$ is now at $v_x = -V$, whereas in the rest frame of the gas it is at zero velocity. The number of atoms in the velocity range $[v_x, v_x + dv_x]$ that hit the grain head-on ($v_x < 0$) equals $-\pi a^2 v_x N(v_x)dv_x$ and each atom imparts in an inelastic collision the momentum mv_x. To obtain the total momentum transfer, one has to integrate over all velocities $v_x < 0$, which leads to the expression

$$\pi a^2 \cdot Nm\left(\frac{m}{2\pi kT}\right)^{1/2} \int_{-\infty}^{0} v_x^2\, e^{-m(v_x+V)^2/2kT}\,dv_x . \tag{7.16}$$

- Suppose the grain moves much slower than sound. Substituting in (7.16) $w = v_x + V$ and taking the momentum difference between front and back, one gets the retarding force on the grain. The term that does not cancel out under this operation equals

$$F = \pi a^2 \cdot 4NmV\left(\frac{m}{2\pi kT}\right)^{1/2} \int_{0}^{\infty} we^{-mw^2/2kT}\,dw ,$$

and therefore,

$$F = \pi a^2 \cdot NmV\left(\frac{8kT}{m\pi}\right)^{1/2} . \tag{7.17}$$

The drag is proportional to the geometrical cross section of the grain, πa^2, and the drift velocity V. With the mean gas velocity $\langle v \rangle = \sqrt{8kT/m\pi}$, the friction coefficient becomes

$$\mu = \pi a^2 \cdot N \cdot m\langle v \rangle .$$

- When the grain moves highly supersonically, there is only a force acting on the front side and the momentum transfer is simply

$$F = \pi a^2 \cdot N m V^2 . \tag{7.18}$$

F is now proportional to the square of the velocity. In elastic collisions, the forces would be larger by about a factor of two.

An astronomically important case of an outer force acting on dust particles is provided by radiation pressure (formulae 2.9 and 2.10, but there the letter F means flux). In case of direct illumination by a star, we may approximately set the cross section for radiation pressure, C^{rp}, equal to the geometrical cross section $\sigma_{\mathrm{geo}} = \pi a^2$, corresponding to an efficiency $Q^{\mathrm{rp}} = 1$. If L_* is the stellar luminosity and r the distance of the grain to the star, the drift velocity is then for subsonic motion

$$V = \frac{L_*}{4\pi c r^2 N m} \left(\frac{m\pi}{8kT}\right)^{1/2} .$$

N, m and T refer to the gas. Evaluating this formula for typical numbers of L_*, N and T, one finds that in stellar environments supersonic drift speeds are easily achieved. However, in many configurations grains are not directly exposed to starlight, but irradiated by infrared photons to which the starlight has been converted through foreground matter. In such circumstances, C^{rp} is much smaller than the geometrical cross section and V accordingly smaller, too.

Around main sequence stars, the gas density is always low and the gas drag unimportant. A grain is attracted towards the star of mass M_* by gravitation and repelled by radiation pressure. From the balance between the two,

$$\frac{GM_*M}{r^2} = \frac{L_*\sigma_{\mathrm{geo}}}{4\pi c r^2} , \tag{7.19}$$

one obtains, neglecting grain charge and magnetic fields, a critical grain radius independent of the distance

$$a_{\mathrm{cr}} = \frac{3L_*}{4\pi c\, G\rho_{\mathrm{gr}} M_*} \tag{7.20}$$

or

$$\frac{a_{\mathrm{cr}}}{\mu m} \simeq 0.24 \cdot \left(\frac{L_*}{L_\odot}\right)\left(\frac{M_*}{M_\odot}\right)^{-1} . \tag{7.21}$$

Grains smaller than a_{cr} are expelled. For instance, a $10\,M_\odot$ star has $a_{\mathrm{cr}} \sim 2\,\mathrm{mm}$, so all particles with sizes of interstellar grains ($\sim 0.1\mu m$) are blown away by radiation pressure. For low mass stars $L_* \propto M_*^{3.5}$ and the critical radius falls with $M_*^{2.5}$. The removal is always rapid as one can show by integrating the outward acceleration $\dot v = 3L_*/16\pi c\rho_{\mathrm{gr}}\, ar^2$ (see 7.19).

7.2.3 Brownian motion of a grain

The grains perform in the interstellar medium a Brownian motion. In equilibrium and in the absence of turbulence, the mean kinetic energy of a gas atom is equal to the mean translatory kinetic energy, E_{kin}, and the mean rotational energy, E_{rot}, of a grain.

7.2.3.1 Translatory Brownian motion

If M and V denote the mass and velocity of the dust particle,

$$\tfrac{3}{2}kT_{gas} = E_{kin} = \tfrac{1}{2}MV^2 ,$$

The kinetic energy of a grain, E_{kin}, is, on average, equally distributed among the three degrees of freedom,

$$MV_x^2 = MV_y^2 = MV_z^2 .$$

The pure translatory Brownian grain velocity,

$$V_{Brown} = \sqrt{\frac{3kT_{gas}}{M}}$$

is typically 10 cm s^{-1} and much smaller than that of gas atoms. We note, however, that the real grain velocity may be a hundred times larger than V_{Brown} because of turbulence in the interstellar medium.

A grain will have lost memory about its present momemtum $\mathbf{p}_0$ after a disorder time t_{dis}. One impinging gas atom of mass $m \ll M$ and velocity v changes the momentum of the grain statistically by mv, and Z atoms by $mv\sqrt{Z}$, as in a random walk. So in equipartition, when $MV^2 = mv^2$, the momentum of the grain is profoundly altered after M/m collisions, i.e., when the mass of the colliding atoms equals the mass of the grain. Therefore, in a gas of number density N,

$$t_{dis} = \frac{M}{N\,mv\,\pi a^2} . \tag{7.22}$$

Figure 7.4 displays a numerical experiment of a two-dimensional Brownian motion assuming inelastic collisions. The parameters are such that a gas atom impacts the grain about once every 1000 seconds. To follow the stochastic evolution, we bin the velocity distribution of the gas atoms $N(v_x)$ of (A.15) into 100 velocity intervals, and likewise for $N(v_y)$. Choosing a time step of 10^3 s, the chance that during one time step the grain is hit by an atom within a certain velocity interval, either from the front or back, is small. The actual occurrence of such an event is prompted by a random number generator. The disorder time according to (7.22) equals 1.5×10^9 s. It is roughly the interval after which there is a change in the direction of the velocity vector of the grain by 90° or in its kinetic energy by 50%.

The computations pertaining to figure 7.4 yield, of course, also the distance r of the grain from its starting point. As predicted by theory, r^2 grows linearly with time t provided t is large compared to the disorder time t_{dis}.

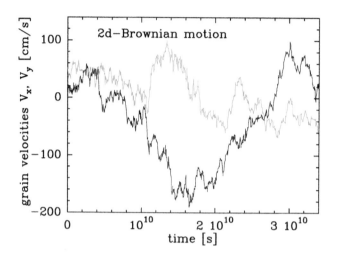

FIGURE 7.4 Numerical experiment of a two-dimensional Brownian motion. Gas parameters: hydrogen atoms at $T = 100$K with number density $10^4 \, \text{cm}^{-3}$. Grain parameters: radius 100Å, mass $M = 1.05 \times 10^{-17}$ g. The figure shows the x and y-component of the velocity; initially $\frac{1}{2}MV_x^2 = kT$ and $V_y = 0$. Because of the large time interval displayed in the plot, the actual jitter is graphically not fully resolved.

7.2.3.2 Rotational Brownian motion

If I and ω are the angular momentum and angular velocity of the grain,

$$\tfrac{3}{2}kT_{\text{gas}} = E_{\text{rot}} = \tfrac{1}{2}I\omega^2 \ .$$

Likewise, E_{rot} is equally distributed over the three degrees of rotational freedom,

$$\omega_x^2 I_x = \omega_y^2 I_y = \omega_z^2 I_z \ , \tag{7.23}$$

where $(\omega_x, \omega_y, \omega_z)$ and (I_x, I_y, I_z) refer to the principal axes of the inertia ellipsoid. Between collisions with gas atoms, the total angular momentum $\mathbf{L} = (I_x\omega_x, I_y\omega_y, I_z\omega_z)$ stays fixed in space, whereas the momentary rotation axis does not and also moves within the grain. The rotational velocity of a dust particle in Brownian motion, $\omega_{\text{Brown}} = \sqrt{3kT_{\text{gas}}/I}$, is typically $3 \times 10^5 \, \text{s}^{-1}$.

The disorder time t_{dis} of equation (7.22) stays practically the same with respect to angular momentum. Now the gas atoms that collide with the spinning grain change its rotation by exerting a torque τ.

As with translatory motion, the real angular velocity of a grain may be orders of magnitude higher than ω_{Brown} because of suprathermal spinup. The process works, for example, when there is a favored site on the grain surface for H_2 formation. H_2 will form there repeatedly, and each time the ejection of the molecule will impart angular momentum to the grain spinning it up.

Another randomizing process, besides collisions with gas atoms, that is always at work is infrared radiation because each photon carries away an angular momentum $\hbar$ in arbitrary direction. Usually the dust is heated by the UV field. In thermal balance, the number of UV photons absorbed per second, N_{UV}, is some 50 times smaller than that of emitted infrared photons, N_{IR}, so N_{UV} may be neglected in the angular momentum balance. After a time t, the grain's angular momentum has changed in a random walk by an amount $\sqrt{N_{IR}t}\,\hbar$. This process may dominate over collisional randomization for small grains.

7.3 Dust in the solar system

7.3.1 Interplanetary dust

Generally, interstellar grains are beyond the reach of spacecraft, because of their distance, so they cannot be brought to Earth and studied in the lab. However, as the Sun moves relative to the local interstellar cloud (it is of low density, $n_H \sim 0.3\,\mathrm{cm}^{-3}$), dust particles of this cloud sweep through the solar system. Some of them were detected by spacecraft at the distance of Jupiter and identified to belong to the local interstellar cloud by the direction and value ($\sim 26\,\mathrm{km/s}$) of their velocity vector [Grü94]. In the process of detection they were destroyed. Their masses, which could also be determined, correspond to $\mu\mathrm{m}$-sized grains, rather atypical for the interstellar medium, as we think. But smaller grains may be prevented from penetrating deep into the solar system by radiation pressure and, if they are charged, by the interplanetary magnetic field.

Further direct evidence about interstellar grains comes from primitive meteorites where one can identify inclusions which distinguish themselves by their isotopic pattern from the solar system. Because the abundance ratios in the isotopic species of the elements C,N,O,Al,Si and the occasionally trapped noble gases Ne and Xe are explainable in terms of nucleosynthesis in AGB stars or supernovae, these subparticles are hypothesized to be of interstellar origin, although they may have undergone considerable reprocessing in the solar nebula.

For example, presolar graphitic grains deviate from the solar isotopic abundance ratio $[^{12}C]/[^{13}C] = 89$ in either direction by more than a factor of ten. Probably most grains in the interstellar medium have a non-solar ratio $[^{12}C]/[^{13}C]$, its value depending on their origin. During planet formation, they are largely destroyed and the isotopic species get mixed. The new grains that condense out and later form asteroids and planets have then solar isotopic abundances. These new grains do not consist of roughly equal amounts of silicate and carbonaceous material, as the grains in interstellar space, instead carbon is very underabundant as witnessed, for instance, by the composition of the Earth.

The most common inclusions in meteorites believed to be older than the solar system are diamond grains. They are extremely small ($\sim$30Å), so that a not negligible fraction of C atoms is at the surface. They are heavily contaminated by H atoms and their average density ρ is considerably lower than in perfect crystals for which $\rho = 3.51\,\mathrm{g\,cm}^{-3}$. Their origin is unclear (carbon stars, supernovae, interstellar shocks).

There are also presolar meteoritic inclusions of graphite and silicon carbide, but they are much bigger (1 to $10\mu m$) than even typical interstellar grains. Such big inclusions can be analyzed individually; the tiny diamonds can not. Interestingly, *meteoritic* presolar grains are almost all C rich, whereas in interplanetary dust particles collected on Earth, one has so far only found presolar silicates.

7.3.2 The Poynting-Robertson effect

Consider a grain of mass m, radius a and geometrical cross section $\sigma_{\mathrm{geo}} = \pi a^2$ circling a star at distance r with frequency ω and velocity $v = \omega r$. The angular momentum of the grain is

$$\ell = mr^2\omega = mrv.$$

Let M_* and L_* be the mass and luminosity of the star. Because there is mostly optical radiation, the absorption coefficient of the grain is $C^{\mathrm{abs}} \simeq \sigma_{\mathrm{geo}}$ and the particle absorbs per unit time the energy

$$\Delta E = \frac{L_*\sigma_{\mathrm{geo}}}{4\pi r^2}.$$

In thermal balance, the same amount is reemitted. Seen from a non-rotating rest frame, the stellar photons that are absorbed travel in radial direction and carry no angular momentum, whereas the emitted photons do because they partake in the circular motion of the grain around the star. If we associate with the absorbed energy ΔE a mass $m_{\mathrm{phot}} = \Delta E/c^2$, the angular momentum of the grain decreases per unit time through emission by

$$\frac{d\ell}{dt} = -m_{\mathrm{phot}}\cdot rv = -\frac{\Delta E}{c^2}\cdot rv = -\frac{L_*\sigma_{\mathrm{geo}}}{4\pi c^2 r^2}\cdot\frac{\ell}{m}.$$

Because for a circular orbit

$$\frac{v^2}{r} = \frac{GM_*}{r^2} \,,$$

we get $v = \sqrt{GM_*/r}$ and

$$\frac{d\ell}{dt} = \frac{\ell}{2r}\frac{dr}{dt} \,.$$

Due to the loss of angular momentum, the distance of the grain to the star shrinks according to

$$\frac{dr}{dt} = -\frac{L_*}{2\pi r} \cdot \frac{\sigma_{\rm geo}}{mc^2} \,. \tag{7.24}$$

When we integrate the equation $dt = -(2\pi mc^2/L_*\sigma_{\rm geo})\, r\, dr$ from some initial radius r to the stellar radius $R_* \ll r$, we find the time $\tau_{\rm PR}$ that it takes a grain to fall into the star,

$$\tau_{\rm PR} = \frac{m\pi c^2}{L_*\sigma_{\rm geo}} \cdot r^2 \,, \tag{7.25}$$

or in more practical units, assuming a density of $\rho_{\rm gr} = 2.5\,{\rm g/cm}^3$ for the grain material,

$$\frac{\tau_{\rm PR}}{\rm yr} = 1700 \left(\frac{a}{\mu{\rm m}}\right)\left(\frac{L_*}{L_\odot}\right)^{-1}\left(\frac{r}{\rm AU}\right)^2 \,. \tag{7.26}$$

The Poynting-Robertson effect is an efficient way to remove dust in the solar system. For instance, during the lifetime of the Sun ($\sim 5 \times 10^9$ yr) only bodies with a diameter greater than 6 m (!) can have survived within the orbit of the Earth. Particles existing today smaller than 6 m must have been replenished from comets or asteroids.

The Poynting-Robertson effect may also be described by an observer in a frame corotating with the grain. In such a reference frame, the stellar photons approach the grain not exactly along the radius vector from the star, but hit it slightly head-on in view of the aberration of light. This phenomenon arises because one has to add the velocity of the photon and the grain according to the rules of special relativity. The photons thus decrease the angular momentum of the grain and force it to spiral into the star. The Poynting-Robertson effect also works when the photons are not absorbed, but isotropically scattered.

7.3.3 Electromagnetic forces on grains: Dust from Io

Spacecraft flying by Jupiter recorded strongly variable dust streams with speeds over 200 km s^{-1}, higher than any virial velocity. The analysis of the time variability in the particle flux yielded as major frequencies the spin period of Jupiter and the orbital period of the innermost Galilean moon Io, making Io the likely source of the fast grains [Hor93].

To understand this ejection of grains into interplanetary space, we have to recall a few facts about Jupiter and its big moons (table 7.1). The giant

TABLE 7.1 Data on Jupiter and the Galilean moons

Jupiter	mass	M_J	=	1.90×10^{30}	g
	equatorial radius	R_J	=	71 400	km
	mean density	$\bar{\rho}_J$	=	1.332	g cm^{-3}
	spin period	P_J	=	9.84	h
	spin frequency	ω_J	=	1.77×10^{-4}	s^{-1}
	surface gravity	g_J	=	2.487×10^4	cm s^{-2}
	magnetic field in	B	=	B_0/L^3	G (L in R_J)
	equatorial plane				(B_0=4.2)
Io	mass	M_{Io}	=	8.94×10^{25}	g
	radius	R_{Io}	=	1 820	km
	mean density	$\bar{\rho}_{Io}$	=	3.53	g cm^{-3}
	surface gravity	g_{Io}	=	180	cm s^{-2}
	distance to Jupiter	r_{Io}	=	5.91	R_J
	eccentricity	e	=	0.004	
	orbital frequency	ω_{Io}	=	4.11×10^{-5}	s^{-1}
	orbital period	P_{Io}	=	1.769	d
Europa	orbital period	P_{Eu}	=	3.551	d
Ganymede	orbital period	P_{Ga}	=	7.155	d

planet Jupiter has a low mean density and is spinning rapidly. Because of its metallic interior (of liquid H_2 and He between 0.2 and 0.78 R_J), it possesses all the ingredients for a strong magnetic field. The field is to first order a dipole with its axis almost parallel to the rotation axis (offset 9.6°) and it corotates with the planet.

Io is in its mass, density and composition similar to our Moon, surprising for a body far from the Sun. As Jupiter is massive and Io close by, tidal forces are strong and one expects

a) inelastic deformations to have forced Io into a synchronous spin so that always the same side faces the planet;

b) an initially elliptical orbit to have become circular because then there would be no more tidal energy dissipation.

However, Europa and Ganymede are in resonant orbits with Io (4:2:1), pulling on it in the same place in the same direction, and that gives Io's orbit its eccentricity, small but big enough to make the satellite geologically active. Io spews out in volcano eruptions silicate material, sulfur compounds and SO_2 gas which condensates, like rain drops in a terrestrial cloud. As the surface gravity on Io is low, the volcano plumes go up high (> 100 km) entering the Jovian magnetosphere where the particles can pick up an electric charge. The commonest ions in the plasma in the Jovian magnetosphere are S^+ and O^+ ions from SO_2 which is supplied by the volcanoes.

To an observer on Io, Jupiter's magnetic field sweeps by at a velocity $v = r_{\text{Io}}(\omega_{\text{J}} - \omega_{\text{Io}}) \simeq 57$ km s^{-1} (table 7.1). A grain of mass m and charge q performs in a homogeneous static magnetic field $\mathbf{B}$ a spiral motion: it travels at constant velocity $v_{\|}$ parallel to $\mathbf{B}$ and gyrates in a plane perpendicular to it with the *Larmor* radius

$$r_{\text{L}} = \frac{\gamma m v_{\perp} c}{qB} \qquad (7.27)$$

where $\gamma = 1/\sqrt{1 - \beta^2}$, $\beta = v/c$, $v = \sqrt{v_{\perp}^2 + v_{\|}^2}$; in our case $\gamma = 1$. So the absolute value of the grain velocity stays constant. But the grains around Io also see an electric field $\mathbf{E}$ (although Jupiter has none, $\mathbf{E}' = 0$) because of the Lorentz transformations

$$\mathbf{E}_{\|} = \mathbf{E}'_{\|}, \qquad \mathbf{E}_{\perp} = \gamma(\mathbf{E}'_{\perp} + \mathbf{v} \times \mathbf{B}'/c)$$

$$\mathbf{B}_{\|} = \mathbf{B}'_{\|}, \qquad \mathbf{B}_{\perp} = \gamma(\mathbf{B}'_{\perp} + \mathbf{v} \times \mathbf{B}'/c) .$$

If the moons are in the (x, y)-plane, Jupiter's magnetic field at the position of Io can be written as $\mathbf{B}' = (0, 0, B_z = B_0/L^3)$, and $\mathbf{E}$ is therefore perpendicular to $\mathbf{B}$ and radially directed away from Jupiter.

The electric potential which the grains acquire follows from the balance between the charging currents by thermal electrons and ions, photoelectron emission, and secondary electrons. It was calculated to be about $+3$ Volt beyond Io's orbit, and -30 V for $r < 5.9R_{\text{J}}$ (where the plasma is cold, $kT \simeq 0.1$ eV).

So the charge of a grain of radius a is after (7.9) $q = 0.01a$ esu. If one compares the electric outward force $F_{\text{el}} = qvB_z/c$ with the gravitational inward force $F_{\text{gr}} = GM_{\text{J}}/r_{\text{Io}}^2$ per unit mass, one finds $F_{\text{el}}/F_{\text{gr}} \simeq 10^{-10}/a^2$. The electric force wins if the particle radius is smaller than 0.1μm.

The dipole field in which a grain gyrates is not homogeneous. But when its changes over the distance of a Larmor radius are small, $|\text{grad } \mathbf{B}| \ll B/r_{\text{L}}$, its motion can be considered as being coupled to the field line. This condition applies and therefore grains smaller than $\sim 0.01\mu$m can leave the magnetosphere and then be detected by a spacecraft.

7.3.4   Shooting stars and less belligerent meteoroids

Extraterrestrial bodies, of a size less than a few meters, that can potentially hit the Earth are called meteoroids. When they do collide, they turn into meteorites if they survive the passage through our atmosphere. Then we can pick them up from the ground and study them in the lab. During the short spell while they are observable by the naked eye, often shining brightly, they are called shooting stars, or meteors (recommended reading [Cep98]). These are renowned for their efficacy in promoting silent wishes.

The optical light from shooting stars is mainly emitted in lines of metallic atoms (Na, Mg, Mg^+, Ca, Ca^+, Fe...) that have been evaporated from the meteoroid and then excited in collisions with atoms of the atmosphere, but one also sees atmospheric N, O and N_2 bands. Optical triangulation and radar measurements, which are possible because the gas becomes partially ionized, are used to determine the trajectories of the bodies. Below we compute the approximate dynamic and thermal evolution of the meteoroids and some of the interesting effects they produce as they impact the Earth.

7.3.4.1 Equations for meteoroid tracks

To sketch the path, brightness and fate of meteoroids, we consider a spherical body of mass M, radius a and mean density ρ_m that enters the (plane) atmosphere of the Earth (density ρ) at a height h with velocity V under an angle α. To push the grain through the atmosphere, one needs according to (7.18) the force $F = c_D \pi a^2 \rho V^2$; the drag coefficient c_D is around one. Energy is therefore dissipated at a rate

$$FV = \pi a^2 \rho V^3$$

at the expense of the kinetic energy $E_k = \frac{1}{2}MV^2$, and $\dot{V} = \partial V/\partial t$ follows from

$$-\pi a^2 \rho V^3 = \dot{E}_k = MV\dot{V} . \tag{7.28}$$

Suppose a constant fraction α_m of the dissipation rate FV is deposited in the meteoroid as heat and the rest, $1 - \alpha_m$, in the atmosphere. Because grains as small as $1\mu m$ can be treated with regard to their *total* radiative power as blackbodies (see figure 6.6), the temperature balance of a grain of specific heat c reads

$$\alpha_m \pi a^2 \rho V^3 = 4\pi a^2 \sigma T^4 + cM\dot{T} .$$

When the meteoroid temperature T reaches (at the surface) the evaporation temperature T_{ev}, ablation starts. T does then not grow any further but stays constant, and the mass M and the radius a decrease. The vaporized atoms are stopped almost momentarily and their loss of kinetic energy per unit time, $\frac{1}{2}MV^2$, is an additional heat source to the atmosphere, besides $(1 - \alpha_m)FV$. Formula (7.28) is thus also correct in case of evaporation. When $T = T_{ev}$, the mass loss by ablation, $\dot{M}$, is determined from

$$\alpha_m \pi a^2 \rho V^3 = 4\pi a^2 \sigma T_{ev}^4 - Q_{ev}\dot{M}$$

where Q_{ev} denotes the energy per gram necessary for the phase transition from solid to gaseous. This leads to the set of equations

$$\dot{V} = -\frac{3}{4\rho_m} \frac{V^2 \rho}{a} \tag{7.29}$$

$$\dot{h} = -V \sin\alpha \tag{7.30}$$

$$\dot{a} = 0 \tag{7.31}$$

$$\dot{T} = \frac{3\alpha_m}{4\rho_m c} \frac{\rho V^3}{a} - \frac{3\sigma}{\rho_m c} \frac{T^4}{a} \tag{7.32}$$

which is valid without ablation, i.e., when the meteoroid is cooling ($\alpha_m \rho V^3 - 4\sigma T^4 < 0$) or heating up, but at a temperature $T < T_{ev}$. Otherwise the last two formula have to be replaced by

$$\dot{a} = -\frac{\alpha_m}{4\rho_m Q_{ev}} \rho V^3 + \frac{\sigma T_{ev}^4}{\rho_m Q_{ev}} \tag{7.33}$$

$$\dot{T} = 0 \,. \tag{7.34}$$

From the straightforward numerical integration of (7.29) to (7.34), we obtain the flight path of the meteoroid, its temperature and, when it exceeds T_{ev}, the evaporation rate $\dot{M}$ or the rate at which its radius shrinks ($\dot{a}$). Its light emission is given by absorption coefficient C_ν^{abs} times Planck function $B_\nu(T_{ev})$.

The calculations displayed in figure 7.5 refer to an initial velocity $V_0 = 30$ km/h and span a wide range of meteoroid masses (from 4×10^{-12} g to 4×10^6 g). The other parameters used for figure 7.5 are: $\alpha = 30°$, $\alpha_m = 0.3$, $\rho_m = 1$ g/cm^3, $T_{ev} = 2200$ K, $Q_{ev} = 6 \times 10^{10}$ erg g^{-1}, $c = 10^7$ erg g^{-1} T^{-1}. The density of the atmosphere, $\rho(h)$, was approximated between 10 and 160 km by $\rho(h) = 6 \times 10^{-4} \exp(-h/7.8 \times 10^5)$ and above 160 km by $\rho(h) = 3 \times 10^{22}/h^{4.8}$ with height h in cm and ρ in g/cm^3.

7.3.4.2 Micrometeorites and meteorites

First we note that braking of the meteoroids sets in only below 300 km (upper left box of figure 7.5). The density ρ and mean molecular weight m at that altitude ($\rho \simeq 4 \times 10^{-14}$ g/cm^3, $m \simeq 22$) imply an average distance between air molecules of $\sim 10\mu$m.

Because the deceleration $\dot{V}$ in (7.29) is proportional to $V^2\rho/a$, particles of small radius a are retarded already high up and arrive in the denser layers with a reduced velocity. Neglecting the time variation of the temperature, $\dot{T}$, formula (7.32) can be used to derive T. It does not depend on the grain size, but is solely determined by the product ρV^3. It turns out that for our particular, but representative example, grains smaller than $\sim 100\mu$m always stay below T_{ev}. They do not evaporate and reach the ground unscathed. These are the micrometeorites that can be collected in Antarctica, on the sea floor or in the stratosphere. Grains smaller than one micron are likely to be removed by radiation pressure or the Poynting-Robertson effect (see 7.26, 7.21) and are therefore not abundant.

Particles greater than $\sim 100\mu$m are vaporized, at least partially. We learn from figure 7.5 that for grains of 1 cm radius gasification sets in ($T = T_{ev}$)

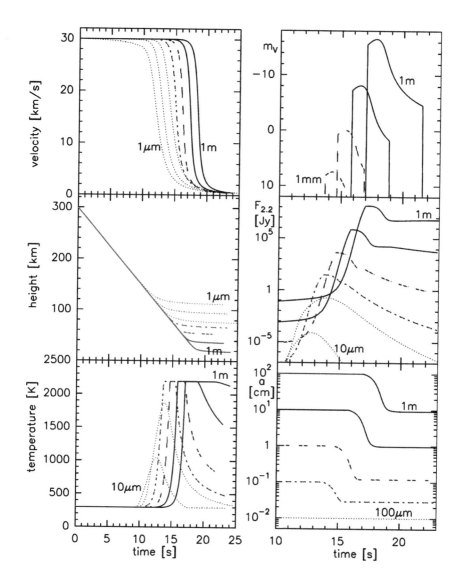

FIGURE 7.5 Velocity V, height h, temperature T, apparent visual magnitude m_V, the 2.2μm flux $F_{2.2}$ and grain radius a as a function of time for meteoroids that enter the Earth's atmosphere at $V_0 = 30$ km/s under an angle of 30°. The initial radii vary between 1μm and 1 m in steps of ten, the smallest and largest radius plotted in each frame are indicated. The meteoroid is o77ptically visible only during the period of ablation when $T = T_{ev}$. The flux $F_{2.2}$ and the magnitude m_V are computed for a distance $D = 1.5\,h$.

at a height of 80 km and lasts for a time t_{ev} of one or two seconds. Smaller grains start evaporating earlier and their luminous phase (the wishing time) is shorter; for bigger grains it is the other way round. While evaporating, the meteors shrink, their velocity plummets to values below 10 km/s, and they produce optical light. In the case depicted in figure 7.5, most of the initial mass is gasified. The tiny rest that has not been ablated ($\sim$0.1% for initial radii greater than 1 cm) continues its passage, but invisibly, in dark flight. When the remnant falls to the ground (rather gently, $V \leq 100$ m/s), it has a characteristic crust formed by solidification during cooling. The ratio of initial to final mass depends sensitively on the initial velocity V_0 and is not far from unity for $V_0 = 10$ km/s.

7.3.4.3   The visual and infrared brightness of meteors

We can approximately compute the brightness of the optical phenomenon if we know the fraction γ of the kinetic energy that is converted into visible line radiation. From a grain of mass M glowing at a distance D, one receives the integrated optical flux

$$F_{opt} = \gamma \tfrac{1}{2} M V^2 \ / \ 4\pi D^2 t_{ev} \ .$$

For $M = 1$g ($a \simeq 1$cm), $D = 100$ km, $V_0 = 30$ km s^{-1}, $t_{ev} = 1$s, and $\gamma \sim 1\%$ (an educated guess), $F_{opt} \simeq 3.6 \times 10^{-5}$ erg cm^{-1} s^{-1} which translates into a visual magnitude of -1^m (see section 10.1.1), as bright as the brightest stars. Such a prominent meteor would certainly catch our attention when we are out at night (and not looking the other way). We can also estimate its near infrared brightness, for example, the 2.2μm flux is simply

$$F_{2.2} = 4\pi \cdot \pi a^2 Q_{2.2} B_{2.2}(T) \ / \ 4\pi D^2 \ .$$

Here no poorly known efficiency is needed (like γ). Again with $D = 100$ km, $T_{ev} = 2200$ K and an absorption efficiency $Q_{2.2} \simeq 1$, one gets a few thousand Jy, like the brightest stellar 2μm sources in the sky.

When one compares the time-integrated purely thermal emission ($\int 4\pi a^2 \times \sigma T^4 \, dt$) with the optical ($\gamma \tfrac{1}{2} M V^2$), taking into account that the grain becomes smaller because of ablation, one finds that the two are about equal for a 1mm-sized grain whereas for bigger meteoroids optical radiation dominates.

7.3.4.4   Catastrophic events

All small bodies in the solar system, with sizes much less than 1 km, are probably produced by erosion or shattering of bigger ones. The lovely shooting stars are therefore just the harmless tail of their menacing big sisters. Indeed, as discussed above, the remains of a meteoroid after vaporization can fall on our head, and the leftover may still be a substantial block if the meteoroid before collision was greater than 1m (these are usually called asteroids). We estimate the frequency with which big asteroids hit the Earth in two ways:

- from historical records: *a)* spectacular events (Tunguska meteorite, 100 yr ago); *b)* regionally devastating impacts (Arizona crater, 5×10^4 yr); *c)* large scale destruction (Nördlingen Ries, 15×10^6 yr); and *d)* global catastrophies leading to the extermination of biological species (Yucatan crater, 65×10^6 yr).

- by extrapolating the number density of the small grains ($< 10\text{m}$) to large sizes following the abundance distribution $n(a) \propto a^{-3.5}$ of (5.9) which seems to be valid not only for interstellar dust but also for the bodies in the solar system. To normalize $n(a)$, we note that the small bodies ($a < 10\text{m}$) have an annual influx to the Earth of $\sim 10^{11}$ g.

In either case, one is forced to the conclusion that something really bad (impact of asteroids $> 1\text{km}$) happens once in a million years. This implies a 10^{-4} chance in a man's lifetime, so the threat is serious and real. Still bigger chunks (≥ 10 km, they do not feel the atmosphere at all) bring us doomsday.

7.4 Grain destruction

7.4.1 Mass balance of gas and dust in the Milky Way

We live in a quasi-steady state where dust formation is balanced by dust destruction, and gas input by gas output. The rates at which this happens are quite uncertain, therefore all numbers below are not precise, and not always internally strictly consistent, but they show the order of magnitude.

As gas and dust are intimately linked, we remark first on the mass balance of *all* interstellar matter in the Milky Way which, of course, consists to 99% of gas. Four important numbers are listed in table 7.2. The star formation rate implies that within an astration time $\tau_{\text{ast}} \sim 10^9$ yr, a short period compared to the Hubble time (13×10^9 yr), all interstellar matter is processed in stars. This is then also the maximum lifetime of a grain. Of course, it might be much shorter if there are other, more efficient destruction mechanisms. 7 The continuous drain of gas in star formation ($5\,\text{M}_\odot\,\text{yr}^{-1}$) must be offset by an equal gas source. It comes mainly from red giants which supply about 80% of the gas ($4\,\text{M}_\odot\,\text{yr}^{-1}$), the rest from planetary nebulae (PN) and supernovae ($1\,\text{M}_\odot\,\text{yr}^{-1}$).

Most of the dust probably forms also in the wind of red giants. As 0.5 to 1% of the matter injected by them into the interstellar medium is in the solid phase, the input rate of dust from red giants is

$$I_{\text{RG}} \sim 0.03\ \text{M}_\odot\ \text{yr}^{-1}\ .$$

TABLE 7.2 Global parameters of the Milky Way relevant to the mass balance of dust

total gas mass	M_{gas}	$\sim$	$5 \times 10^9 \, M_\odot$
total dust mass	M_{dust}	$\sim$	$3 \times 10^7 \, M_\odot$
star formation rate	τ_{SFR}	$\sim$	$5 \, M_\odot \, \mathrm{yr}^{-1}$
supernova rate	τ_{SN}	$\sim$	$0.03 \, \mathrm{yr}^{-1}$

The observational footing for this dust-to-gas injection ratio (0.5 to 1%) rests on the determination of the gas and dust mass of the giant envelope and the velocity with which matter escapes.

Planetary nebulae are thought to produce little or no dust, but supernovae of type II (from high mass stars) are very efficient. Their ejecta are massive ($\sim 10 \, M_\odot$) and a large fraction consists of heavy elements ($\sim 3 \, M_\odot$) which is all potential dust material. However, some of it will be locked up in CO, H_2O or otherwise and will be unable to form grains. If one supernova explosion creates $\sim 0.3 \, M_\odot$ of dust and if there are three events per century (table 7.2), the SN dust input rate becomes

$$I_{\mathrm{SN}} \sim 0.01 \, M_\odot \, \mathrm{yr}^{-1} \, .$$

The total input rate in the Milky Way is then

$$I_{\mathrm{dust}} = I_{\mathrm{SN}} + I_{\mathrm{RG}} \sim 0.04 \, M_\odot \, \mathrm{yr}^{-1} \, .$$

It should equal the total dust destruction rate,

$$I_{\mathrm{dust}} = D_{\mathrm{dust}} \, .$$

D_{dust} is a purely theoretical quantity and not amenable to observations. It can be written as the ratio of total dust mass over mean lifetime of a grain,

$$D_{\mathrm{dust}} = \frac{M_{\mathrm{dust}}}{t_{\mathrm{dust}}} \, . \tag{7.35}$$

Fortunately, M_{dust} can be estimated from the total far infrared or millimeter emission of the Milky Way (table 7.2). Putting $D_{\mathrm{dust}} = I_{\mathrm{dust}} = 0.04 \, M_\odot \, \mathrm{yr}^{-1}$ yields $t_{\mathrm{dust}} = 7.5 \times 10^8$ yr. This number is about equal to the maximum lifetime of a grain set by star formation. It implies that a grain, as the spiral waves sweep over it, is cycled ~ 10 times between cloud and intercloud medium.

Theory, however, predicts a lifetime substantially shorter than 7.5×10^8 yr (because of sputtering and destruction in shocks) and thus a destruction rate $D_{\mathrm{dust}} > I_{\mathrm{dust}}$. Because creation must balance annihilation, an additional input term, I_{add}, is postulated that accounts for the growth of grain mass in the interstellar medium. Speculations favor accretion in dark clouds.

7.4.2 Destruction processes

Grains are destroyed in various processes (recommended reading [Sea87], [McK89], [Dwe92], [Tie94], [Jon96]) such as

- *Evaporation by radiative heating.* The solid is heated up to the condensation temperature. This happens very close to a star, to a certain extent in HII regions, also in the diffuse medium when the grains are extremely small. By far the greatest sink is star formation and the loss rate of dust associated with astration equals $M_{dust}/\tau_{ast} \sim 0.03$ $M_{\odot}$ yr^{-1}.

- *Sputtering.* Atoms can be ejected from a grain surface by the impact of an energetic gas particle. The projectile transfers its kinetic energy, E_{proj}, to target atoms in the grain which can then leave the soild provided E_{proj} is a few times larger than their binding energy (5 to 10 eV). At high energies ($E_{proj} \gtrsim 10$ keV), the target atoms generate secondary atoms that can also escape. The relevant gas particles for sputtering are ions, electrons are inefficient and neutrals rare. Therefore, the collisional cross section is determined by the grain charge.

- *Chemical sputtering.* At low particle energies, near the threshold, the projectile may form an unstable molecule with a surface atom. When it disintegrates, atom and projectile are ejected. This erosion process is relevant for carbonaceous material.

- *Grain-grain collisions.*

  - *Evaporation.* The critical relative grain velocities are around 20 km s^{-1} when the kinetic energy exceeds the binding energy (~ 10 eV per atom) by a factor ~ 10. The grains acquire such high velocities through acceleration in a magnetic field behind a supernova shock.

  - *Shattering.* The grains are smashed into smaller units; the total solid mass stays constant. Critical velocities are an order of magnitude smaller than for evaporation (~ 2 km s^{-1}), critical energies even two orders (it is easier to break a vase than to gasify it). Shattering is believed to be the major mechanism determining the grain size distribution.

- *Photodesorption* where grain atoms are ejected by photons.

7.4.2.1 Destruction in shocks

Besides star formation, grains are mostly destroyed in shocks associated with supernova remnants (SNR). The observational evidence comes from clouds of high velocity ($v > 100$ km/s) in which more than half of the silicon and iron are in the gas phase, whereas in normal clouds only a fraction of order 1% of these atoms is in the gas. High-velocity clouds are interpreted to be fragments of an expanding and shocked supernova shell. The mechanism of grain destruction depends on the type of shock:

- In a fast shock, cooling is slow. The shock is adiabatic or non-radiative and has a moderate density jump of four. Sputtering is thermal because the velocities of the gas atoms are thermal. A shock with $v \sim 300\,\mathrm{km/s}$ corresponds to a temperature of about $3 \times 10^6\,\mathrm{K}$. If one determines the rate of impinging protons using (8.1) and adopts a gas density $n = 1\,\mathrm{cm}^{-3}$, one finds that at least small grains are likely to be eroded.

- In a low-velocity shock, the cooling time is smaller than the expansion time of the supernova remnant. The shock is radiative and the density jump is much greater than four. A charged grain of mass m will gyrate in the magnetic field $\mathbf{B}$. As $B_\parallel$, the component of $\mathbf{B}$ parallel to the shock front, is compressed, the grain velocity v_{gr} increases because its magnetic moment $\mu = mv_{gr}^2/2B_\parallel$ is conserved and the grain is accelerated (betatron acceleration). The collisional velocities between the grain and the gas atoms are now nonthermal and nonthermal sputtering dominates. The final velocity v_{gr} is limited by drag forces from the gas. The deceleration is proportional to the inverse of the grain radius, so big grains become faster than small ones and are more easily destroyed; small grains may survive.

Dust destruction through grain-grain collisions is thought to be much less efficient than sputtering.

To estimate the importance of shock destruction, we turn to the theory of supernova blasts. During the first (Sedov) stage, the remnant of mass M_{SNR} and velocity v is adiabatic and its kinetic energy $E = \frac{1}{2}M_{SNR}v^2$ conserved; at later times, the momentum vM_{SNR} is constant. According to [McK89], the shock becomes radiative when $v \le 200(n^2/E_{51})^{1/14}\,\mathrm{km/s}$; E_{51} is the energy expressed in units of $10^{51}\,\mathrm{erg}$. In a gas of density $n = 1\,\mathrm{cm}^{-3}$ and with the standard explosion energy $E_{51} = 1$, one finds for the mass of the supernova shell $M_{SNR} \sim 10^3\,\mathrm{M_\odot}$ at the onset of the radiative phase. A supernova rate $\tau_{SN} = 0.03\,\mathrm{yr}^{-1}$ then implies that all interstellar material of the Milky Way is processed in a time

$$t_{SNR} = \frac{M_{gas}}{M_{SNR}\,\tau_{SN}} \sim 2 \times 10^8\,\mathrm{yr}\;.$$

7.5 Grain formation

We derive approximate equations that describe the nucleation of monomers into large clusters. The examples refer to water because its properties are experimentally well established, but the physics apply, with some modifications, also to interstellar grains. There, however, material constants are poorly known.

7.5.1 Evaporation temperature and vapor pressure

When two phases, like liquid and gas, are in equilibrium at temperature T, the vapor pressure P changes with T according to the *Clausius-Clapeyron* equation,

$$\frac{dP}{dT} = \frac{Q}{(V_g - V_f)\,T} \,, \tag{7.36}$$

where V_g and V_f are the volume per mol in the gas and fluid phase, respectively, and Q is the heat necessary to evaporate 1 mol. We remember that 1 mol consists of

$$L = 6.02 \times 10^{23}$$

molecules. L is Loschmidt's number, and the gas constant R, the Boltzmann constant k and L are related through

$$k = \frac{R}{L} \,. \tag{7.37}$$

The Clausius-Clapeyron equation is with obvious modifications also valid with respect to sublimation, which is the phase transition between solid and gas, and thus applies to the evaporation of interstellar grains and their mantles. Equation (7.36) is derived in a thought experiment from a Carnot cycle. Because $V_f \ll V_g$, we can approximate (7.36) by

$$\frac{d\ln P}{dT} = \frac{Q}{RT^2} \,, \tag{7.38}$$

which has the solution

$$P = P_0\, e^{-T_0/T} \,, \qquad\qquad T_0 = Q/R \,. \tag{7.39}$$

The affinity to the Boltzmann distribution is not only formal, but physical because atoms in the gas are in an energetically higher state than atoms in a solid. The term $e^{-T_0/T}$ corresponds to $e^{-E_b/kT}$ in (8.14). One gets close to equation (7.39) by equating the evaporation rate R_{evap} of (8.14) to the accretion rate of (8.2). Equation (7.39) is very general and does not say anything about the details of the phase transition, therefore, P_0 and T_0 have to be determined experimentally. Laboratory data for water ice from $0\,°C$ down to $-98\,°\,C$ are displayed in figure 7.6. They can be neatly fit over the whole range by equation (7.39) with properly chosen constants.

One gram of H_2O, in whatever form, has $N = 3.22 \times 10^{22}$ molecules. If it is ice, it takes 79.4 cal to liquify it at $0\,°C$, then 100 cal to heat it up the water to $100\,°\,C$ and another 539.1 cal to vaporize the water. In cooling the vapor by $\Delta T = 100\,K$ from 100 to $0\,°\,C$, assuming six degrees of freedom per molecule, one gains $3Nk\Delta T = 1.37 \times 10^9$ erg. Because

$$1\,\mathrm{cal} = 4.184 \times 10^7\,\mathrm{erg} \,,$$

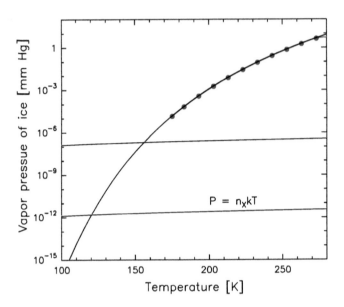

FIGURE 7.6 Vapor pressure of ice. Experimental points (dots) from [Wea77], the solid curve is a fit according to (7.39) with $T_0 = 6170\,\mathrm{K}$ and $P_0 = 4.0 \times 10^{13}\,\mathrm{dyn/cm^2}$ or $3.0 \times 10^{10}\,\mathrm{mm}$ of Hg column. The fit is extrapolated to lower temperatures. The lower horizontal line shows the partial gas pressure, $P = n_X kT$, for a number density $n_X = 0.1\,\mathrm{cm^{-3}}$. Such an H_2O pressure prevails in interstellar space when the total gas density $n_H = 10^6\,\mathrm{cm^{-3}}$ and the water abundance (in the gas phase) is 10^{-7}. The upper horizontal line is for $n_X = 10^5\,\mathrm{cm^{-3}}$.

the total sublimation energy per H_2O molecule at $0\,^\circ$C is therefore $kT = 8.64 \times 10^{-13}\,\mathrm{erg}$ corresponding to a temperature of $6260\,\mathrm{K}$. This agrees nicely with the fit parameter $T_0 = 6170\,\mathrm{K}$ in figure 7.6.

The evaporation temperature T_{evap} in interstellar space of a species X frozen out on grain mantles follows by setting its partial pressure in the gas phase, $n_X kT$, equal to the evaporation pressure,

$$P_0\, e^{-T_0/T_{\mathrm{evap}}} = n_X kT_{\mathrm{evap}} . \tag{7.40}$$

Because of the exponential dependence of the vapor pressure on temperature, the result is rather insensitive to the gas density. At hydrogen densities of order $\sim 10^6\,\mathrm{cm^{-3}}$, water ice evaporates at $T \sim 120\,\mathrm{K}$ as one can read off from figure 7.6 assuming an H_2O abundance of 10^{-7}. Ammonia and methane go into the gas phase already at about $90\,\mathrm{K}$ and $70\,\mathrm{K}$, respectively. A rise in density from 10^6 to $10^{12}\,\mathrm{cm^{-3}}$ increases the evaporation pressure for H_2O, NH_3 and CH_4 by $\sim 30\,\mathrm{K}$.

Typical evaporation temperatures of astronomically important species are:

- H_2 (3K), CO (20K), CO_2 (60K), CH_4 (70), NH_3 (80), CH_3OH (100K), H_2O (120), silicates (1500K), graphite (2500K).

7.5.2 Vapor pressure of small grains

Condensation or evaporation of small interstellar grains have the peculiarity that the transition surface between solid and gas is not flat, but curved. This has fundamental consequences. For illustration, consider a rain drop; it has the advantage that its properties are known and that it is a perfect sphere. Not to be a sphere, i.e., to have (for the same mass) a surface larger than necessary, would be energetically disadvantageous because to create an area dA of surface takes the work

$$dW = \zeta \, dA \tag{7.41}$$

where $\zeta \simeq 75 \, \mathrm{erg/cm^2}$ is the surface tension of water. By equating the $p \, dV$ work to the work required to create new surface, one finds the internal pressure p inside a drop of radius r,

$$p = \frac{2\zeta}{r} \; . \tag{7.42}$$

When molecules leave the rain drop, its surface area shrinks by an amount dA thereby creating the energy $dW = \zeta \, dA$. Evaporation is therefore easier in grains of small radius r and their vapor pressure p_r is higher than that over a flat surface, p_∞.

To compute p_r, one performs an isothermal reversible cycle of four steps. Below, we write down the work W_i that is being done in each step; the formulae for W_i are elementary.

1. In vessel 1 is a small liquid sphere of radius r at temperature T in equilibrium with its vapor of pressure p_r (figure 7.7). The volumes per mol of gas and fluid are $V_{g,r}$ and V_f, respectively. We evaporate the droplet, but in such a way that while evaporating we inject with a syringe into the sphere against the internal pressure $p = 2\zeta/r$ of (7.42) and the outer pressure p_r the same amount of liquid as is being lost to the gas, so all the time $r = \mathrm{const}$. When 1 mol has been gasified,

$$W_1 = p_r V_{g,r} - \left(p_r + \frac{2\zeta}{r}\right) V_f \; .$$

2. Isothermal expansion from pressure p_r to p_∞ of the 1 mol of gas that has been vaporized in step 1,

$$W_2 = RT \ln(p_r/p_\infty) \; .$$

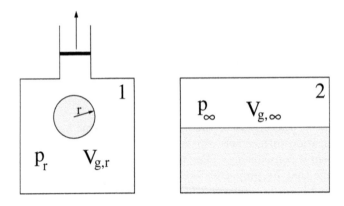

FIGURE 7.7 A reversible cycle employing two vessels to find the vapor pressure of a drop of radius r. The shaded area marks the fluid; see text.

3. The 1 mol of vapor is pressed into vessel 2 where the liquid has a plane surface and is under its vapor pressure p_∞,

$$W_3 = -RT .$$

4. To complete the cycle, we take from vessel 2 with a syringe 1 mol of liquid,

$$W_4 = V_f \, p_\infty .$$

As the cycle was isothermal ($dT = 0$), the sum $\sum W_i$ must be zero. Let v_0 denote the volume of one molecule in the liquid phase. Because $RT = pV$ and because the pressure in the drop, $2\zeta/r$, is much bigger than $p_r - p_\infty$, we obtain

$$\ln \frac{p_r}{p_\infty} = \frac{2\zeta v_0}{kTr} . \qquad (7.43)$$

This tells us that the vapor pressure $p_r = p_\infty \cdot e^{2\zeta v_0/kTr}$ equals p_∞ for big spheres, but increases exponentially when the radius becomes very small. Condensation of the first seed grains is thus possible only when the partial pressure p exceeds p_∞, or when the saturation parameter s (capital S for the entropy) defined by

$$s = \frac{p}{p_\infty} \qquad (7.44)$$

is greater than one. The vapor is then said to be supersaturated. At pressure p_r given by (7.43), a sphere of diameter $2r$ is in equilibrium with the vapor, a bigger drop will grow indefinitely, and a smaller one will evaporate, so r is the critical cluster radius.

7.5.3 Critical saturation

To estimate the critical saturation parameter $s_{cr} > 1$ at which clusters are created, we calculate the work W needed to form in a reversible cycle a drop of radius r. The drop consists of n atoms and is formed in a vessel with vapor at pressure p and temperature T. The work in each of the four steps is denoted by W_i, so $W = \sum W_i$.

1. Remove n gas atoms from the vessel,

$$W_1 = -nkT .$$

2. Isothermally expand these n atoms from p to p_∞,

$$W_2 = -nkT \ln(p/p_\infty) .$$

3. Press the n atoms into another vessel, also at temperature T, but containing a liquid of flat surface under vapor pressure p_∞,

$$W_3 = nkT .$$

4. Form there from the liquid a drop of radius r (and bring it back to the first vessel),

$$W_4 = 4\pi\zeta r^2 .$$

For the sum we get

$$W = -nkT \ln(p/p_\infty) + 4\pi\zeta r^2 = \tfrac{1}{3}\zeta \cdot 4\pi r^2 . \tag{7.45}$$

The term $-nkT \ln(p/p_\infty)$ is the potential energy of the drop. It is proportional to the number of molecules n because the molecules possess short range forces and therefore only connect to their nearest neighbors. For long range forces, the term would be proportional to n^2. The other term $4\pi\zeta r^2$, which goes as $n^{2/3}$, introduces a correction because atoms on the surface are attracted only from one side.

As the energy of the system stays constant over the cycle, one has according to the first law of thermodynamics to subtract from the system the heat $Q = W$. Therefore, during formation of the droplet the entropy of the system falls by (see A.36)

$$S = Q/T = 4\pi\zeta r^2/3T .$$

S decreases because a liquid represents a state of higher order than a gas. Although the entropy of a macroscopic system can only increase, on a microscopic level all processes are reversible and S fluctuates. Formation of a seed grain comes about by such an entropy fluctuation. As the entropy is after (A.28) equal to Boltzmann constant k times logarithm of the number Ω of states, $S = k \ln \Omega$, the probability $P = 1/\Omega$ for the formation of a seed is given by $e^{-S/k}$. When we insert the saturation parameter $s = p/p_\infty$ of (7.44), simple algebra yields

$$P = \frac{1}{\Omega} = e^{-S/k} = \exp\left(-\frac{4\pi\zeta r^2}{3kT}\right) = \exp\left(-\frac{16\pi\zeta^3 v_0^2}{3k^3T^3 \ln^2 s}\right). \quad (7.46)$$

For water ($\zeta = 75\,\mathrm{erg/cm^2}$, $v_0 = 3.0 \times 10^{-23}\,\mathrm{cm^3}$) at temperature $T = 275\,\mathrm{K}$, the exponent $S/k \sim 116/\ln^2 s$. Of course, rain drops are not relevant astronomically, but the same formula yields for carbon compounds ($\zeta \sim 1000\,\mathrm{erg/cm^2}$, $v_0 = 9 \times 10^{-24}\,\mathrm{cm^3}$) at a condensation temperature of $1000\,\mathrm{K}$ a value $S/k \sim 500/\ln^2 s$.

The probability P in (7.46) is most sensitive to ζ/T and, if ζ/T is fixed, to s changing from practically impossible to highly likely in a narrow interval Δs. It is exactly this property which allows to estimate the critical value s_{cr}:

> If N, v and σ denote the number density, mean velocity and collisional cross section of the vapor atoms, there are per second $N^2 v\sigma$ atomic collisions each leading with a probability P to the formation of a seed. So $N^2 v\sigma P$ is the rate at which seeds form and the condition
>
> $$N^2 v\, \sigma\, P \sim 1$$
>
> yields s_{cr}. The outcome of the simple calculation depends entirely on the exponent in (7.46) which has to be of order one; the factor $N^2 v\sigma$ has very little influence. One finds typical values of the critical saturation parameter for water at room temperature around 10 which implies in view of (7.43) a critical seed radius r_{cr} of a few Angstrom.

7.5.4 Time-dependent homogeneous nucleation

Let us study the creation of seed grains in a kinetic picture. If the gas has only one kind of atoms or molecules of mass m, one speaks of homogeneous nucleation, in contrast to heterogeneous nucleation when different molecular species or ions are present. A small cluster or droplet of n molecules is an n-mer with concentration c_n. The number density of the gas molecules (they are monomers) is therefore denoted c_1. A cluster has a radius r_n and a surface area $A_n = 4\pi r_n^2$. When v_0 is the volume of one molecule, n and r_n are related through

$$n\, v_0 = \frac{4\pi}{3} r_n^3 .$$

Consider a gas of constant temperature T and pressure $p = c_1 kT$ with saturation parameter $s = p/p_\infty > 1$ after (7.44). If atoms impinge on a drop at a rate $\beta_n A_n$ and evaporate from it at a rate $\alpha_n A_n$, the concentration c_n changes with time like

$$\dot{c}_n = -c_n\left[\beta_n A_n + \alpha_n A_n\right] + c_{n-1}\beta_{n-1}A_{n-1} + c_{n+1}\alpha_{n+1}A_{n+1} . \quad (7.47)$$

This is a very general equation; all the physics is contained in the coefficients for evaporation and accretion, α_n and β_n. If we assume that atoms are added to the grain at a rate $\pi r_n^2 \langle v \rangle c_1$ and leave at a rate $\pi r_n^2 \langle v \rangle p_n/kT$, where $\langle v \rangle$ is the mean velocity of (8.1) and p_n/kT from (7.43), then

$$\alpha_n = \frac{p_n}{\sqrt{2\pi m kT}} , \qquad \beta = \frac{p}{\sqrt{2\pi m kT}} . \qquad (7.48)$$

In this case, β is constant for all n and the sticking coefficient is one; p_n is the vapor pressure of an n-mer according to (7.43). One defines a particle flux or current from n-mers to $(n+1)$-mers by

$$J_n = c_n \beta A_n - c_{n+1} A_{n+1} \alpha_{n+1} . \qquad (7.49)$$

It allows to transform equation (7.47) into the deceptively simple looking form

$$\dot{c}_n = J_{n-1} - J_n .$$

Equation (7.47) disregards collisions between clusters; only molecules impinge on a drop; but that is a valid assumption as one can easily verify.

7.5.5   Steady-state nucleation

When the system is in phase equilibrium, detailed balance holds. We will flag equilibrium concentrations by the superscript 0. In equilibrium, $J_n = 0$ and

$$c_n^0 \beta A_n = c_{n+1}^0 \alpha_{n+1} A_{n+1} .$$

The last relation enables us to write also for non-equilibrium conditions

$$J_n = -c_n^0 \beta A_n \left[\frac{c_{n+1}}{c_{n+1}^0} - \frac{c_n}{c_n^0} \right] . \qquad (7.50)$$

The equilibrium concentrations c_n^0 are given by the Boltzmann formula

$$c_n^0 = c_1^0 \, e^{-\Delta G_n/kT} , \qquad (7.51)$$

where ΔG_n is the work to be expended for creating an n-mer out of monomers (see 7.45),

$$\Delta G_n = 4\pi \zeta r_n^2 - n kT \ln(p_n/p_\infty) . \qquad (7.52)$$

In view of the discussion in section A.3.5 on the equilibrium conditions of the state functions, ΔG_n is for a system at constant temperature and pressure the difference in free enthalpy. The function ΔG_n is depicted in figure 7.8; it has its maximum at n_* where $\partial \Delta G_n/\partial n = 0$. All values at n_* will in the following be marked by an asterisk. Clusters of size smaller than n_* are inside a potential well and fight an up-hill battle when growing. They have to overcome the barrier of height ΔG_*. When they have climbed the

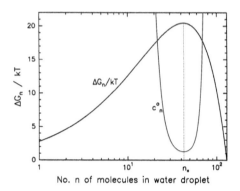

FIGURE 7.8 The full curve shows $\Delta G_n/kT$ where $\Delta G_n = 4\pi\zeta r_n^2 - nkT \ln p/p_\infty$ is the change in free enthalpy when n water molecules condense into a drop of radius r_n (see 7.52). For clusters of size n_*, the difference in the Gibbs potential between gas and droplet, ΔG_n, has its maximum and $\partial\Delta G_n/\partial n = 0$. The dashed curve depicts the equilibrium concentration c_n^0 of n-mers on an arbitrary linear scale; again the minimum is at $n = n_*$. In these plots, $T = 300\,\mathrm{K}$, $s = p/p_\infty = 4$ and the surface tension $\zeta = 75\,\mathrm{erg/cm^2}$.

barrier, their size is n_* and for $n > n_*$, they will grow unrestrictedly. So n_* is the critical size. Figure 7.9 plots the dependence of n_* on the saturation parameter s.

In a steady state (superscript s), nothing changes with time and $\partial/\partial t = 0$. The conditions for steady state are less stringent than for equilibrium because J_n need not vanish, it only has to be a positive constant,

$$J_n = J^s = \mathrm{const} > 0 .$$

Under this condition, we get from (7.50)

$$\frac{J^s}{c_n^0 \beta A_n} = \frac{c_n^s}{c_n^0} - \frac{c_{n+1}^s}{c_{n+1}^0} \simeq -\frac{\partial}{\partial n}\left(\frac{c_n}{c_n^0}\right) . \tag{7.53}$$

To realize a steady state in a thought experiment, one has to invoke a Maxwellian demon who removes large clusters with $n \geq L > n_*$ by gasifying them. The demon supplies the boundary condition for the cluster distribution,

$$c_n^s = 0 \qquad \text{for } n \geq L .$$

One does not have to be very particular about L; it just has to be somewhat greater than n_*. As a second boundary condition, we impose

$$c_1^s/c_1^0 = 1$$

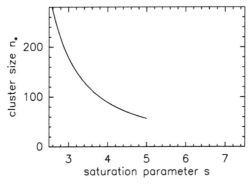

FIGURE 7.9 The dependence of the critical cluster size n_*, defined in figure 7.8, on the saturation parameter $s = p/p_\infty$ for water vapor at 273 K.

which means that only a very small fraction of the total mass is in clusters. Consequently, by performing a sum over (7.53),

$$J^s \sum_{n=1}^{L-1} \frac{1}{c_n^0 \beta A_n} = \frac{c_1^s}{c_1^0} - \frac{c_L^s}{c_L^0} = 1 , \tag{7.54}$$

we can directly compute the steady state nucleation rate J^s. The latter may also be approximated analytically from the relations

$$1 \simeq \frac{J^s}{\beta} \int_1^L \frac{dn}{A_n c_n^0} \simeq \frac{J^s}{\beta c_1^0} \int_1^L \frac{e^{\Delta G_n/kT}}{A_n} \, dn \simeq \frac{J^s}{\beta A_* c_1^0} \int_1^L e^{\Delta G_n/kT} \, dn$$

which follow from (7.54). Because ΔG_n has a fairly sharp peak at n_*, the surface area $A_n = 4\pi r_n^2$ can be taken out from under the integral and replaced by A_*. A Taylor expansion of ΔG_n around the maximum n_*,

$$\Delta G_n \simeq \Delta G_* + \frac{1}{2} \frac{\partial^2 \Delta G_n}{\partial n^2} (n - n_*)^2 ,$$

with

$$\Delta G_* = \frac{4\pi\zeta}{3} r_*^2 = \frac{16\pi\zeta^3}{3} \left(\frac{v_0}{kT \ln s} \right)^2 \tag{7.55}$$

and

$$\frac{\partial^2 \Delta G_n}{\partial n^2} = -\frac{\zeta v_0^2}{2\pi r_*^4}$$

yields, upon replacing the integration limits $(1, L)$ by $(0, \infty)$ to obtain the standard integral (B.5), the steady state nucleation rate J^s. The number of clusters that grow per unit time and volume from size n to $n + 1$ is then independent of n and one readily finds

$$J^s = v_0 \left(\frac{2\zeta}{\pi m} \right)^{1/2} \left(c_1^0 \right)^2 \exp \left[-\frac{4\mu^3}{27 \ln^2 s} \right] . \tag{7.56}$$

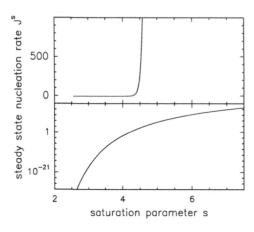

FIGURE 7.10 Variation of steady state nucleation rate J^s [$s^{-1}\,cm^{-3}$] with saturation parameter $s = p/p_\infty$ for H_2O vapor at $T = 273\,K$ on a linear (***top***) and logarithmic scale (***bottom***). Above $s \simeq 4.7$, nucleation is catastrophic.

The exponent in square brackets equals $-\Delta G_*/kT$ and μ is defined by

$$\mu = \frac{4\pi\zeta a_0^2}{kT} \qquad (7.57)$$

where a_0 denotes the radius of a monomer, so $v_0 = 4\pi a_0^3/3$. The expression in (7.56) is quite a statisfactory approximation to the sum in (7.54). The magnitude of J^s is, of course, entirely determined by the exponent. Figure 7.10 illustrates the immense change of J^s, over many powers of ten, with the saturation parameter s.

Figure 7.11 shows the deviation of the equilibrium from the steady state concentrations, the latter being calculated from

$$c_n^s = c_n^0 \, J^s \sum_{i=n}^{L-1} \frac{1}{c_i^0 \beta A_i} \, .$$

For small n-mers, equilibrium and steady state concentrations are the same. At the critical cluster size n_*, the ratio c_n^s/c_n^0 is about one half. For large n, the ratio c_n^s/c_n^0 tends to zero.

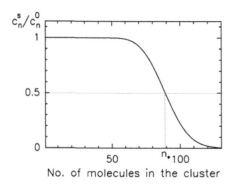

FIGURE 7.11 The ratio of the cluster abundance in a steady state to the cluster abundance in equilibrium as a function of cluster size n for water vapor at $T = 273\,K$ and saturation parameter $s = p/p_\infty = 4$.

7.5.6 Solutions to time-dependent homogeneous nucleation

7.5.6.1 The evolution towards the steady state

The time-dependent system of equations for nucleation presented in (7.47) can easily be solved when the coefficients for evaporation and accretion, α_n and β_n, are as simple as in (7.48). Let us write the time derivative of the concentration of n-mers as

$$\dot{c}_n = \frac{c_n - c_n^{\text{old}}}{\tau} ,$$

where c_n^{old} is the old value of the concentration a small time step τ ago; we want to find the new one, c_n. Equation (7.47) can immediately be brought into the form

$$A_n c_{n-1} + B_n c_n + C_n c_{n+1} + D_n = 0 , \qquad (7.58)$$

where A_n, B_n, C_n, D_n are known coefficients (so for the moment, A_n is not the surface area of an n-mer). Putting

$$c_n = \gamma_n\, c_{n-1} + \delta_n , \qquad (7.59)$$

and inserting into (7.58) yields

$$\gamma_n = -\frac{A_n}{B_n + \gamma_{n+1} C_n} , \qquad \delta_n = -\frac{\delta_{n+1} C_n + D_n}{B_n + \gamma_{n+1} C_n} .$$

We determine the concentrations c_n in the size range

$$g \leq n \leq L ,$$

with $1 \ll g \ll n_*$ (as far as possible) and $L > n_*$. The boundary conditions are suggested by figure 7.11: At the upper end, we take $c_L = 0$ because $c_L/c_L^0 = 0$. Therefore, $\gamma_L = \delta_L = 0$, and this allows to compute γ_n and δ_n for $n = L - 1, \ldots, g + 1$. At the lower end, the abundance of clusters is very close to equilibrium and we put $c_g = c_g^0$. By choosing g considerably greater than one, we avoid using the enthalpy ΔG_n of (7.52) for the smallest clusters where it cannot be correct. We then find c_n for $n = g + 1, \ldots, L - 1$ from (7.59).

Figure 7.12 presents a numerical experiment for water vapor at constant temperature $T = 263\,\text{K}$ and saturation parameter $s = 4.9$. These two values imply a monomer concentration $c_1^0 = 3.5 \times 10^{17}\ \text{cm}^{-3}$ and a critical size $n_* = 71$. The computations are performed for clusters consisting from $g = 12$ to $L = 120$ molecules. At time $t = 0$, the smallest grains are in equilibrium, $c_n/c_n^0 = 1$ for $n \leq g = 12$, and bigger clusters are absent ($c_n = 0$ for $n > g$). The fluxes J_n, as defined in (7.50), converge towards their steady state values

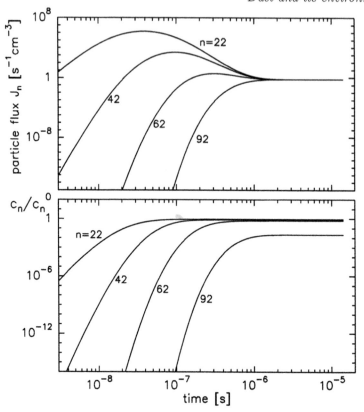

FIGURE 7.12 Formation of water drops out of vapor at 263 K supersaturated by a factor $s = 4.9$. **Bottom:** Time evolution of selected cluster concentration. **Top:** Evolution of the particle fluxes J_n of (7.49).

J^s (see 7.56) after a relaxation time τ_{rel}, which can be shown by an analysis of equation (7.53) to be of order

$$\tau_{rel} \sim \frac{r_*^2 kT}{4\,\beta\,\zeta v_0^2}. \tag{7.60}$$

In the particular example of figure 7.12, $\tau_{rel} \sim 2 \times 10^{-7}$ s and $J^s \simeq 1\,\mathrm{s}^{-1}\,\mathrm{cm}^{-3}$.

7.5.6.2 Time-dependent nucleation using steady-state fluxes

As the relaxation time τ_{rel} in (7.60) is short compared to the time it takes to grow big stable drops, the steady state flux J^s represents a good approximation to all fluxes J_n after the time τ_{rel} when transient effects have died out. The J_n may then be replaced by J^s and this greatly simplifies the further analysis in which we study the nucleation in a cooling gas, a situation that prevails in the outward flowing wind of a mass loss giant.

When the partial gas pressure of the condensible component, $P = c_1 kT$, reaches the vapor pressure P_∞ over a flat surface, the saturation parameter s equals one. At this instant, the gas temperature is denoted by T_e and the time t is set to zero. As the gas cools further, s increases. At first, the steady state nucleation rate J^s is negligible because all clusters are below the critical size n_* (see figure 7.8) and the time it takes to form one critical cluster, roughly given by the inverse of J^s, is unrealistically long. Only when the saturation parameter approaches the critical value s_{cr} at time t_{cr} do clusters become bigger than the critical size n_* and catastrophic nucleation sets in.

Let $N(t)$ be the number of monomers in clusters that were formed at time t_{cr} and and let $R(t)$ be their radius, so $4\pi R^3/3 = N v_0$. For $t \geq t_{cr}$, these clusters grow at a rate

$$\dot{R} = \tfrac{1}{3}\pi a_0^3 \langle v \rangle c_1(t) \tag{7.61}$$

where a_0 and $\langle v \rangle = \sqrt{8kT/\pi m}$ are the radius and mean velocity of a monomer. The rate at which monomers are depleted through steady state nucleation is therefore

$$\dot{c}_1(t) = -J^s(t) N(t) \tag{7.62}$$

which leads to a monomer depletion since time zero of

$$c_1(t) - c_1(0) = -\int_0^t J^s(t') N(t') \, dt' . \tag{7.63}$$

If the cooling is due to an adiabatic expansion, $T/c_1^{\kappa-1}$ is a constant, where $\kappa = C_p/C_v$ is the ratio of specific heats. One then has to add to the density decrease $\dot{c}_1(t)$ on the right side of (7.62) the term $\gamma c_1 \dot{T}/T$ where $\gamma = 1/(\kappa-1)$. The time dependence of the saturation parameter s follows via P_∞ as given by (7.39),

$$s(t) = \frac{T(t)}{T_e} e^{T_0/T(t) - T_0/T_e} .$$

To illustrate the formation of grains, we numerically integrate equations (7.61) and (7.62) assuming an adiabatic expansion with a cooling rate

$$T = T_e - wt$$

with $T_e = 292\,\mathrm{K}$ and $w = 100\,\mathrm{K\,s^{-1}}$. Results are plotted in figure 7.13.

Until the time of crtitical supersaturation, t_{cr}, grains do not form and the decrease of the gas density and temperature and the rise of the saturation parameter are due only to the adiabatic expansion at the prescribed rate. Then at $t_{cr} = 0.247\,\mathrm{s}$, nucleation rises dramatically in a spike depriving the gas of its atoms at an accelerated pace. At that moment, the saturation parameter s is near its critical value of 4.7, the temperature has fallen to $268\,\mathrm{K}$, the parameters $\mu = 9.7$ and $\Lambda \sim 6 \times 10^6$ (see 7.57 and 7.64). When the spike in J^s appears, clusters grow beyond n_* according to (7.61) whence

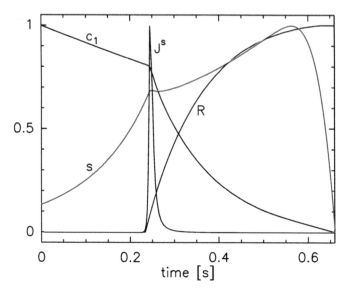

FIGURE 7.13 Nucleation in an adiabatically cooling water vapor. At $t = 0$, the gas is at $T_e = 292$ K and just saturated ($s(0) = 1$) with a monomer concentration $c_1(0) = 6.6 \times 10^{17}$ cm^{-3}. The gas expands adiabatically cooling by 100 K every second. We plot as a function of time the saturation parameter s, the number density of the gas molecules c_1, the steady state nucleation rate J^s and the radius R of the drops. All four variables are normalized at their maximum and plotted on a linear scale. These maxima occur at 7.3 for s, 10^3 s^{-1} cm^{-3} for J^s and 280μm for R.

they are stable. The radius at which their growth levels off (at $t \simeq 0.65$ s) represents the typical final grain size. The saturation continues to increase after t_{cr} because of cooling, but eventually falls for lack of gas atoms.

7.5.6.3 Similarity relations

The number of variables that enters the nucleation calculations is quite large ($v_0, m, c_1, \zeta, T, \langle v \rangle, s, p_0, T_0$), but, as shown in [Yam77], one can drastically reduce the parameter space to two variables: μ from (7.57) and Λ defined in (7.64). Therefore, the models presented here for the condensation of water near room temperature can be carried over to grain formation in late type stars ($T \sim 1000$ K, $c_1 \sim 10^9$ cm^{-3}, $\zeta \sim 10^3$ dyn/cm^2) as long as μ and Λ are comparable.

As the gas temperature changes only a little during the brief period of catastrophic nucleation when $T \sim T_{cr}$, it may be approximated by a linear function in time,

$$T(t) = T_{cr} - at .$$

When one inserts this $T(t)$ into the evaporation pressure $p_\infty = p_0\, e^{T_0/T}$ of (7.39), one finds for the time dependence of the saturation parameter

$$s(t) = \frac{c_1(t)}{c_1(0)}\, e^{t/\tau_{\rm sat}}$$

where

$$\frac{1}{\tau_{\rm sat}} \simeq \frac{T_0}{T_{\rm cr}} \cdot \left|\frac{d\ln T}{dt}\right| .$$

The cooling rate $d\ln T/dt$ is evaluated at $T_{\rm cr}$. By defining

$$y = c_1(t)\,/\,c_1(0)$$
$$x = t\,/\,\tau_{\rm sat}$$
$$\Lambda = \tau_{\rm sat} \cdot c_1(0)\, \pi a_0^2\, \langle v\rangle \tag{7.64}$$
$$\rho = \frac{3R}{a_0\,\Lambda} , \tag{7.65}$$

the nucleation formulae (7.61) and (7.63) can be transformed into the new equations

$$\frac{d\rho}{dx} = y \tag{7.66}$$

$$1 - y = \frac{\Lambda^4}{81}\left(\frac{\mu}{\pi}\right)^{1/2} \int_0^x y^2 \rho(x') \exp\left\{-\frac{4\mu^3}{27(x'+\ln y)^2)}\right\} dx' . \tag{7.67}$$

They contain only the two independent parameters μ and Λ. The first, μ, is fixed by the ratio of surface tension over temperature and thus incorporates grain properties. The second, Λ, is determined by the conditions in the environment, such as gas density c_1 and cooling rate $\dot{T}$, but it also reflects the physics of the dust via the constant T_0 in the expression for the evaporation temperature of the grain material (see 7.39). The final grain size is largely influenced by Λ (see definition of ρ in 7.65).

The quantitative results of homogeneous nucleation theory (recommended reading [Fed66], [Abr74]) outlined above are, however, fairly speculative because of the intrinsically exponential behavior of the process of grain formation and a number of dubious assumptions, oversimplifications and unsolved problems. For example, the smallest clusters are not spheres and their surface tension is unknown. Moreover, nucleation in the wind of giants is not homogeneous but proceeds in a vast network of chemical reactions; and stellar photons, so far neglected, are present that create new reaction channels and are not in equilibrium with the gas.

8

Grain surfaces

In the interstellar medium, gas atoms and molecules continually collide with dust grains. They may either rebounce from the surface or stick. Such collisions can have great astronomical consequences:

- They transfer energy from gas to dust and back which affects the temperature of the components.

- They deplete gas species as they freeze out on the grains. Such species may then not be observable in the gas any more, whereas otherwise they would be strong emitters and important coolants.

- The depleted gas atoms or molecules are deposited in ice layers that change the optical properties of the grains. Prominent infrared absorption bands appear.

- In the accreted gas, new molecules form, notably molecular hydrogen, but also complex molecules.

8.1 Gas accretion on grains

Consider a gas of temperature T_{gas} containing different species of atoms or molecules, like H, H_2, CO, NH_3 and others. Let the particles of species i have a mass m_i, a number density density n_i and mean velocity (see A.17)

$$v_i = \sqrt{\frac{8kT_{\text{gas}}}{\pi m_i}} .$$ (8.1)

Relative to the gas, the dust grains can be considered to be at rest. One spherical grain of geometrical cross section πa^2 accretes the gas species i at a rate

$$R_{\text{acc}} = \pi a^2 n_i v_i \eta_i$$ (8.2)

where η_i denotes the sticking probability. When the grains have a size distribution $n(a) \propto a^{-3.5}$ with lower and upper limit a_- and a_+, and $a_+ \gg a_-$, their total cross section F for gas capture per cm^3 is (see section 5.5.1)

$$F = \frac{3 n_H m_H R_d}{4 \rho_{\text{gr}} \sqrt{a_- a_+}} .$$

Here R_d is the dust-to-gas mass ratio, ρ_{gr} the density of the grain material, $n_H = n(HI) + 2n(H_2)$ the total number density of hydrogen, either in atomic or molecular form. The total accretion rate per cm^3 of species i is therefore $R_{acc}^{tot} = F n_i v_i \eta_i$. The mean lifetime τ_{acc} of a gas atom (or molecule) before it is swallowed by dust, defined by $\tau_{acc} R_{acc}^{tot} = n_i$, thus becomes

$$\tau_{acc} = \frac{1}{F v_i \eta_i} = \frac{4\rho_{gr}}{3 m_H R_d \eta_i} \frac{\sqrt{a_- a_+}}{n_H v_i} . \tag{8.3}$$

Inserting reasonable numbers ($R_d = 0.007$, $\rho_{gr} = 2.5$ g/cm^3, $a_- = 100$ Å, $a_+ = 3000$ Å, $v_i = 0.3$ km/s) and neglecting processes that can remove the adsorbed gas again from the surface (section 8.2), one finds that an atom with a sticking probability of one ($\eta_i = 1$) freezes out after a time

$$\tau_{acc} \sim \frac{2 \times 10^9}{n_H} \text{ yr} . \tag{8.4}$$

τ_{acc} is shorter than the lifetime of molecular clouds ($\tau_{cloud} \sim 10^7 \ldots 10^8$ yr) which always have densities above 10^2 cm^{-3}. If the gas density is high ($n_H > 10^4$ cm^{-3}), as in clumps, depletion and mantle formation proceed more quickly ($\tau_{acc} \leq 10^5$ yr) than the dynamical processes which are characterized by the free-fall time scale ($t_{ff} \sim 2 \times 10^7/\sqrt{n_H}$ yr).

The growth rate da/dt of the grain radius should be independent of a, because an impinging atom does not feel the grain curvature; it only sees the local environment of the impact point. Therefore, the accreted matter goes primarily onto small dust particles as they have the larger total surface area. Large grains hardly grow any bigger by accretion, although the mean grain size, which is the average over the total size distribution, increases.

8.1.1 Physical adsorption and chemisorption

8.1.1.1 The van der Waals potential

To describe the collision between an atom on the grain surface and an approaching gas atom, we first derive their interaction potential. Both atoms are electrically neutral, but have dipole moments: let $\mathbf{p}_1$ refer to the surface and $\mathbf{p}_2$ to the gas atom.

A point charge q at position $\mathbf{r}_0 = \mathbf{0}$ has in an external electrostatic potential ϕ, given by some fixed far away charges, the electrostatic potential energy $U = q\phi(\mathbf{r}_0)$. If instead of the point charge q there is, localized around $\mathbf{r}_0$, a charge distribution $\rho(\mathbf{r})$, the energy becomes

$$U = \int \rho(\mathbf{r}) \phi(\mathbf{r}) \, dV .$$

Expanding $\phi(\mathbf{r})$ to first order around $\mathbf{r}_0$,

$$\phi(\mathbf{r}) = \phi(\mathbf{r}_0) + \mathbf{r} \cdot \nabla \phi(\mathbf{r}_0) = \phi(\mathbf{r}_0) - \mathbf{r} \cdot \mathbf{E}(\mathbf{r}_0) ,$$

and substituting the expansion into the above volume integral for U, we obtain

$$U = q\,\phi(\mathbf{r_0}) - \mathbf{p} \cdot \mathbf{E}(\mathbf{r_0}) ,$$

where we have used the general definition (1.2) of a dipole $\mathbf{p}$. In our case of colliding atoms, the effective charge q vanishes. Suppose the field $\mathbf{E}(\mathbf{r})$ is due to the dipole $\mathbf{p_1}$ on the grain surface; $\mathbf{E}(\mathbf{r})$ is then described by (3.13). Denoting by $\mathbf{r}$ the vector from $\mathbf{p_1}$ to $\mathbf{p_2}$, we get

$$U(r) = \frac{\mathbf{p_1} \cdot \mathbf{p_2}}{r^3} - \frac{3(\mathbf{p_1} \cdot \mathbf{r})(\mathbf{p_2} \cdot \mathbf{r})}{r^5} . \tag{8.5}$$

Let us further assume that the dipole $\mathbf{p_2}$ of the gas atom is induced by $\mathbf{p_1}$ of the surface atom. The strength of the former depends then on its polarizability α (see 3.13 and 1.8). For $\mathbf{p_1} \parallel \mathbf{r}$,

$$p_2 = \frac{2\alpha p_1}{r^3} ,$$

the dipole moments are parallel ($\mathbf{p_1} \cdot \mathbf{p_2} = p_1 p_2$) so that

$$U(r) = -\frac{4\alpha p_1^2}{r^6} . \tag{8.6}$$

By and large, polar molecules of large dipole moment, like H_2O, are more tightly bound and have a higher evaporation temperature than apolar ones where it is small or zero (CO, N_2), but, as mentioned in section 5.2.3, the induced dipole moment (CO_2) can be more important than the permanent one.

8.1.1.2 The full potential including repulsion

The absolute value of the van der Waals potential increases rapidly at short distances, much quicker than for monopoles. But when the two dipoles are very close, a repulsive interaction sets in because the electrons tend to overlap and *Pauli's* exclusion principle forbids them to occupy the same quantum state. This potential bears a more empirical character and changes with distance even more abruptly, like $1/r^{12}$. The net result of the combination between repulsion and attraction is

$$U(r) = 4D \left[\left(\frac{\sigma}{r}\right)^{12} - \left(\frac{\sigma}{r}\right)^6 \right] \tag{8.7}$$

with $D\sigma^6 = \alpha p_1^2$. The parameter D defines the strength of the potential and σ its range. $U(r)$ becomes infinite for $r \to 0$, changes sign at $s = \sigma$, reaches its minimum value $-D$ at $r_0 = \sqrt[6]{2}\sigma$ and remains negative as it approaches zero for $r \to \infty$ (figure 8.1).

In a collision of an atom with the grain surface, one has to include the contribution from all force centers, i.e., atoms on the grain surface. On a

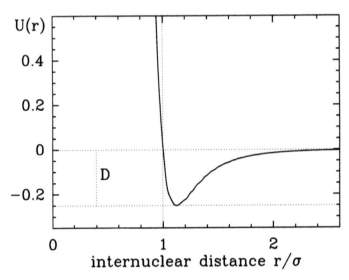

FIGURE 8.1 The potential $U(R)$ between dipoles after (8.7); here with $D = 0.25$ and $\sigma = 0.5$. The curve is relevant for physical adsorption. The minimum is at $r_0 = \sqrt[6]{2}\sigma$ where $U(r_0) = -D$.

regular surface, the potential attains minima at various locations privileged by symmetry. For a simple or body centered cubic lattice, these minima are at the mid-points between four neighboring surface atoms (see figure 5.2).

8.1.1.3   Physical and chemical binding

Physical adsorption is weak; it needs typically only 0.1 eV to remove the atom or molecule from the grain surface, corresponding to 1000 K if the binding energy E_b is expressed as a temperature $T = E_b/k$. The exact value depends on the composition and structure of the surface and on the type of the gas species.

A much tighter coupling of a gas atom colliding with the grain surface is possible through chemical binding or chemisorption. It involves a profound change in the electron structure of the binding partners. The interaction is therefore much stronger (from 0.5 to 5 eV) and the range of the chemical potential shorter than in the case of physical adsorption. As a precondition for chemisorption, the surface must contain chemically active sites and the gas species should not consist of saturated molecules, like H_2 or H_2O, but of atoms, particularly hydrogen.

Chemically active sites will have disappeared once the grain surface is covered by physical adsorption by one monolayer of, for example, H_2 or H_2O. Therefore, the mean binding energy of atoms and molecules in an ice mantle which is several Å or more thick is less than 0.5 eV.

8.1.2 The sticking probability

Let us estimate the probability η that an impinging gas atom (subscript g) stays on the grain surface. For the gas atom to stick, it must transfer to the dust particle in the collision an energy ΔE_s greater than its kinetic energy E_k far from the surface; otherwise it will rebound. For simplicity, we consider the interaction in one dimension along the x-axis with just one surface atom (subscript s) of mass m_s which we imagine to be attached to a spring of force constant κ, so that it oscillates at frequency

$$\omega_s = \sqrt{\kappa/m_s} \ .$$

For the interactive potential $U(r)$ we use (8.7), where $r = x_g - x_s > 0$ is the distance between the two atoms. The force associated with the potential, $F(r) = -U'(r)$, is attractive and negative when r is large and repulsive and positive when r is small. The gas atom passes in the collision first through the attractive potential where its kinetic energy increases, then encounters the steeply rising repulsive part (see figure 8.1) and, without a transfer of energy ($\Delta E_s = 0$), rebounces at the distance of closest approach, $r_{\min}$, where $U(r_{\min}) = E_k$, and then returns to infinity. If the surface atom has its equilibrium position at $x = 0$, such a simple mechanical system is governed by the equations of motion

$$m_s \ddot{x}_s = -\kappa \, x_s - F \tag{8.8}$$
$$m_g \ddot{x}_g = F \ . \tag{8.9}$$

The potential $U(r)$ may be approximated near its minimum at $r_0 = \sqrt[6]{2}\sigma$ (see figure 8.1) by a parabola, $U(r) = -D + \frac{1}{2}\omega_0^2(r - r_0)^2$ where ω_0^{-1} can be regarded as the collision time. Under the simplifying assumption of a quadratic interaction potential everywhere, not just around r_0, the transferred energy ΔE_s can be determined analytically [Wat75],

$$\Delta E_s = \frac{m_g}{m_s}(D + E_k) \cdot
\begin{cases}
4 & : \quad \omega_0 \gg \omega_s \\
\pi^2/4 & : \quad \omega_0 = \omega_s \\
\dfrac{2\,\omega_0^4}{(\omega_s^2 - \omega_0^2)^2} & : \quad \omega_0 \ll \omega_s
\end{cases} \tag{8.10}$$

Because under most interstellar conditions $E_k \ll D$, we learn from (8.10) that the sticking efficiency is largely determined by the the binding energy for physical adsorption, D, and by the mass ratio of gas to surface atom. Sticking coefficients are typically between 0.1 and 1.

Numerical solutions to (8.8) and (8.9) with the full potential (8.7) are presented in figure 8.2 for the following parameters: The vibrational frequency of the surface atom is equal to the collisional frequency, $\omega_s = \omega_0 = 1.58 \times 10^{14}\,\mathrm{s}^{-1}$, binding energy $D = 0.05\,\mathrm{eV}$, $\sigma = 0.2\,\text{Å}$, kinetic energy of the gas atom $E_k = \frac{1}{2}kT_g$ with $T_g = 50\,\mathrm{K}$, kinetic energy of the surface atom $\frac{1}{2}kT_s$

with $T_s = 20\,\mathrm{K}$, mass of gas atom $m_g = m_H$. Note how the elongation of the surface atom increases when the gas atom looses energy in the collision. The numbers which one computes for $\Delta E_s/E_k$ from the time evolution qualitatively agree with formula (8.10) and confirm that the gas atom sticks when $\Delta E_s/E_k$ is well above one and rebounds otherwise.

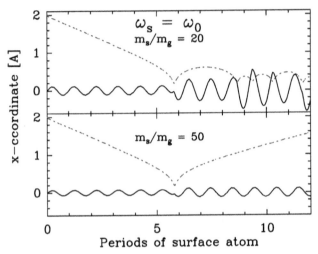

FIGURE 8.2 Two examples of a 1-d collision between a gas and a surface atom (subscript g and s; see text for parameters). **Solid lines**: vibration of surface atom, 20 times enlarged; **dash-dots**: evolution of gas atom. The time is in units ω_s^{-1}. **Top:** The gas atom sticks; here $m_g = m_H$, $m_s/m_g = 20$ and $\Delta E_s/E_k = 3.0$. **Bottom:** It rebounces; here $m_s/m_g = 50$ and $\Delta E_s/E_k = 1.2$.

8.2 Mobility of atoms on grain surfaces

The more or less evenly spaced atoms on the surface of a grain act as force centers and the surface potential has a semi-regular hilly structure. A physically adsorbed gas atom or molecule finds itself in a potential well where it is surrounded by a barrier of height U_0. It is not totally immobile but can leave its present location by

- hopping over the barrier after thermal excitation;
- quantum mechanically tunneling under the barrier;
- evaporating from the grain;
- photodesorption.

8.2.1 Thermal hopping

Let us make an educated guess how long it takes until the adsorbed atom jumps over the barrier. Its height, U_0, is typically three times smaller than the binding energy E_b of the gas atom: $E_b \sim 3\,U_0$. Let P_1 and P_2 be the probabilities that the atom is at an energy level below or above U_0, respectively. Usually, $P_2 \ll 1$, so $P_1 \simeq 1$, and, of course, $P_1 + P_2 = 1$. For a Boltzmann distribution, $P_2/P_1 \sim \exp(-U_0/kT)$ where T is the grain temperature. Let the atom change its energy levels at a frequency ν_0. In one second, it will be $P_2\nu_0$ times above U_0 and, on the average, it gets there after $(P_2\nu_0)^{-1}$ seconds. The timescale for thermal hopping is therefore

$$t_{\mathrm{hop}} \sim \nu_0^{-1}\, e^{U_0/kT} \ . \tag{8.11}$$

We determine the frequency ν_0 classically. Let a be the distance between surface atoms (or surface sites or local minima in the potential curve). The adsorbed atom of mass m moves in the interval from $-a/2$ to $a/2$ in the harmonic potential $U(x) = (4U_0/a^2)x^2$; its velocity equals

$$v = \sqrt{2(U_0 - U(x))/m} \ .$$

Elementary integration of $dt = dx/v$ from $-a/2$ to $a/2$ yields

$$\nu_0 = \sqrt{4U_0/m\pi^2 a^2} \simeq \sqrt{4E_b/3m\pi^2 a^2} \ .$$

Typical values for ν_0 are a few times $10^{12}\,\mathrm{s}^{-1}$. Of course, relevant in the estimate (8.11) is not ν_0, but the exponent. For $U_0 < kT$, thermal hopping happens momentarily ($t_{\mathrm{hop}} \sim \nu_0^{-1} \sim 10^{-12}\,\mathrm{s}$), otherwise t_{hop} increases exponentially. Binding temperatures, E_b/k, are of order $400\,\mathrm{K}$ for H and H_2, and $800\,\mathrm{K}$ for C,N,O. A carbon atom on a $10\,\mathrm{K}$ cold grain, for example, makes one hop per second. But these estimates are very rough. The adsorption energy, E_b/k, depends, of course, also on the type of surface, whether it is made of graphite, silicate, water ice, CO ice or something else.

8.2.1.1 Scanning time in thermal hopping

An atom hopping from one site to the next has no preferential direction; its path is a random walk (see figures 7.2 and 7.3). If it takes N steps of constant length a, the root mean square deviation from the starting position is $\sigma = a\sqrt{N}$.

Consider the trajectory of an atom on a quadratic lattice, as in figure 5.4, but with $b = a$. In N random steps of length a, the atom makes close to $N/2$ steps in x-direction and $N/2$ in y-direction. Statistically, it ends up at a point with coordinates $|x| \sim |y| \sim a\sqrt{N/2}$. Of all the sites it has visited in N steps, only about $\sqrt{N}$ have different x-values and $\sqrt{N}$ different y-values. As there are $\sqrt{N}$ y-coordinates per accessed x-coordinate, to first order, the number of different sites which the atom visits in N steps is itself of order N, and not much smaller, despite the zig-zag path.

A grain of radius r and grid constant a has $N_{\text{surf}} \simeq 4\pi(r/a)^2$ sites. There-fore, to screen them all, an atom needs the time

$$t_{\text{scan}} \sim N_{\text{surf}} \, t_{\text{hop}} \,. \qquad (8.12)$$

A typical number for N_{surf} is 10^6, which implies in the example above that a carbon atom scans the surface of a cold grain in one year. But a time scale acquires an astrophysical meaning only when it is put into a relation with another one. Therefore, t_{scan} should be compared with the time for evaporation, photodesorption, cloud destruction and the like.

8.2.2 Evaporation

Replacing in equation (8.11) U_0 by the binding energy E_b leads to the evap-oration timescale

$$t_{\text{evap}} \sim \nu_0^{-1} e^{E_b/kT} \,. \qquad (8.13)$$

For example, when an ice covered grain is warmed up to $20\,\text{K}$, the CO ice disappears in 10 years (again, a factor of ten does not worry us) assuming an adsorption energy per CO molecule, E_b/k, of $1000\,\text{K}$.

Because T stands in (8.13) in the exponent, the evaporation time is radically shortened for very small dust particles whose temperature fluctuates in a hard radiation field and shows occasional spikes (see figure 6.11).

If a grain is covered by an ice mantle that has N_{surf} surface atoms bound with an energy E_b, the number of atoms evaporating from it per second equals

$$R_{\text{evap}} \sim N_{\text{surf}} \, \nu_0 \, e^{-E_b/kT} \,. \qquad (8.14)$$

8.2.3 Tunneling

Because of its exponential dependence on grain temperature in (8.11), thermal hopping does not work when the grain is cold. However, an atom may then still have the chance to quantum mechanically tunnel through the potential barrier, U_0, as discussed in section A.8. According to (A.108), the tunneling time equals

$$t_{\text{tun}} \sim \nu_0^{-1} \exp\left(\frac{2a}{\hbar} \sqrt{2mU_0} \right) \,.$$

It is very sensitive to the mass, m, of the atom and tunneling can only be important for the lightest atoms, in practice for hydrogen or deuterium. The mobility of other species is restricted to thermal hopping.

Let us evaluate the tunneling time for a hydrogen atom ($m = 1.67 \times 10^{-24}\,\text{g}$) physically adsorbed on an ice layer with a spacing $2a = 2\text{Å}$ between force centers and vibrating at a frequency $\nu_0 = 3 \, 10^{12} \, \mathrm{s}^{-1}$. When the the potential barrier U_0/k is low, say $100 \, \mathrm{K}$, one obtains $t_{\mathrm{tun}} \simeq 10^{-11} \, \mathrm{s}$. The atom is then very mobile.

8.2.3.1 Scanning time for tunneling

For an *ideal* surface, tunneling to a site n grid spacings away is not a random walk, but the well is reached after a time $t_n \simeq n \, t_{\mathrm{tun}}$. A real grain has defects on its surface, i.e., sites surrounded by a high potential which the atom cannot penetrate and where it is deflected in its motion, thus leading to a random walk. The defect sites invariably arise because the grain is bombarded by soft X-rays and cosmic rays.

Suppose the grid constant of the atomic lattice is of unit length. Let ℓ be the mean free path before the atom is scattered. The time to reach a site n spacings away increases from $n \, t_{\mathrm{tun}}$ without scattering to $(n^2/\ell) \, t_{\mathrm{tun}}$ when defects are taken into account because one needs $(n/\ell)^2$ steps of length ℓ. If the surface contains N_{surf} sites altogether, its linear dimension is of order $N_{\mathrm{surf}}^{1/2}$ and the mobile particle covers in a random walk such a distance in the scanning time

$$t_{\mathrm{scan}} \sim \frac{N_{\mathrm{surf}}}{\ell} \, t_{\mathrm{tun}} \, . \tag{8.15}$$

Inserting numbers for a hydrogen atom on a grain of $1000 \, \text{Å}$ radius, we find $t_{\mathrm{scan}} \lesssim 10^{-6}/\ell \, \mathrm{s}$, so scanning is quick.

8.2.4 Photodesorption

The absortion of a UV photon by a grain can lead to the ejection of physically adsorbed atoms or molecules. Indeed, ice mantles are found only towards molecular clouds which are well shielded from hard radiation ($A_{\mathrm{V}} \geq 4$ mag). At the transition zone between ice-coated and bare grains, the rates at which a molecular species is accreted onto a grain (see 8.2) and photodesorbed should be equal,

$$\pi a^2 \, n v \eta \; = \; \pi a^2 \, 4\pi \int_{\mathrm{UV}} \frac{J_\nu}{h\nu} \, d\nu \times Y \, e^{-\tau_{\mathrm{UV}}} \, .$$

The integral times $4\pi^2 a^2$ gives the number of absorbed UV photons per second, τ_{UV} is the shielding optical thickness, and Y the yield for photodesorption per UV photon.

The yield is poorly constrained, theoretically and experimentally, but with will and force one can squeeze out a number. For example, in the case of CO we know from observations that $\tau_{\mathrm{UV}} \geq 6 \ldots 8$. At the edge of a cold molecular clump, the CO number density $n \sim 0.1 \, \mathrm{cm}^{-3}$, the mean molecular velocity $v \sim 10^4 \, \mathrm{km \, s}^{-1}$ and the sticking probability $\eta \simeq 0.3$. If we then insert for the mean intensity, J_ν, the interstellar radiation field of figure 6.1, the integral gives $3 \times 10^8 \, \mathrm{s}^{-1} \, \mathrm{ster}^{-1}$ for $\lambda < 0.3\mu\mathrm{m}$, and we obtain for the yield $Y \sim 10^{-4}$.

In this number joggling, our estimate for Y might easily be off by a factor of 100. Nevertheless, we now know its order of magnitude; we have bracketed Y: it lies somewhere in the range between 10^{-6} and 10^{-2}. The little exercise also nicely illustrates how such diverse things like dust opacity, radiation field, molecular gas abundance, grain surface physics all come into play in one particular process: here photodesorption.

8.3   Grain surface chemistry

Gas phase chemistry can explain the abundances of many molecules, ions, atoms and their isotopes in the various phases of the interstellar medium, however, it fails to account for the existence of the most common species (H_2) and there are several complex key molecules, like CH_3OH or H_2CO, which one finds hard to make in the gas. These molecules are thought to form on grains (recommended reading [Dul84], [Leq05], [Tie05]).

8.3.1   Chemical reactions in the gas

We can better appreciate the particularities and theoretical difficulties of grain surface chemistry, if we first recall a few essentials of gas reactions. To theoretically determine the concentration of gas species (like H, H^+, H_2, HCO, CO, ...), one writes down for *each* species the balance between *all* processes in which it is created and destroyed, and then solves the whole set of equations. For example, in the (charge transfer) reaction

$$H^+ + O \;\rightarrow\; O^+ + H \,,$$

which is one of thousands in the full set, the number densities of H^+ and O decrease and those of O^+ and H increase per second by the amount

$$n(H^+)\, n(O)\, k_{O,H^+} \,.$$

The change in number densities is equal to the product of the densities of the species on the left times the rate coefficient, k_{O,H^+}, of the particular reaction. There are analogous expressions when one reactant is a molecule (O + CH $\rightarrow$ CO + H), or when partners fuse (CO + S $\rightarrow$ OCS), or when they dissociate (H + $H_2O_2 \rightarrow H_2O$ + OH). When photons or cosmic rays are involved, for example, in the ionization of hydrogen,

$$H^+ + h\nu \;\rightarrow\; H^+ + e \,,$$

one writes the rate at which H is eliminated by radiation as $n(H)\, k_{rad}$. Here k_{rad} incorporates the energy density of the photons and their spectral distribution.

In a steady state, the densities of all species ($n(\text{H}^+), n(\text{O}), \text{n}(\text{O}^+), n(\text{H}),...$) are constant in time. Sometimes the environment changes before the chemistry reaches equilibrium because some reactions are slow. Then, as well when one wants to explicitly follow the chemical evolution, time-dependent calculations are necessary.

Let $\sigma(u)$ be the cross section of the reaction A + B → C for partners with relative velocity u. Then, by definition,

$$n(\text{A})\,n(\text{B})\,u\sigma(u)$$

gives the number of reactions per s and cm^3. The rate coefficient, $k(T)$, is the mean of $u\sigma(u)$ weighted over the Maxwellian velocity distribution $f(u)$ (see A.14) and depends only on the gas temperature,

$$k(T) \;=\; \int_0^\infty u\sigma(u) f(u)\,du \; .$$

The rate coefficients include all the physics and can, ideally, be measured in the lab or calculated quantum mechanically. If one knows them for all reactions, one can feed the set of equations into a computer and it will supply one with a solution to the abundance problem for any environmental conditions in terms of density, temperature, cosmic ray flux and radiation field. If the resulting gas composition is in agreement with observations, one claims to understand what is going on. To contrast it below with grain surface reactions, we summarize the salient points of gas phase chemistry:

- Reactions in the gas are the result of a collision between two partners; three-body collisions do not happen at interstellar densities.

- Before the collision, at distances much greater than the square root of the cross section $\sigma(u)$:

    the reactants are free (unperturbed);
    they are in an exactly definable quantum mechanical configuration;
    they move on straight lines in three dimensions with known velocity.

- One number, the rate coefficient $k(T)$, precisely describes the efficiency of the reaction. It does not change during the chemical evolution of the cloud (provided T stays constant).

8.3.2 Chemical reactions on dust

Next consider a chemical reaction on a grain surface, say, A + B → C. There are two possibilities as to how it can proceed:

- Langmuir-Hinshelwood mechanism:

    – At least one particle of species A and one of species B are (physically) adsorbed simultaneously.

- Either A or B, or both, are mobile on the surface by way of tunneling or thermal hopping.
- A and B find each other before one of them is desorbed or otherwise chemically transformed.
- Upon encounter, the particular reaction takes place.

- Eley-Rideal mechanism:

  - Only one species is adsorbed on the surface, say B.
  - A *gas* particle of species A hits the grain at the spot where B is located. No surface mobility is required.
  - The reaction A + B → C takes place.

There is no general rule regarding the fate of the product C. It may stay on the surface or be ejected into the gas, with a certain kinetic energy and possibly in an excited state. H_2 is, however, always ejected and reaction products that are heavy or have a large binding energy tend to stay on the surface. They accumulate there and form a mantle around the grain.

The mantle material may slowly be further chemically transformed, on a timescale $\sim 10^7$ yr, through the occasional interaction with cosmic rays or UV photons. Or it may be released when the grain is heated above the evaporation temperature of the ice species, as happens in regions of massive star formation.

8.3.2.1   Timescales replace rate coefficients

The processes that lead to a surface reaction are diverse and have large inherent uncertainties. Therefore one uses timescales to quantify them. They are crude and can principally not be made accurate in view of the immense variety of surface conditions which always remain unknown in their details. Take, as an example out of many, the reaction

$$H + CO \rightarrow HCO .$$

First, there is the timescale t_{ar} of arrival of an H atom or a CO molecule on the surface. It is defined as the inverse of the accretion rate and depends obviously on the composition of the gas. It is believed that in dense clouds hydrogen is almost entirely molecular and carbon bound in CO, with only small traces of atomic hydrogen or carbon. The dependence of t_{ar} on the gas composition demonstrates that gas and surface chemistry are not independent but linked. To quantify the amount of H and CO particles on the surface, t_{ar} has to be compared with the timescales for photodesorption and evaporation, t_{evap} and t_{des}, as these are the processes that compete with accretion.

Next there is the timescale for migration, t_{mig}, which includes hopping and tunneling, possibly also diffusion in the mantle, and determines how quickly H and CO get together. The trajectory of an atom on the surface and the speed at which it moves are sensitive to the grain temperature and the details of the surface and are not amenable to exact computations. At present, the closest approach to reality is to assume in a model a certain surface structure and to follow the path of an atom with a Monte Carlo technique.

The particular reaction $H + CO \rightarrow HCO$ has an activation barrier E_a, and therefore, when H and CO finally meet, a new timescale enters: the one for chemical reaction. In perfect analogy to (8.13), it is defined by

$$t_{\text{chem}} \sim \nu_0^{-1} e^{E_a/kT} . \tag{8.16}$$

Summarizing, in grain surface chemistry, the (ideally) precise rate coefficients, $k(T)$, of the gas have to be replaced by very approximate timescales. For each $k(T)$, one needs several timescales ($t_{\text{ar}}, t_{\text{tun}}, t_{\text{hop}}, t_{\text{evap}}, t_{\text{chem}}, \ldots$) that furthermore change during the chemical evolution.

8.3.2.2 The crux and power of surface chemistry

There are four reasons why chemistry on a grain is much more complicated than in the gas:

1. The reactants get together in two steps. First, they collide with the grain and then they move over its surface in an erratic manner.

2. The reactants are never free, but coupled to the solid, not strongly ($E_b \lesssim 0.5\,\text{eV}$) but unpredictably.

3. The solid participates in the reaction as a third body, while its exact properties remain unknown.

4. The timescales change during the chemical evolution, not only t_{ar} as gas components freeze out, but all others too because the surface changes as new layers are deposited.

   (For the Eley-Rideal mechanism, item 1 does not apply and under item 2, only one reactant is coupled.)

The use of timescales in surface chemistry renders model results less reliable, but this does not mar its importance:

– The grain is the third partner that acts as a catalyst. It can remove energy, take care of angular momentum conservation or lower the activation barrier (when there is one).

– Encounters between reactants can be more frequent. In the gas, the reactants search for each other in $3d$-space, on a grain in $2d$-space, and that is easier. If the partners meet more frequently, the chance for a reaction with an activation barrier is enhanced (see 8.16). Therefore, some reactions are much faster on a grain.

Despite intrinsic theoretical difficulties, current models of surface chemistry in dense clouds have yielded encouraging results. For example, one can now explain how CH_3OH and H_2CO form on dust (figure 8.3) by repetitively adding H to CO; or how CO_2, whose gas phase abundance is very low, is made in mantles from $CO + O \rightarrow CO_2$ (figure 8.4); or how water is synthesized by reactions between H_2 and OH.

8.3.3 The formation of H_2 in diffuse clouds

All surface reactions take place in dark clouds where the accreted gas particles are well protected against photodesorption, but for one exception, and that concerns molecular hydrogen. The only way to make it in relevant quantities is by accretion of H atoms on grain surfaces in diffuse clouds. Radiative association $H + H \rightarrow H_2 + h\nu$ is strongly forbidden, three-body reactions require excessive densities $n \geq 10^{12}\,\mathrm{cm}^{-3}$, and other reactions are negligible (however, they may have been important in the early universe).

The major destruction route of H_2 in diffuse clouds is photo-dissociation. Absorption of photons with wavelengths between 912 and 1108Å leads to electronic transitions $X\,^1\Sigma_g^+ \rightarrow B\,^1\Sigma_u$ (Lyman band) and $X\,^1\Sigma_g^+ \rightarrow C\,^1\Pi_u$ (Werner band). They are immediately followed by electric dipole decay ($A \sim 10^8\,\mathrm{s}^{-1}$) to the electronic ground state. If the final vibrational quantum number v is greater than 14 – this happens in about one out of seven cases – the state is unbound and the molecule dissociates.

Molecular hydrogen is observed in the diffuse medium if the ultraviolet radiation field is not too strong. Its photo-dissociation can then be calculated and must be balanced by H_2 formation. The reaction rate R (per cm^3 and second) to create H_2 on grain surfaces is obviously proportional to the number of H atoms impinging on a grain ($n(\mathrm{HI})\,\sigma_{\mathrm{gr}}\langle v \rangle$) times the number density of grains, n_{gr},

$$R = \eta\, n(\mathrm{HI})\, \sigma_{\mathrm{gr}}\, \langle v \rangle\, n_{\mathrm{gr}} \ .$$

The efficiency η includes the probability for sticking, finding another hydrogen atom by tunneling over the grain surface, molecule formation and desorption. $n(\mathrm{HI})$ is the density of hydrogen atoms, $\langle v \rangle$ from (A.17) their mean velocity and $\sigma_{\mathrm{gr}} = \pi a^2$ the geometrical cross section of spherical grains of radius a. With a gas-to-dust of ratio 150 and a bulk density of $2.5\,\mathrm{g/cm^3}$,

$$n_{\mathrm{gr}} \simeq 1.5 \times 10^{-27}\, \frac{n_{\mathrm{H}}}{a^3} \ ,$$

where $n_{\mathrm{H}} = n(\mathrm{HI}) + n(\mathrm{H_2})$ is the total hydrogen density. Therefore

$$R \simeq 7 \times 10^{-23}\, \eta\, n(\mathrm{HI})\, n_{\mathrm{H}}\, \frac{T^{1/2}}{a} \ . \tag{8.17}$$

In the diffuse medium, $n(\text{HI}) \simeq n_\text{H}$. The smallest grains should get the highest weight (see 5.12), so $a \sim 10^{-6}$ cm seems appropriate. If $\eta \sim 0.2$,

$$R \sim 10^{-17} n_\text{H}^2 T^{1/2} \quad \text{cm}^{-3} \text{ s}^{-1} \,,$$

and this should indicate the right range. The chemical reaction $\text{H} + \text{H} \to \text{H}_2$ on the grain surface leads to the ejection of the molecule. On leaving, it is unclear how much of the formation energy of 4.5eV is transmitted to the dust particle and how much is carried away by the H_2 molecule in the form of excited rot-vib levels.

8.4 Ice mantles

Ice mantles form in cold and dense molecular clouds as a result of gas accretion onto refractory cores. The attachment of the ice molecules to the core and to each other is via physical adsorption. The mantles consist of a mixture of molecules (H_2O, CH_3OH, NH_3, CO, CO_2, CH_4, O_2, N_2, ...) in varying abundance ratios, but water usually dominates (70%). Some of the molecules were, at least, partially, made in reactions on the grain surface, like NH_3 or H_2O, others probably entirely, for instance, CO_2 or CH_3OH (section 8.3). Therefore, when methanol evaporates in a warm environment (star forming region), its gas phase abundance jumps up by two or three orders of magnitude.

An impressive example of absorption by ice mantles is presented in figure 8.3. The infrared bands allow us to identify molecules or, at least, certain characteristic groups within them, for example, OH in water molecules. They thus help us unravel the chemical composition of the grain mantle. For the refractory cores, this is only possible in silicate material (section 9.3). The features are obviously grain size independent.

In a simplified picture, the vibrational fundamental frequency ω in a characteristic group that consists of two molecules of mass m_1 and m_2 which are connected by a spring of force constant κ equals

$$\omega = 2\pi\nu = \sqrt{\kappa/\mu} \tag{8.18}$$

where the reduced mass is

$$\mu = \frac{m_1 m_2}{m_1 + m_2} \,. \tag{8.19}$$

The frequency is high when the molecules are strongly coupled and the masses of the involved atoms small. For example, the stretching mode of the light CH molecule occurs at $\sim 3\mu$m; for the heavier CN, the resonance is at

~10μm. Likewise, in a single C–C bond, the stretching band lies at ~10μm, in a triple C≡C bond at ~5μm.

The spectrum of a vibrational transition of a diatomic gas molecule, like CO, splits into rotational lines which generally group into an R-branch and a P-branch corresponding to rotational transitions with $\Delta j = 1$ and $\Delta j = -1$, respectively. Molecules in an ice mantle are squeezed in by other molecules and cannot rotate freely, so these branches are suppressed (see figure 8.4). The center frequency and band shape are then influenced by the composition of the matrix into which the molecule is embedded.

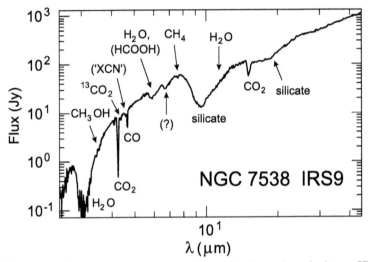

FIGURE 8.3 The infrared spectrum towards the infrared object IRS9 in the star forming region NGC 7538 [Whi96]. The resonances at 10 and 18μm are due to refractory silicate cores. The other features come from molecules that have frozen out onto refractory grains and are now in ice mantles.

The amount of ice can be estimated for each mantle constituent from the strength of the absorption feature. Take, for example, the CO_2 band at 4.27μm (figure 8.3). One measures in the lab the optical depth integrated over the band, $\int \tau_\lambda \, d\lambda$, which a CO_2 ice layer of thickness d produces. The column density of CO_2 molecules in the layer is $\rho d/m$, where ρ is the ice density and m the molecular mass. For *one* CO_2 molecule (see [Sch99] for full list of mantle components), one finds

$$\int \tau_\lambda \, d\lambda = 7.6 \times 10^{-17} \text{ cm .}$$

Integration over an observed CO_2 band yields then the solid CO_2 column density. With respect to water ice, the abundance is usually quite large

$([CO_2]/[H_2O] \sim 20\%)$; in the gas phase, however, CO_2 is rare (figure 8.4). One therefore has to conclude that the molecule is formed on the grain.

That CO_2 is absent in the gas but often abundant as a solid is also evident from the remarkable spectra in figure 8.4. We also see there that the CO gas-to-solid ratio varies among the sources and that one can detect the two phases of CO simultaneously. Towards GL4176, the very broad R- and P-band to the left and right of the center wavelength arise from absorptions in the gas. Towards NGC7538, the absorbing molecules are in a solid state and the individual rotational lines have disappeared.

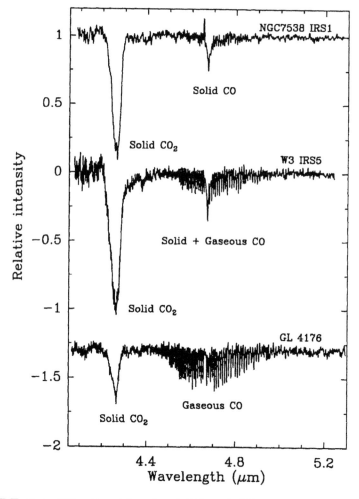

FIGURE 8.4 Vibrational bands of CO and CO_2 towards three sources [Dis98]. The CO $v = 1 \rightarrow 0$ rot-vib transition is centered at $\lambda_c \simeq 4.66 \mu m$.

FIGURE 8.2

9

PAHs and spectral features of dust

Gas atoms or molecules emit and absorb narrow lines, from the UV to the microwave region, by which they can be uniquely identified. Because of their sharpness, the lines also contain kinematic information via the Doppler shift in the radial velocity. Observing several transitions simultaneously allows to infer the conditions of excitation, mainly gas temperature and density.

The spectroscopic information from dust, on the other hand, is poor. Being a solid, each grain emits very many lines, but they overlap to an amorphous continuum. Moreover, the emission (not the absorption) is restricted to infrared wavelengths. A fortunate exception are the smallest grains, PAHs, which show very characteristic bands. Resonances are also observed from silicate material and ice mantles. These features are the topic of the present chapter. Because of the strong disturbance of any oscillator in a solid, the transitions are broad and do not allow to extract velocity information. *Precise* chemical allocation is also not possible, but individual functional groups partaking in the transition can be identified and exciting conditions may occasionally be derived.

9.1 Polycyclic Aromatic Hydrocarbons

Polycyclic Aromatic Hydrocarbons, in short PAHs, are planar molecules, made of a number of benzene rings, usually with hydrogen atoms bound to the edges. The smallest consist of barely more than a dozen atoms, so they are intermediate between molecules and grains and constitute a peculiar and interesting species of their own (recommended reading: [Omo86], [All89], chapter 6 of [Tie05]).

The first observation of PAHs, without proper identification, dates back to the early 1970's when one found in the near and mid infrared a number of hitherto unknown features [Gil73]. Although their overall spectral signature was subsequently observed in very diverse objects (HII regions, reflection nebulae, planetary nebulae, cirrus clouds, evolved stars, external galaxies, especially star burst nuclei), their proper identification took place only in the mid eighties.

9.1.1   Microcanonic emission of PAHs

Consider a PAH of total energy E which is the sum of the kinetic and potential energy of all atoms. We approximate the latter by harmonic oscillators. The accessible states are in a narrow range from E to $E+\delta E$. Figure 9.1 illustrates the locus of the states in phase space when the system has only one degree of freedom ($f=1$). There are then two coordinates, position q and momentum p, and all states lie in an annulus around an ellipse. The oscillating particle is most likely to be found near the greatest elongation, which is at the left and right edge of the ellipse. There the velocity is small and, for a given Δq, one finds the highest density of cells. For $f > 1$, the situation is multi-dimensional, but otherwise analogous.

The population of the cells in phase space is statistical and the probability to find the i-th oscillator of a PAH in quantum state v is in view of (A.33)

$$P_{i,v} = \rho(E; i, v) / \rho(E) .$$

$\rho(E)$ is the density of states at energy E, and $\rho(E; i, v)$ is the density of states at energy E subject to the condition that oscillator i is in quantum state v. When one extends the expression (A.27) for $\rho(E)$ to the case when the oscillators are not identical, but each has its own frequency ν_i, and further refines the expression by adding to E the zero-point energy E_0 of all excited oscillators, one obtains [Whi63]

$$\rho(E) \simeq \frac{(E + aE_0)^{f-1}}{(f - 1)! \prod\limits_{j=1}^{f} h\nu_j} \tag{9.1}$$

with

$$a \simeq 0.87 + 0.079 \times \ln(E/E_0), \qquad E_0 = \tfrac{1}{2} \sum_j h\nu_j .$$

The fudge factor, a, is of order unity. $\rho(E; i, v)$ is also calculated after (9.1), but now one has to exclude oscillator i. There is one degree of freedom less and E, as well as the zero-point energy E_0, have to be diminished accordingly. This gives

$$\rho(E; i, v) = \left[(E - vh\nu_i) + a(E_0 - \tfrac{1}{2}h\nu_i) \right]^{f-2} \left[(f - 2)! \prod\limits_{j\neq i} h\nu_j \right]^{-1} .$$

Hence the relative population of level v is

$$P_{i,v} = \frac{\rho(E; i, v)}{\rho(E)} = (f - 1)\,\xi \left[1 - \left(v + \frac{a}{2} \right) \xi \right]^{f-2}, \qquad \xi = \frac{h\nu_i}{E + aE_0} . \tag{9.2}$$

The i-th oscillator emits in the transition $v \to v-1$ the power

$$\varepsilon_{\mathrm{mic}}(E; i, v) = P_{i,v}\, h\nu_i \cdot A^i_{v,v-1} ,$$

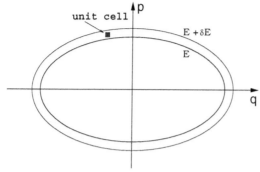

FIGURE 9.1 The phase space accessible to a one-dimensional oscillator with energy between E and $E + \delta E$ is according to equation (6.2) given by the shaded annulus. The ellipses correspond to total (kinetic plus potential) energy E and $E + \delta E$, respectively. The unit phase cell has the size $\delta q \, \delta p = \hbar$.

and the total microcanonic emission from oscillator i is therefore

$$\varepsilon_{\mathrm{mic}}(E; i) \;=\; h\nu_i \, A_{1,0}^i \sum_{v \geq 1} v \, P_{i,v} \,. \tag{9.3}$$

9.1.2 An example: anthracene

9.1.2.1 The vibrational modes of anthracene

The PAH anthracene, $C_{14}H_{10}$, has a simple and fairly symmetric structure (see insert in figure 9.2) and is therefore easy to handle theoretically. It is built up of three benzene rings; all bonds at its edges are saturated by hydrogen atoms. There are altogether $f = 3 \times (14 + 10) - 6 = 66$ degrees of freedom. The vibrational modes can be computed if one knows the potential in the molecule. In the model on which figure 9.2 is based seven vibrational types are considered. Besides the four obvious ones,

- CC stretching
- CH stretching
- CC out-of-plane bending
- CH out-of-plane bending,

three mixed modes in which several atoms take part in the oscillation are also included. These are CCC and CCH bending as well as CCCC torsion. The frequencies of the 66 oscillators by which the molecule is replaced are found by introducing normal coordinates (section A.5.1). Despite the simplicity of the model, the achieved accuracy is good, and the theoretical frequencies agree well with laboratory data.

Some vibrational modes lie very close together because of the symmetry of the molecule. It is clear that atoms in similar force fields must swing in a

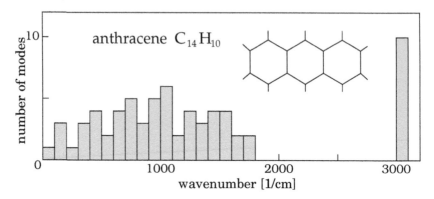

FIGURE 9.2 The polycyclic aromatic hydrocarbon anthracene has 66 degrees of freedom. They correspond to 66 oscillators (modes), each having its eigenfrequency. The wavenumbers are tabulated in [Whi78] and grouped here into intervals of $100\,\mathrm{cm}^{-1}$ width.

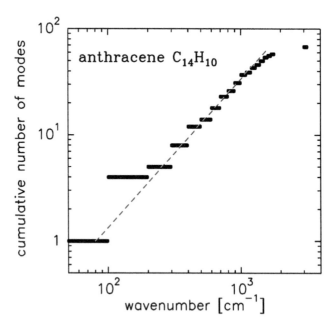

FIGURE 9.3 The cumulative number of modes in anthracene. Due to the small number of atoms, the increase is only roughly proportional to ν^2 (dashed line, see 6.41).

similar manner. For instance, all CH stretches of the ten hydrogen atoms in anthracene are crowded between 3030 and 3040 cm⁻¹. For reasons of symmetry, at most four of these CH-frequencies can be different, but these four really are distinct because they correspond to sites where the surroundings of the H-atoms have different geometries. The distribution of the number of modes over energy is displayed in figure 9.2. It contains several noteworthy features:

- The modes strongly cluster around 3035 cm⁻¹. This range is completely isolated and due to CH stretching. When κ is the force constant of CH binding and m the mass of an H atom, the frequency is roughly $\omega = \sqrt{\kappa/m}$ (see 8.18). Hydrogen atoms are also involved in the modes around 1000 cm⁻¹ which are due to a CH bend.

- The rest of the histogram is due to the skeleton of carbon atoms and surprisingly smooth.

- In section 6.4, we gave the Debye temperatures, $\theta_z = 950\,\mathrm{K}$ and $\theta_{xy} = 2500\,\mathrm{K}$, for a graphite plane. Excluding the CH bend at 3000 cm⁻¹, the high frequency cutoff, which by definition is the Debye frequency, is at 1800 cm⁻¹ and corresponds nicely to θ_{xy} (see 6.32).

- The longest wavelengths lie in the far infrared. The lowest energy is 73 cm⁻¹ (137 μm) and again in qualitative agreement with the estimate for the minimum frequency ν_{min} after (6.44) if one uses $\theta_z = 950\,\mathrm{K}$.

9.1.2.2 Microcanonic vs. thermal level population

As described in section 9.1.1, we compute the microcanonic level population of anthracene, which is always the statistically correct way, and check whether the assumption of thermalization, which when fulfilled makes life so much easier, is permitted.

The total energy E of a thermalized PAH is equal to the sum of the mean energies of all N oscillators (see A.12),

$$E = \sum_{i=1}^{N} \frac{h\nu_i}{e^{h\nu_i/kT} - 1} .$$

For anthracene, $N = 66$ and the ν_i are known. When one calculates from this equation T after the molecule has absorbed a stellar photon of typical energy $E = 4\,\mathrm{eV}$, one finds $T = 1325\,\mathrm{K}$. Figure 9.4 demonstrates that in this case, one may indeed use the Boltzmann formulae (dotted lines) for the level population, and thus also for line emission. For higher photon energies, things only improve.

We also learn from the microcanonic results of level populations that those oscillators in the PAH which correspond to the major bands (table 9.1) are

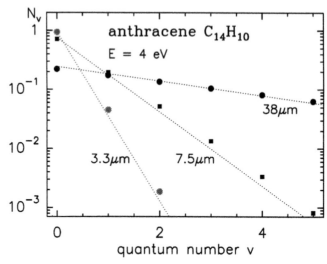

FIGURE 9.4 Level population N_v in anthracene for three oscillators (their wavelengths are indicated) after absorption of a photon with $E = 4\,\text{eV}$ from the ground state. Dots show the microcanonic fractional population as approximated by (9.2); always $\sum N_v = 1$. The dotted lines, which only have a meaning at integral quantum numbers v, give the thermal results at a temperature $T = 1325\,\text{K}$.

mostly found in the lowest quantum numbers, $v \leq 1$. This implies that effects produced by the anharmonicity of the oscillator potential are, to first order, not important. However, if one looks closer, they have to be taken into account. Because of anharmonicity, the spacing between the upper levels decreases and the lines from higher v-states are shifted towards the red. The best example is presented by the C–H stretches at $3.28\mu\text{m}$ ($v = 1 \to 0$, see table 9.1) and at $3.42\mu\text{m}$ ($v = 2 \to 1$). The shift is sufficiently large so that the $v = 2 \to 1$ line can occasionally be seen well detached from the $v = 1 \to 0$ transition. The $v = 2 \to 1$ line is always weaker because of low level population, but note that the Einstein A-coefficient increases proportionally to v. If the shift due to anharmonicity is small, as for the C–H out-of-plane bend at $11.3\mu\text{m}$, the higher v-lines blend with the ground transition producing an overall broadening on the red side of the band.

9.1.3 Photo-excitation of PAHs

In interstellar space, PAHs are almost all of the time in their electronic ground state. Collisions with gas atoms are rare, and the gas temperature is too low anyhow to thermalize the PAH, so vibrational modes are also in their lowest level. Only occasionally do PAHs get excited by the absorption of a single ultraviolet photon upon which they emit their characteristic infrared lines.

For a neutral PAH molecule, the electronic ground state is the singlet S_0. Emission of infrared lines from the PAH molecule is initiated by absorption of a UV photon of energy E_{UV}, say, once a day. Afterwards the molecule is highly excited and in a new electronic singlet state S_i. Then various and complicated things happen, very fast, some on time scales of 10^{-12} s. But by and large, the outcome is that the ultraviolet energy is thermalized, i.e., distributed over all available vibrational modes, and then degraded into infrared photons, mainly by emission in the fundamental modes (see table 9.1). The emission lasts typically for a few seconds.

Before the IR photons are emitted, immediately after the UV photon has been absorbed, the PAH may change from the electronic state S_i without radiative loss to another singlet state. Or it may undergo intersystem crossing to a triplet state. In this case, because the triplet system has a metastable ground level T_1, some of the energy E_{UV} will eventually go into phosphorescence from the triplet T_1 to S_0 and thus be lost for the IR features.

For a singly photo-ionized PAH, the excitation scheme is similar; only the electronic ground state is then a doublet because of the unpaired electron. A PAH may also be negative or bear multiple charges. The degree and kind of ionization depends on the radiation field, the gas temperature, the electron density and, of course, on the UV cross sections of the PAH; some of the basic ideas are outlined in section 7.1. The ionization details may not be decisive for evaluating the IR emission of PAHs, but probably have an influence on the cutoff wavelength and thus on the conditions for excitation (see below).

9.1.4 Cutoff wavelength for electronic excitation

There is a minimum photon energy for electronic excitation of a PAH corresponding to a cutoff wavelength λ_{cut}. Light of wavelengths greater than λ_{cut} is simply not absorbed and thus cannot produce the infrared resonances. For PAHs of circular shape and size $\langle a \rangle$, one finds empirically, in the lab, the relation

$$\lambda_{cut} \simeq 1630 + 370\sqrt{N_C} \simeq 1630 + 450\,\langle a \rangle \qquad [\text{Å}] . \qquad (9.4)$$

Here $\langle a \rangle$ is measured in Å and N_C is the number of carbon atoms in the molecule. Small PAHs ($N_C < 50$) are only excited by UV radiation, larger ones also by optical photons.

One can understand the existence of the threshold by considering a free electron in a linear metallic molecule of length L. The electron is trapped in the molecule and its energy levels are given by (A.98),

$$E_n = \frac{h^2 n^2}{8 m_e L^2} , \qquad n = 1, 2, \ldots .$$

The phase space available to electrons with energy $E \le E_n$ is $hn/2$, so if there are s free electrons, all levels up to $n = s$ are populated; a denser population is forbidden by Pauli's exclusion principle. A transition with $\Delta n = 1$ has therefore the energy

$$h\nu_{\text{cut}} = \left. \frac{dE_n}{dn} \right|_{n=s} \Delta n = \frac{h^2 s}{4m_e L^2} = \frac{h^2}{4m_e(\frac{L}{s})L} .$$

If the atoms are spaced at intervals of 2Å and if there is one free electron per atom, then $L/s = 2\,\text{Å}$ and

$$\lambda_{\text{cut}} = \frac{4m_e(\frac{L}{s})Lc}{h} \sim 330 \times L \qquad [\text{Å}] . \tag{9.5}$$

This qualitative and simplified model of a linear metallic grain already yields the right order of magnitude for the cutoff wavelength of PAHs.

9.1.5   Photo-destruction and ionization

Because the PAHs are so small, absorption of a UV photon can lead to dissociation. In a strong and hard radiation field, PAHs will not exist. Because of the anharmonicity of the potential of a real oscillator, an atom detaches itself from the PAH above a certain vibrational quantum number v_{dis}. The chance $P(E)$ of finding in a PAH with f degrees of freedom and total energy E the oscillator i in a state $v \ge v_{\text{dis}}$ is given by equation (9.2), and $P(E)$ is proportional to the dissociation rate $k_{\text{dis}}(E)$ of the molecule. In slight modification of (9.2), one therefore writes

$$k_{\text{dis}}(E) = \nu_0 \left(1 - \frac{E_0}{E} \right)^{f-1} . \tag{9.6}$$

ν_0 has the dimension of a frequency; E_0 may be loosely interpreted as the bond dissociation energy. The two factors, ν_0 and E_0, can be determined experimentally. For example, with respect to hydrogen loss, $E_0 = 2.8\,\text{eV}$ and $\nu_0 = 10^{16}\,\text{s}^{-1}$; with respect to C_2H_2 loss, $E_0 = 2.9\,\text{eV}$, $\nu_0 = 10^{15}\,\text{s}^{-1}$ [Joc94]. If hydrogen atoms come off, the aromatic cycles are unscathed; if however acetylene evaporates, the carbon rings are broken.

After UV absorption, the excited PAH is hot and has a certain chance to evaporate. When relaxation through emission of IR photons is faster than dissociation, the PAH is stable. The infrared emission rate of PAHs, which is of order $10^2\,\text{s}^{-1}$, can thus be considered as the critical dissociation rate, k_{cr}. When $k_{\text{dis}}(E) > k_{\text{cr}}$, evaporation will occur. k_{cr} is associated after (9.6) with a critical internal energy E_{cr}. Because E_{cr} is considerably greater than E_0, series expansion of the function $\ln(1 - x)$ gives approximately

$$\ln k_{\text{cr}} \simeq \ln \nu_0 - (f - 1) \cdot \frac{E_0}{E_{\text{cr}}} .$$

As one would expect, there is a rough linear relationship between the degrees of freedom of the PAH and the critical energy. Because k_{cr} enters logarithmically, an exact value is not required. The critical temperature T_{cr} can be found from E_{cr} via equation (A.12),

$$E_{cr} = \sum_{i=1}^{f} \frac{h\nu_i}{e^{h\nu_i/kT_{cr}} - 1} .$$

Ionization after absorption of an energetic UV photon is another reaction channel, besides dissociation and infrared emission, and it has the overall effect to stabilize the molecule against destruction. Big PAHs are easier to ionize than small ones; the ionization potential of benzene is 9.24 eV and declines to ~6 eV for molecules with 100 carbon atoms.

For the astronomically relevant question under what conditions do PAHs survive in a harsh radiation field, one has to balance their destruction rate to the rate at which they are built up by carbon accretion. In the interstellar radiation field (figure 6.1), we estimate that PAHs with only 30 carbon atoms can survive; near an O star (distance $r = 10^{16}$ cm, $L = 10^5 L_\odot$), N_C should be above 200.

As the physical processes connected with PAH destruction are most complicated, one may be tempted to use a purely phenomenological approach and simply assume that PAHs evaporate if the fraction of PAHs above some critical temperature T_{cr} exceeds a certain value f_{evap}, i.e., when

$$\int_{T_{cr}}^{\infty} P(T) \, dT \geq f_{evap} . \tag{9.7}$$

The probability function $P(T)$ of (6.45) has to be computed anyway for the evaluation of PAH emission. $f_{evap} \sim 10^{-8}$ and $T_{cr} \sim 2500$ K do not seem to be unreasonable numbers and one may use such an approach to decide whether PAHs evaporate or not.

9.1.6 Cross sections and line profiles of PAHs

A PAH has several hundred vibrational modes, but one detects typically only half a dozen major resonances around the frequencies where the strong modes cluster. Table 9.1 compiles their vibrational type, approximate center wavelength and integrated line strength. One observes (generally mild) variations in the position of the features, their shape and strength. They occur from one source to another, and even within a source. The likely reasons are local changes in the environment, but genuine differences in the composition of the PAHs are also conceivable. There are also minor resonances, not listed in table 9.1, which are more variable in intensity. They present other vibrational modes, but may also refer to transitions between higher levels, like the

TABLE 9.1 Approximate numbers for strong PAH resonances. The cross sections integrated over the band, $\sigma_{\text{int}} = \int \sigma(\lambda)\,d\lambda$, are per hydrogen atom for C–H vibrational modes, and per carbon atom for C–C modes. In the column "Vibrational type", mono and trio refers to hydrogen atom position 1 and 3 in figure 9.5, respectively.

Center wavelength λ_0 [μm]	Damping constant γ [$10^{12}\,\text{s}^{-1}$]	Integrated cross section σ_{int} [$10^{-26}\,\text{cm}^3$]	Vibrational type
3.3	20	12	C–H stretch
6.2	14	14	C–C stretch
7.7	22	51	C–C stretch and C–H bend in-plane
8.6	6	27	C–H bend in-plane
11.2	4	41	C–H bend out-of-plane, mono
12.8	3.5	47	C–H bend out-of-plane, trio

3.42μm band due to the $v = 2 \rightarrow 1$ transition of the oscillator with the ground wavelength at 3.3μm (see section 9.1.2.2).

We distinguish between two classes of bands. One involves hydrogen atoms, which are all located at the periphery of the PAH, the other only carbon atoms. The strength of the emission is therefore proportional either to N_{H} or N_{C}, the number of hydrogen an d carbon atoms, respectively. As the total amount of absorbed (and thus also emitted) energy is determined by the carbon skeleton, a PAH with hydrogen atoms will have weaker C–C bands than one without (and the same N_{C}).

We approximate the line shapes by Lorentzian profiles (see 1.100),

$$\sigma_\nu = N\,\sigma_{\text{int}}\,\frac{\nu_0^2}{c}\,\frac{\gamma\,\nu^2}{\pi^2\,[\nu^2 - \nu_0^2]^2 + [\gamma\nu/2]^2}\,, \tag{9.8}$$

where

$$\sigma_{\text{int}} = \int \sigma(\lambda)\,d\lambda$$

is the cross section of the particular band integrated over wavelength, $\nu_0 = c/\lambda_0$ is the center frequency, and N the number of C or H atoms, respectively. The band results from an overlap of many lines with peak frequencies clustering around ν_0; its width is determined by the damping constant γ.

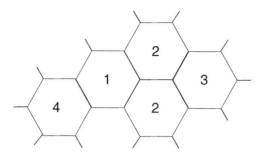

FIGURE 9.5 The hydrogen atoms in a PAH are attached to the edge of the carbon skeleton. The exact frequency of the C–H out-of-plane bend depends on whether the hydrogen atoms are isolated (1, mono positions), or whether there are 2 (duos), 3 (trios) or 4 (quartets) adjacent H atoms. Due to the coupling between the vibrating H atoms, the center wavelength increases from approximately 11.2μm to 13.6μm (see table 9.1).

9.2   ERE and DIBs

In reflection nebulae, where the stellar light is scattered by dust, which gives them their name, one observes in the nebular light a strong and very broad bump. It is named **E**xtended **R**ed **E**mission, short ERE, as it lies somewhere between 5500 and 9500Å. As with PAHs, its center and width are variable, within one source and from one to another. Because the polarization is much lower in the bump than outside, the bump cannot be caused by scattering, and because the wavelength is so small, thermal emission is excluded. Another process is required, and photo-luminescence is invoked. Assuming that one optical or UV photon yields one ERE photon, about 10% of the stellar light is needed to produce the ERE phenomenon. Therefore, the carriers have to be very abundant and the particles small. PAHs, or clusters of them, are a candidate; first, they do absorb 10% or 20% of the starlight; second, the transition (in a neutral PAH) from the metastable electronic triplet T_1 to the singlet S_0 (see section 9.1.3) presents such a luminescence process; and, third, one sees extended red emission and PAH features together, for example, in the Red Rectangle (figure 9.6). Alternatively, silicon nano-particles are also discussed as the source for the extended red emission.

Another class of spectral features are the **d**iffuse **i**nterstellar **b**ands, or DIBs. They are only seen in absorption. One finds them as weak dips in observations of reddened stars when the spectral resolution is much higher than what is usually used in photometry (see section 10.1.1). The features are narrow and their width is typically only a few Angstrom. Altogether there are more than a hundred of them, mainly between 0.4μm and 0.9μm, of variable strength,

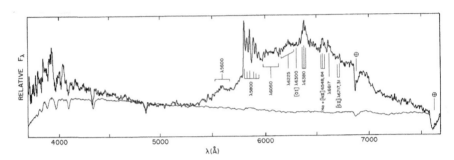

FIGURE 9.6 Extended red emission (ERE) between 5400 and 7500Å in the Red Rectangle, a reflection nebula around the star HD 44179 [Sch80]. The broad bump is due to photo-luminescence, not to scattering by dust, whereas the narrow lines are from the scattered stellar spectrum.

shape and width. Although discovered 80 years ago, the identification of the DIBs is still unclear. The favorite explanation is electronic transitions in large molecules (more than 10 atoms). Their oscillator strength is high and there would be no abundance problem if the molecules contained mainly carbon, so that they too might arise from PAHs.

9.3 The silicate bands at 10μm and 18μm

9.3.1 The strength of the resonances

The most prominent and intensively studied resonance is the one around 10μm (often called the 9.7μm feature) due to an Si–O stretching vibration in silicates. It is the strongest indicator of the element silicon in grains. Towards star forming regions, it is observed in absorption, towards oxygen-rich mass loss giants it is seen in absorption and occasionally in emission. At around 18μm, there is a related feature due to an O–Si–O bending mode which is weaker and more difficult to observe from the ground because of atmospheric attenuation. In these two bands, silicates dominate dust absorption (figures 5.10 and 10.7) and the optical depth (or absorption coefficient) ratio is

$$\frac{\tau_{18}}{\tau_{9.7}} = \frac{K_{18}}{K_{9.7}} \simeq 0.35 . \tag{9.9}$$

To derive the ratio of the visual to the 9.7μm optical depth, $\tau_V/\tau_{9.7}$, one measures the extinction through a cloud that lies in front of a carbon star. In these stars, all oxygen is bound in CO and silicate grains are absent. The stars have shells of carbon dust that produce a smooth 10μm emission background with a temperature of $\sim 1000\,$K and without the 9.7μm feature. On the average,

$$\frac{\tau_V}{\tau_{9.7}} = 18 \pm 2 . \tag{9.10}$$

The ratio $\tau_V/\tau_{9.7}$ can be used to determine the visual extinction A_V from the 10μm absorption, provided there are no hot grains along the line of sight that emit themselves at 9.7μm. This method still works at high extinction when stellar photometry is no longer possible. For example, from the depression of the 10μm feature in figure 8.3, we estimate that the object IRS9 in NGC 7538 suffers about 35 mag of visual extinction corresponding to $\tau_{9.7} \simeq 2$.

When one converts the depth of the depression at 9.7 and 18μm into an optical thickness under the assumption of pure absorption, one sometimes ends up with a ratio $\tau_{18}/\tau_{9.7}$ that is considerably smaller than 0.35 and thus probably false (see 9.9). The likely explanation for the reduced ratio is the presence of dust which is warm enough to radiate at 18μm, but not at 9.7μm.

9.3.2 How the bands change with temperature and grain size

The band profiles are not uniform but vary from one source to another and within a source. They depend on several parameters such as

- the color (spectral slope) of the background at 10μm and 18μm
- the optical depth, τ_ν, of the foreground
- the size (figure 9.7) and shape (figure 3.4) of the grains as well as their degree of porosity
- the mineralogical structure of the silicate material (figure 9.8) and its degree of crystallinity
- the temperature of the silicate grains (figure 9.7)
- radiative transfer effects when there is emission and absorption.

The dust temperature influences the band shape only when the feature is in emission; the spectrum of the background source and the optical thickness of the foreground cloud affect the profile when the feature is in absorption. Figure 9.7 illustrates how the band changes with grain size and grain temperature assuming pure silicate dust.

- In absorption (lower right frame), the observed flux is proportional to background intensity, $I_{b\nu}$, times $\exp(-\tau_\nu)$. For small particles (radii $a \leq 0.1\mu$m), the feature has its minimum at $\lambda_{min} = 9.5\mu$m. As they become bigger, λ_{min} increases; but up to $a = 1\mu$m only mildly, and the band shape is conserved. For particles with 3.1μm radius, $\lambda_{min} = 11.5\mu$m and the resonance has almost disappeared; when they are larger still, as in accretion disks, the extinction turns grey.

- In emission, the observed flux is proportional to absorption coefficient
  times Planck function (remaining three frames).

  - When the grains are small ($a \lesssim 0.1\mu$m) and hot ($T \geq 800$ K), the
    radiation peaks at $\lambda_{\max} = 9.5\mu$m. As they get cooler, $\lambda_{\max}$ shifts
    towards longer wavelengths, by almost 2μm when the temperature
    is down to 200 K. If the silicate particles are colder than 130 K (not
    shown in the figure), the 10μm bump is swamped altogether by the
    steeply rising Planck function.
  - Up to radii of 1μm, there is not much change in the band shape, but
    when $2\pi a \geq \lambda$, the resonance broadens and eventually disappears.
    The grains emit then like blackbodies.

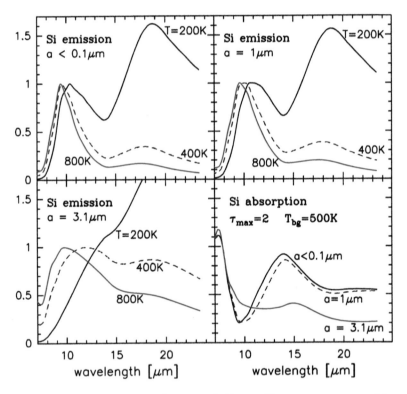

FIGURE 9.7 Variations in the 10 and 18μm silicate resonances. ***Absorp-***
tion profiles: (bottom right) against a 500 K blackbody for various grain
radii a. Fluxes are set to one at 8μm. The maximum optical thickness in the
interval from 8μm to 13μm is for all lines two. ***Emission profiles:*** (other
boxes) grain temperatures T = 200 K, 400 K and 800 K, grain radii $a \leq 0.1\mu$m,
1μm and 3.1μm. Curves are normalized such that their maximum in the interval from 8μm to 13μm equals one. Optical constants are from figure 5.10.

9.4 Crystalline silicates

9.4.1 Where they are found and how they form

Interstellar silicates are amorphous, but in several classes of objects (comets, stellar disks, shells of AGB stars, planetary nebulae) a not negligible fraction ($\sim$10%) is crystalline (recommended reading [Mol05]). The 10μm band is then not smooth but ragged, and various new features appear longward of 20μm (figure 9.8). They are also seen in the lab (figure 9.9). In amorphous materials, the spikes blend and are not discernible.

Only silicates without iron (similar to the minerals enstatite, $MgSiO_3$, and forsterite, Mg_2SiO_4, see section 5.3.4) are observed as crystals, although the mixed types (similar to bronzite, $(Mg,Fe)SiO_3$, and olivine, $(Mg,Fe)_2SiO_4$) certainly exist in interstellar dust but, strangely, they never seem to be crystalline. Here are clues to the formation and destruction of crystals.

Methamorphosis can occur between the amorphous and crystalline state in either direction. The crystalline form has a lower potential energy, but to get there from the amorphous state, a barrier, E_{bar}, has to be surmounted. A mechanical analog are pebbles in a box which only after shaking settle to a more ordered (and denser) configuration. The activation energy may be supplied to the whole grain by warming it up, or locally, for example, by electron irradiation (at least, in the lab). Cosmic rays, on the other hand, introduce lattice defects and tend to destroy the order of the grid. Interestingly, crystalline silicates have not been found in the general interstellar medium.

If the crystalline silicates condense as such in the wind of AGB stars or in planetary disks, the grains probably form with an energy above E_{bar} and are initially glassy; i.e., the atoms are poorly ordered but possess enough mobility to settle to the crystalline state. As the dust cools, the mobility rapidly decreases. The timescale for crystallization, t_{cryst}, is known to be a very steep function of temperature. At 1100 K, t_{cryst} is of order minutes, at $T < 900$ K of order many years and the atomic sites are then effectively frozen in. Therefore, condensation of crystals only works if t_{cryst} is shorter than the cooling time. The energetic situation is similar to thermal hopping (section 8.2.1). Assuming $t_{cryst} = 10^2$ s, $T = 10^3$ K and a characteristic collisional frequency $\nu_0 \sim 10^{13}$ s^{-1}, equation (8.11) yields $U_0 = E_{bar}$ around 3 eV. The atoms probably reach their final position not in one big leap, but in several smaller hops, the same way that the pebbles in a box arrange themselves through a sequence of weaker jostles, but that does not qualitatively change the timescale considerations.

Because narrow spikes extend over a broad wavelength range (figure 9.8), one can use their intensity ratios to estimate the dust temperature, T_d, from

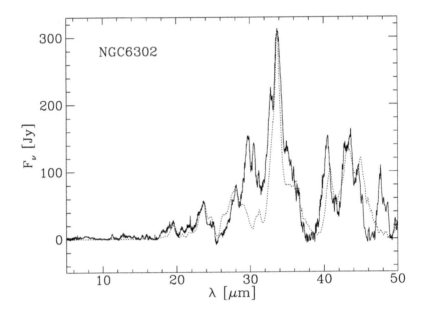

FIGURE 9.8 The spectrum of the planetary nebula NGC 6302 (solid line) with many resonances of crystalline silicate after subtraction of the continuum. The dots show a fit with forsterite and enstatite at about 70 K (from [Mol01]).

formula (6.20), provided all the preconditions listed in section 6.2.4 are fulfilled. Determining T_d from spikes will give results superior to those obtained from flux ratios in the continuum only if the ratio of the mass absorption coefficient, $K(\lambda_1)/K(\lambda_2)$, which refers to all dust (amorphous and crystalline), is better known in the resonances than outside.

9.4.2   Thermal expansion of grains

Grains, like all bodies, become bigger when they are warmed up. This phenomenon cannot be explained if the atoms move in a potential, V, that is strictly quadratic. An anharmonicity term is needed, and V has to be expanded to the next higher order. Let a_0 be the equilibrium spacing between two atoms at zero temperature. If x denotes the displacement from a_0, we write

$$V(x) = \tfrac{1}{2}\kappa x^2 - g x^3 .$$

The term gx^3 takes into account the mutual repulsion ($g > 0$), while κ corresponds to the force constant of the spring with which the atoms are coupled. As the temperature rises, the atoms swing more violently about their equilibrium position while their mean seperation increases. Classically, the average

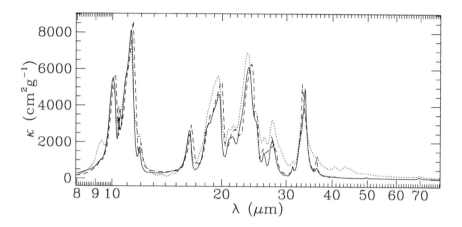

FIGURE 9.9 Various lab measurements of the mass absorption coefficient of forsterite compiled by [Mol02] and scaled so that they match.

displacement at temperature T becomes

$$\langle x \rangle = \frac{\int_{\infty}^{\infty} x e^{-\beta V(x)} dx}{\int_{\infty}^{\infty} e^{-\beta V(x)} dx} = \frac{3g}{\kappa^2} kT . \tag{9.11}$$

Here $\exp[-\beta V(x)]$ is the Boltzmann factor, $\beta = 1/kT$, and we assumed in the integration that the anharmonicity term, gx^3, is small compared to kT and then used (B.5) and (B.9). Obviously, $\langle x \rangle = 0$ when $g = 0$ or $T = 0$.

Let a be the mean separation between two atoms at temperature $T > 0$, then $\langle x \rangle = a - a_0$ and the expansion coefficient becomes

$$\alpha = \frac{1}{a_0} \frac{d\langle x \rangle}{dT} = \frac{3kg}{a_0 \kappa^2} . \tag{9.12}$$

A solid has, of course, a continuous vibrational spectrum (frequencies from 0 to ν_D) and it can be shown that α follows the same temperature dependence as the specific heat, so $\alpha \propto T^3$ for $T \to 0$ and $\alpha = $ const at high T.

The expansion between atoms in a crystal is not uniform. Temperature changes generally affect the proportions in the base of the unit cell. For example, the size and shape of the $[SiO_4]$ tetrahedron in silicates is rather insensitive to heating because the Si–O bond is strong whereas the cations do expand (the Mg–O bond is weak).

9.4.3   The frequency shift of a resonance in grain heating

In section A.3, we argued that in an expansion of a grain the derivative of the free energy, F, with respect to the volume V at constant temperature does not change (see A.39). In equation (A.40), we expressed the free energy of the gas through the partition function Z and wrote $F = -kT \ln Z$. For a solid where the atoms are not free, we must add the potential energy Φ, therefore

$$F = \Phi - kT \ln Z .$$

Replacing $\partial/\partial V$ in (A.39) by the derivative with respect to the mean atomic position, a, at temperature T gives

$$0 = \left(\frac{\partial \Phi}{\partial a} \right)_T - kT \left(\frac{\partial \ln Z}{\partial a} \right)_T .$$

The partition function of an harmonic oscillator of frequency ω equals $Z = e^{-\frac{1}{2}\beta\hbar\omega}/(1 - e^{-\beta\hbar\omega})$ and its mean energy $\langle E \rangle = \frac{1}{2}\hbar\omega + \hbar\omega/(e^{\beta\hbar\omega} - 1)$ (see A.11 and A.12). Because $\partial\Phi/\partial a = \kappa(a - a_0) = \kappa a_0 \alpha T$, we get

$$0 = \kappa a_0 \alpha T + \frac{\langle E \rangle}{\omega} \frac{\partial \omega}{\partial a}$$

The frequency depends on the equilibrium separation, a, which is a function of T. With

$$\frac{d\omega}{da} = \frac{d\omega}{dT} \frac{dT}{da} = \frac{1}{a_0 \, \alpha(T)} \frac{d\omega}{dT}$$

one finds

$$\frac{d\omega}{dT} = -\frac{\omega}{\langle E \rangle} \kappa \alpha^2 a_0^2 T = -\frac{\omega}{\langle E \rangle} \frac{9k^2 g^2}{\kappa^3} T . \tag{9.13}$$

This simplistic formula can help us understand what is being measured in the lab and what should then also happen in space. For example, the 69μm peak in forsterite shifts by about $\Delta\lambda = +0.5\mu$m when T is raised from 200 to 300 K (figure 9.10). If one adopts for this mineral a standard expansion coefficient $\alpha = 10^{-5}$ K^{-1}, an atomic separation $a = 4$Å, an oscillator mass $m = 20m_{\rm H}$, and if we set the mean oscillator energy $\langle E \rangle$ to $\frac{1}{2}\hbar\omega$, appropriate at low temperatures ($\hbar\omega \simeq kT$ at 200 K), and write for the force constant $\kappa \simeq m\omega^2$ (section 1.2.1), one obtains for $\Delta\lambda$ a similar value, at least, of the right order of magnitude. According to (9.13), the shift in ω increases with temperature and is very much reduced at shorter wavelengths. Both predictions are qualitatively borne out in figure 9.7.

Note that in the displacement of the center wavelength of the 10μm resonance illustrated in figure 9.7, the optical constants of the grain material stayed fixed, whereas here they change as the oscillator frequencies vary with temperature.

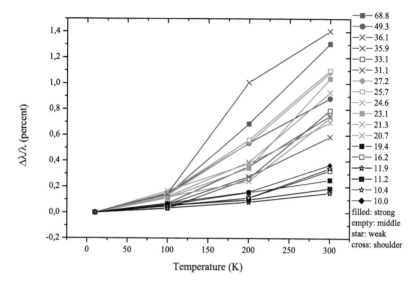

FIGURE 9.10 Lab data of peak shifts in forsterite bands as a function temperature [Koi06]. Wavelengths on the right.

9.5 The feature at 3.4µm

Another common absorption band lies at 3.4µm (figure 9.11, see [Pen02] for description of chemistry). It is generally weak and to better analyze it, one subtracts from the spectrum a reddened blackbody as the baseline. The difference between baseline and band can be converted into an optical depth assuming pure extinction proportional to $e^{-\tau}$. Typical values are $\tau_V/\tau_{3.4\mu m} \sim 250$.

- The band can be decomposed into subfeatures, altogether four, which are attributed to aliphatic (the carbon atoms do not belong to a ring) C–H stretches of the CH_2 and CH_3 group. There is further splitting due to different vibrational types: symmetric and anti-symmetric.

- The absolute and relative strengths of the subfeatures suggest a CH_2/CH_3 ratio of ~2.5 and an abundance of carbon in these aliphatic compounds of ~2.5%.

- The feature seems to arise only in the diffuse medium and to be absent in molecular clouds (where the nearby water ice resonance of OH stretching at 3.1 µm is produced, see figure 8.3). The reason why aliphatic groups are destroyed in dense clouds, but unscathed in diffuse ones is unclear.

- As the light in the feature is unpolarized, when the 10μm silicate band is polarized, the organic material of the 3.4μm resonance cannot reside in mantles around "normal" silicate cores. The carriers possibly cover the cores of very small, fast rotating and therefore unaligned particles.

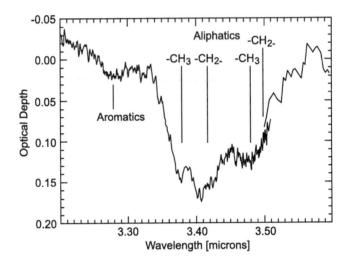

FIGURE 9.11 The optical thickness in the 3.4μm band towards the source GCS 3 in the Galactic Center (figure 12 in [Chi00]). The feature at 3.28μm is attributed to the C–H stretch in aromatic hydrocarbons (PAHs).

10

Interstellar reddening and dust models

10.1 Reddening by interstellar grains

Our knowledge of interstellar dust rests on five observational pillars:

- **Interstellar extinction.** The interstellar extinction or reddening curve specifies how dust weakens the light from background stars as a function of wavelength.

- **Dust emission.** We receive this from all kinds of objects: proto–stars, old stars with shells, interstellar clouds and whole galaxies.

- **Infrared resonances.** These are features seen in absorption and emission which allow to identify chemical components of the dust material.

- **Polarization.** It refers to light from stars behind dust clouds, but also to scattered radiation and dust emission.

- **Scattered starlight.** Examples are reflection nebulae, the diffuse galactic light or the zodiacal light of the solar system.

In this section, we are concerned with interstellar extinction and the clues it provides to dust properties.

10.1.1 Stellar photometry

The interstellar extinction curve is obtained from stellar photometry. The wavelength resolution in such measurements is usually poor, typically $\lambda/\Delta\lambda \simeq 10$, but the broadness of the observational bands enhances the sensitivity. The standard photometric system is due to *H.L. Johnson* and *W.W. Morgan* and rooted in optical astronomy. It was later expanded into the infrared; there the choice of wavelengths was dictated by the transmission of the atmosphere of the Earth. The observational bands are designated by letters: U,B,V,R,I, ... Table 10.1 lists their approximate center wavelengths λ_c and the conversion factors, w_λ, for translating magnitudes into Jansky and back; by definition,

$$1 \text{ Jy } = 10^{-23} \text{ erg cm}^{-2} \text{ s}^{-1} \text{ Hz}^{-1}.$$

The precise effective observing wavelength follows only after folding the source spectrum with the transmission of the instrument and the atmosphere.

TABLE 10.1 The standard photometric system. The center wavelengths λ_c and the conversion factors w_λ between magnitudes and Jansky after (10.1) are averages gleaned from the literature. The last column gives conversion factors for a blackbody of 9500 K; see text.

Band	Historical meaning	λ_c [μm]	w_λ	w_λ (bb)
U	ultraviolet	0.365	1810	2486
B	blue	0.44	4260	2927
V	visual	0.55	3640	3084
R	red	0.70	3100	2855
I	infrared	0.90	2500	2364
J		1.25	1635	1635
H		1.65	1090	1116
K		2.2	665	715
L		3.7	277	294
M		4.8	164	184
N		10	37	46
Q		20	10	12

In photometry, especially at shorter wavelengths, it is customary to express the brightness of an object in apparent magnitudes. These are logarithmic quantities and they were appropriate units in the days when the human eye was the only detector because according to the *psycho-physical rule* of *W. Weber* and *G.T. Fechner*: the subjective impression of the eye changes proportionally to the logarithm of the physical flux. But to the pride of many astronomers, magnitudes are still in use today. As they are laden with five millenia of history, only he who has thoroughly studied the five thousand year period can fully appreciate their scientific depth.

The *apparent magnitude* m_λ at wavelength λ is related to the flux F_λ through

$$m_\lambda = 2.5 \, \log_{10} \left(\frac{w_\lambda}{F_\lambda} \right) \tag{10.1}$$

where w_λ sets the zero point. If F_λ is expressed in Jy, it can be converted into apparent magnitude if one uses the values of w_λ in table 10.1. The formula is simple, the difficulty lies in the calibration factors w_λ. The numbers in the literature scatter by about 10% and, furthermore, they do not refer to identical center wavelengths because of the use of different filters.

After (10.1), a step of one up in magnitude means a factor of $10^{0.4} \simeq 2.512$ down in observed flux. The brightest star of all, Sirius, has a visual apparent magnitude $m_V = -1.58$ mag; the weakest stars we can detect with our eye under optimal conditions have $m_V \simeq 6.5$ mag.

Whereas the apparent magnitude m_λ at wavelength λ depends on the stellar distance D, the *absolute magnitude* M_λ, defined by

$$m_\lambda \; - \; M_\lambda \;\; = \;\; 5 \, \log_{10} \left(\frac{D}{\mathrm{pc}} \right) \; - \; 5 \; , \tag{10.2}$$

does not and thus measures the intrinsic brightness of a star. At a distance of 10 pc, apparent and absolute magnitude are, by definition, equal. When there is intervening dust, one adds to the right-hand side of equation (10.2) a term A_λ to account for the weakening of starlight through interstellar extinction,

$$m_\lambda \; - \; M_\lambda \;\; = \;\; 5 \, \log_{10} \left(\frac{D}{\mathrm{pc}} \right) \; - \; 5 \; + \; A_\lambda \; . \tag{10.3}$$

A_λ is related to the optical depth τ_λ (see definition 11.14) through

$$A_\lambda \;\; = \;\; 1.086 \, \tau_\lambda \; .$$

The strange conversion factor $2.5/\ln 10 \simeq 1.086$ implies that an increase of the optical depth by $1/1.086$ raises the magnitude by one. The term $m - M$ in (10.2) is the *distance modulus*. So ingenious astronomers have contrived a means to express length in stellar magnitudes.

The magnitude difference between two bands is called *color*. It is equivalent to the flux ratio at the corresponding wavelengths and thus determines the gradient in the spectral energy distribution. By definition, the colors of main sequence A0 stars are zero. The most famous A0V star is α Lyr (= Vega, $D = 8.1$ pc, $m_V = 0.03$ mag, $L = 54 \, L_\odot$). Such stars have an effective surface temperature of about 9500 K and they can be approximated in the infrared by a blackbody. Indeed, if we calibrate the emission of a blackbody at 9500 K at the J band (1.25μm) and put all colors to zero, we find the conversion factors $w_\lambda(\mathrm{bb})$ listed in table 10.1. In the infrared, they are not very different from the w_λ (adjacent column), however, the discrepancies are large at U, B and V because of the many absorption lines in the stellar spectrum that depress the emission relative to a blackbody.

10.1.2 The interstellar extinction curve

10.1.2.1 Its definition and the standard color excess $E_{\mathrm{B-V}}$

The interstellar extinction or reddening curve is obtained from photometry at various wavelengths λ on two stars of identical spectral type and luminosity class, one of which is reddened (star No. 1), whereas the other is not (star No. 2). When dust emission along the line of sight is negligible, one receives from the stars the flux

$$F_i(\lambda) \;\; = \;\; \frac{L(\lambda)}{4\pi D_i^2} \, e^{-\tau_i(\lambda)} \qquad i = 1, 2 \; .$$

$L(\lambda)$ and $\tau_i(\lambda)$ are the spectral luminosity and extinction optical thickness which is due to absorption plus scattering. From equation (10.1), we get

$$m_i(\lambda) = 1.086 \times \left[-\ln L(\lambda) + \tau_i(\lambda) + \ln(4\pi D_i^2) + w(\lambda) \right],$$

where $w(\lambda)$ is the constant of equation (10.1) given in table 10.1. Because $\tau_2(\lambda) = 0$, we write simply $\tau(\lambda)$ instead of $\tau_1(\lambda)$. The difference in magnitude $\Delta m(\lambda) = m_1(\lambda) - m_2(\lambda)$ between the two stars then becomes

$$\Delta m(\lambda) = 1.086 \times \left[\tau(\lambda) + 2\ln(D_1/D_2) \right].$$

The difference in Δm at two wavelengths, λ and λ', gives the *color excess*

$$E(\lambda, \lambda') = \Delta m(\lambda) - \Delta m(\lambda') = 1.086 \times \left[\tau(\lambda) - \tau(\lambda') \right],$$

which does not contain the distance D anymore. At $\lambda = 0.44 \mu m$ and $\lambda' = 0.55 \mu m$, the center wavelengths of the B and V band, the color excess is denoted by

$$E_{B-V} = E(B, V)$$

and called standard color excess. The band symbol is also used for the apparent magnitude, for example, $V = m_V$ or $B = m_B$. The intrinsic color is denoted $(B - V)_0$ in contrast to the observed color $B - V$, which includes the effect of the selective weakening by interstellar dust. With this notation, we can write

$$E_{B-V} = (B - V) - (B - V)_0 \qquad (10.4)$$

and likewise for any other pair of wavelengths.

Fixing in the color excess $E(\lambda, \lambda')$ the wavelength λ' at $0.55 \mu m$ and normalizing $E(\lambda, V)$ by E_{B-V}, one arrives at the extinction curve in its traditional form,

$$\text{Ext}(\lambda) = \frac{E(\lambda, V)}{E_{B-V}} = \frac{A_\lambda - A_V}{A_B - A_V} = \frac{\tau_\lambda - \tau_V}{\tau_B - \tau_V}. \qquad (10.5)$$

A_λ is the extinction in magnitudes at wavelength λ. The normalization of the extinction curve by E_{B-V} is important because it allows a comparison of the wavelength dependence of extinction towards stars which suffer different amounts of reddening. Obviously,

$$\text{Ext}(B) = 1 \qquad \text{Ext}(V) = 0.$$

The quantity

$$R_V = \frac{A_V}{E_{B-V}} = -\text{Ext}(\lambda = \infty) \qquad (10.6)$$

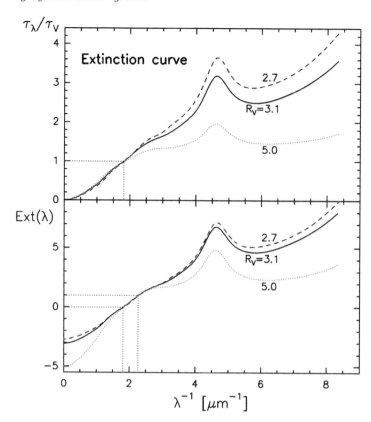

FIGURE 10.1 The observed interstellar extinction curve in the form τ_λ/τ_V and as $\text{Ext}(\lambda)$ according to equation (10.5). The ratio of total over selective extinction, $R_V = 3.1$, refers to the diffuse medium, $R_V = 5$ to the edges of molecular clouds. Whereas τ_λ/τ_V is normalized at V, $\text{Ext}(\lambda)$ is zero at V and equals one at B.

is called *ratio of total over selective extinction*. The visual extinction in magnitudes A_V, which also appears in the distance modulus of (10.3), is the most commonly used quantity to characterize the opaqueness of a cloud, but it is not directly observable. To obtain A_V from photometry, one measures, for example, E_{B-V}, assumes an R_V-value and then uses (10.6).

One can alternatively express the reddening curve through

$$\frac{\tau_\lambda}{\tau_V} = \frac{\text{Ext}(\lambda)}{R_V} + 1 \,. \tag{10.7}$$

τ_λ/τ_V and $\text{Ext}(\lambda)$ are mathematically equivalent. Still other forms, containing the same information, are sometimes more suitable for displaying certain trends, for example, A_λ/A_J or $(A_\lambda - A_J)/(A_V - A_J)$.

Because $\tau_\lambda/\tau_V = K_\lambda^{ext}/K_V^{ext}$, the reddening curve depends only on grain properties and it tells us how the mass extinction coefficient, K_λ^{ext}, varies with wavelength.

10.1.2.2 Properties of the reddening curve

The interstellar extinction curve (figure 10.1) is observationally determined from about 0.1 to 10μm. At longer wavelengths, the optical depth is too small to measure the weakening of background light; moreover, the dust begins to emit itself. We summarize the salient features:

- In the diffuse interstellar medium, Ext(λ) of equation (10.5) is at optical and infrared wavelengths fairly uniform over all directions in the sky and R_V equals 3.1; deviations occur only in the UV.

- Towards clouds, the shape of the curve Ext(λ) varies from one source to another, even in the infrared. The reduction in ultraviolet extinction (relative to V) for large R_V (see figure 10.1) suggests that the small grains have disappeared.

- If one plots instead of Ext(λ) the ratio τ_λ/τ_V then, to first order, *all* extinction curves look alike for $\lambda \geq 0.55\mu$m; at shorter wavelengths, they can differ substantially. The curves are largely (but not totally) fixed over the *entire* wavelength range by the one parameter R_V of equation (10.6). The observed variations of R_V range from 2.1 to well over five, the larger values being found towards molecular, the smaller ones towards high latitude clouds.

- The only resonance in the extinction curve is the broad bump at 4.6μm^{-1} or 2175Å which is always well fit by a Drude profile. Its central position stays constant from one star to another to better than 1% implying that the underlying particles are in the Rayleigh limit (sizes < 100Å). Usually one assumes that they are graphitic (section 4.5.1). The width of the bump varies ($\Delta\lambda^{-1} \simeq 1 \pm 0.2\mu$m^{-1}) and there is also scatter in its strength which is correlated with R_V in the sense that the resonance is weak in clouds where R_V is large. A plausible explanation is coagulation of dust particles in clouds as a result of which the small grains disappear.

  The 4.6μm^{-1} feature is due to absorption, not scattering (figure 10.4), and it is not accompanied by an increase in polarization (section 4.3). Therefore the grains responsible for it are not aligned.

- The reddening curve informs us about dust extinction; it does not tell us which fraction is due to scattering and which to absorption.

- Piecewise analytical approximations to the extinction curves as a function of R_V and wavelength are given in [Car89].

10.1.3 Two-color-diagrams

In two-color-diagrams (TCDs), one plots one color against another. TCDs are a simple and efficient tool to separate distinct or identify similar astronomical objects. As an example of a TCD, we show in figure 10.2 the UBV diagram of unreddened main sequence stars. The position in the UBV diagram of blackbodies with temperatures from 3500 K to 40 000 K is always well above the main sequence. So compared to a blackbody of the same color B–V, a stellar photosphere is much weaker at U. The reason for this behavior is that photo-ionization of the $n = 2$ level in hydrogen falls into the U band. As this is the dominant process for the photospheric opacity at this wavelength, it greatly suppresses the stellar flux. The characteristic wiggle in the main sequence line in figure 10.2 reflects the variation of this suppression (called Balmer decrement) with effective temperature.

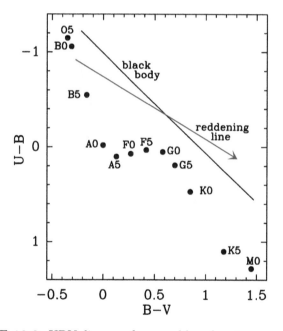

FIGURE 10.2 UBV diagram for unreddened main sequence stars (fat line) and for blackbodies (thin line). The arrow shows the direction and distance by which a star is displaced from the main sequence under the standard reddening law with $R_V = 3.1$ (figure 10.1) and $A_V = 5$ mag of foreground extinction. The main sequence locations of a few spectral types are indicated (after [All73]). The lower tip of the blackbody line refers to a temperature of 3500 K, the upper to 40 000 K. The blackbody and reddening curve have been calculated. The figure is therefore quite accurate and may be used to find the position of a main sequence star that suffers a certain amount of extinction A_V.

Interstellar reddening with the standard extinction law ($R_V = 3.1$) shifts any star in the UBV diagram from the main sequence in the direction parallel to the arrow of figure 10.2. The length of the shift is proportional to the foreground A_V, and in figure 10.2 it corresponds to $A_V = 5$ mag. In case of another extinction law, say, with $R_V = 5$, the slope and length of the arrow would be different (for the same A_V). If we observe an object somewhere in the right part of figure 10.2 and somehow know that it is a main sequence star, we can determine the amount of extinction by moving it up, parallel to the arrow, until it reaches the main sequence. This dereddening is unique as long as the star is of type B6 or earlier so that the reddening path does not intersect the wiggle of the main sequence curve.

10.1.4 Spectral indices

Akin to a color is the spectral index, commonly defined in the infrared as

$$\alpha = \frac{d \log(\lambda F_\lambda)}{d \log \lambda} . \tag{10.8}$$

So a spectral index depends on wavelength. In practice, α is calculated not as a derivative, but as the slope of λF_λ between two wavelengths λ_1 and λ_2,

$$\alpha = \frac{\log(\lambda_2 F_{\lambda_2}) - \log(\lambda_1 F_{\lambda_1})}{\log \lambda_2 - \log \lambda_1} . \tag{10.9}$$

For a blackbody obscured by a foreground extinction $\tau(\lambda)$, one has in the limit of high temperatures

$$\alpha = -3 - \frac{\tau(\lambda_2) - \tau(\lambda_1)}{\ln(\lambda_2/\lambda_1)} . \tag{10.10}$$

An important example is the spectral index between 2.2 and 20μm which serves as a classifier for protostellar objects. In this case and for standard dust, which is depicted by the MRN-line in figure 10.9,

$$\alpha \rightarrow -3 + A_V \times 0.036 \qquad \text{for } T \rightarrow \infty .$$

Of course, other wavelength intervals fulfill the same purpose; for instance, one also uses the index between 2.2 and 5μm. Figure 10.3 demonstrates how the indices change as a function of source temperature and foreground extinction.

10.1.5 The mass absorption coefficient

The reddening curve yields only a normalized extinction coefficient, for instance, K_λ/K_V. To obtain the absolute value of the cross section per gram of interstellar matter, K_λ, one has to know K_V. For this purpose, one measures

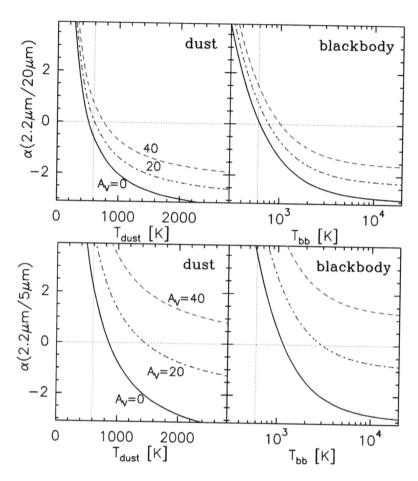

FIGURE 10.3 *Top:* The spectral index α in the wavelength interval from 2.2 to 20μm after (10.9) of a source at temperature T that is observed through a foreground of visual extinction A_V. In the right box, the object is a blackbody, or to first approximation a star. In the left, it is an optically thin dust cloud with the standard mixture of silicate and carbon grains (section 10.2). Because α is sensitive to both T and A_V, it is used in the classification of protostars. Without extinction, α of a blackbody approaches –3 at high temperatures. The weak dotted lines are for comparison between left and right boxes (note the different temperature ranges). *Bottom:* As above, but for spectral index α between 2.2 and 5μm.

towards a star the standard color excess E_{B-V} and the total hydrogen column density N_H to which both atomic and molecular hydrogen contribute,

$$N_H = N(HI) + 2N(H_2) .$$

The observations are performed in the far UV from a satellite and consist of absorption measurements towards early type stars in the $Ly\alpha$ line of HI and the Werner and Lyman bands of H_2. The average value over a large sample of stars is

$$N_H \simeq 5.8 \times 10^{21} E_{B-V} \quad cm^{-2} \tag{10.11}$$

or with (10.6) and $R_V = 3.1$,

$$N_H \simeq 1.9 \times 10^{21} A_V \quad cm^{-2} . \tag{10.12}$$

As $A_V = 1.086 \times N_H K_V$, one finds $K_V \simeq 4.9 \times 10^{-22}$ cm^2 per H-atom and a mass extinction coefficient

$$K_V \simeq 200 \text{ cm}^2 \quad \text{per g of interstellar matter.} \tag{10.13}$$

This value is easy to reminisce. In combination with the interstellar reddening curve, we know K_λ in the range from about 0.1 to $10\mu m$.

10.1.5.1   The dust-to-gas mass ratio

Equation (10.12) also allows to determine a credible dust-to-gas mass ratio, R_d, if one makes very reasonable assumptions about the composition of the grains, their shape and size. The amount of dust relative to gas follows then from the condition (10.11) that a hydrogen column density of 5.8×10^{21} cm^{-2} produces a standard color excess E_{B-V} of 1 mag. Dust models compatible with the interstellar reddening curve and constrained by cosmic abundances (section 5.3.5) yield

$$R_d = \frac{M_{dust}}{M_{gas}} \simeq 0.7 \dots 1.4\% . \tag{10.14}$$

The lower number may be appropriate for the diffuse medium, the higher for dense clouds where grains are ice-coated. Determining the gas mass from interstellar lines (HI or CO) and the dust mass from millimeter dust emission (to which it is directly proportional) leads to similar values, although the scatter is larger.

It follows from equations (10.12) and (10.14) that for a bulk dust density of 2.5 g cm^{-3} the total dust volume in a column of 1 cm^2 cross section with $A_V = 1$ mag equals 1.2×10^{-5} cm^{-3}. Considering the uncertainties, this number compares well with V_{dust} in (2.63) which was derived in a totally different way from the Kramers-Kronig relation.

On the average, the visual light is weakened in the galactic plane by about two magnitudes per kpc, of course, with great excursions in various directions. Using (10.12), this implies a mean gas density $n_H \sim 1$ cm^{-3} or a mean dust density $\rho_d \sim 3 \times 10^{-4}$ M$_\odot$ pc^3.

How dusty is the interstellar medium? Very dusty! If the atmosphere of the Earth, which has a density $\rho \sim 10^{-3}$ g/cm^3, had the same relative dust content, the optical depth $\tau = \ell \rho K_V$ would become unity over a distance of only $\ell = 10$ cm. We would not see our feet!

10.2 Dust models

We are now in a position to present a tentative, but complete model of interstellar dust with respect to its interaction with light. A dozen rival dust models can be found in the literature. They are all capable of explaining the basic data on extinction, polarization, scattering and infrared resonances and, if employed in radiative transfer calculations, yield successful matches to observed spectra.

All major dust models basically agree on the solid state abundances, the approximate grain sizes and the minimum number of dust components. Unfortunately, such consensus on important issues does not insure against common error. All models need

- big grains consisting mainly of silicate and carbon for the extinction curve longward of 1 μm;

- small graphitic particles for the extinction bump at 2200Å (figure 10.1);

- PAHs, or something similar to explain the near and mid IR resonances;

- very small grains undergoing temperature fluctuations to account for the hot emission of Cirrus clouds or reflection nebulae.

Differences among models show up primarily in the details of the chemical composition (mixing of components, coating, degree of fluffiness, fraction of built in refractory organic compounds). All models can be refined by varying the shape, size and optical constants of the grains. This allows to further enhance the agreement with observational data, but does not necessarily bring new insight. A more promising attempt is to increase the internal consistency of the models. Steps in this direction have been undertaken with respect to the smallest particles (PAHs and vsg) by *computing* grain properties, such as enthalpy and vibrational modes ([Dra01] and [Li01]).

10.2.1   Description of the model components

The model proposed here is similar to most others and certainly not superior. Its advantage, if it has any, lies in its relative simplicity. It uses published sets of optical constants, without modifying them at certain wavelengths to improve fits. The grains have the simplest possible structure and shape: except for PAHs, they are uncoated compact spheres. Therefore, to account for the interstellar polarization, some of the spheres have to be turned into cylinders (section 4.3). Altogether, we use only 12 types of dust particles, which include PAHs and very small grains (vsg). One particle type differs from another either by its size or chemical composition.

- **Big grains.** Most of the solid matter in the interstellar medium resides in big grains (also called MRN grains, section 5.5). There are three chemically distinct components (see section 5.3):

  - *Silicates* (Si); their abundance in the solid phase $[Si]/[H] = 3 \times 10^{-5}$ and they account for $\sim 60\%$ of the total dust mass.
  - *Amorphous carbon* (aC); solid abundance $[C]/[H] = 1.6 \times 10^{-4}$.
  - *Graphitic particles*; solid abundance $[C]/[H] \simeq 9 \times 10^{-5}$.

  The aC grains have four radii from 150Å to 1200Å, the Si grains from 300Å to 2400Å, both with an $a^{-3.5}$ size distribution (section 5.5). The graphitic particles, which make up about one third of solid carbon, are all small ($a \simeq 100$Å). We list them among the big grains because their temperature is still constant in most radiation fields.

  The bulk density of all dust materials is $\sim 2.5\,\text{g/cm}^3$. Absorption and scattering cross sections are calculated from Mie theory employing optical constants from figure 5.10, 5.11 and 5.12.

- **Very small grains (vsg).** A not negligible fraction of the solid matter resides in particles with dimensions less than 100Å. They form a population of very small grains (vsg) which may be regarded as the small-size extension of the big grains. If the $n(a) \propto a^{-3.5}$ size distribution of the big grains extends all the way down, the vsg make up more than 10% of the total dust mass. This may be an overestimate, and we assume, as suggested by their contribution to the total flux of the interstellar radiation field, a mass fraction of about 5%.

  Their optical cross sections are still calculated from Mie theory, although the concept of a continuous solid medium may not be fully valid. We usually assume only a graphitic carbon component, and no very small silicates. The main reason why one has to include very small grains is their time-variable temperature which leads to very "hot" emission (sections 6.5 and 6.6).

- **PAHs.** Finally, we add PAHs to the dust mixture. Although there must be a plethora of different species, we consider only two types:

  - small ones with $N_C = 50$ carbon atoms and
  - big ones with $N_C = 300$,

each type containing $Y_C^{PAH} = 3\%$ to 5% of the mass of all solid carbon.

The temperature of the PAHs also fluctuates violently and has to be calculated according to the recipes of section 6.5.

In estimating PAH properties, we widely follow [Sch93]. For the hydrogenation parameter, $f_{H/C}$, which is the ratio of hydrogen to carbon atoms, we put for $N_C > 30$

$$f_{H/C} = \frac{2.8}{\sqrt{N_C}} .$$
(10.15)

$f_{H/C}$ is thus proportional to the number of carbon atoms at the periphery of the PAH, where the H atoms are found, over the total number N_C. The formula gives $f_{H/C} = 0.40$ for $N_C = 50$, and $f_{H/C} = 0.16$ for $N_C = 300$. In a strong radiation field one may have to lower the hydrogenation parameter.

For the absorption cross section of a single PAH, C^{PAH}, we disregard differences between neutral and ionized species, although ionized PAHs, which can be excited by less energetic photons, have greater infrared absorption coefficients than neutral ones [Li01].

At wavelengths shorter than λ_{cut}, which is usually in the UV (see 9.4), we put the cross section of a single PAH with N_C carbon atoms to

$$C^{PAH} = 7 \times 10^{-18} \, N_C \, \text{cm}^2 .$$

It does not depend on frequency and is proportional to the number of carbon atoms, irrespective of their geometrical arrangement. The cross sections in the resonances are taken from table 9.1.

In the far infrared, at $\lambda > 25\mu$m, one expects a continuum (see in figure 9.2 the smooth frequency distribution in the modes even of a small PAH) and C_λ should fall off like λ^{-m} with m between one and two. However, this continuum is usually swamped by the emission from big grains and not observable.

The absorption coefficient of PAHs per gram of interstellar matter is displayed in figure 10.8.

   If helium accounts for 28% of the gas mass, the above cited abundances for all three dust components (big grains, vsg, PAHs) imply a total dust-to-gas mass ratio in the diffuse medium of $R_d = 1 : 140$.

10.2.2   Extinction and scattering of the dust model

10.2.2.1   The extinction curve

What kind of reddening does our dust model produce? Figure 10.4 reassures
us that, despite its simplicity, it more or less fits the observed extinction curve
of the diffuse medium. The deficit in the far UV is irrelevant in all radiative
transfer calculations of infrared sources where the observer does not receive
far UV photons anyway. The figure also displays the individual contributions
of the chemical components.

- In the visual ($\lambda^{-1} = 1.82\mu m^{-1}$), absorption is due to amorphous carbon,
  silicates account for only 20%. But silicates are responsible there for 80%
  of the scattering, or of 50% of the total extinction.

- Small graphite grains are irrelevant in the visual, but they produce the
  $4.6\mu m^{-1}$ hump via absorption and become important again in the far
  UV at $\lambda^{-1} > 8\mu m^{-1}$. If we replaced them by amorphous carbon, there
  would otherwise be little change.

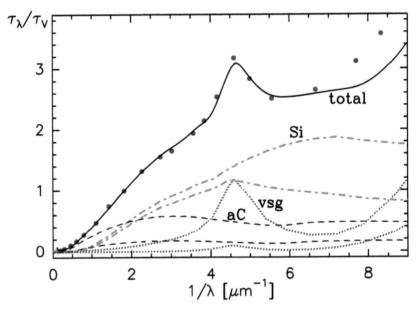

FIGURE 10.4 The normalized extinction optical depth, τ_λ/τ_V, of our dust
model (solid line, without PAHs) in comparison to the observed interstellar
extinction curve (dots). The two dash-dot curves show the contribution of the
silicates (Si); the upper one refers to extinction, the lower to pure scattering.
Likewise for amorphous carbon (aC, broken) and graphite particles (labeled
vsg, dotted) which consist of small and very small grains (vsg).

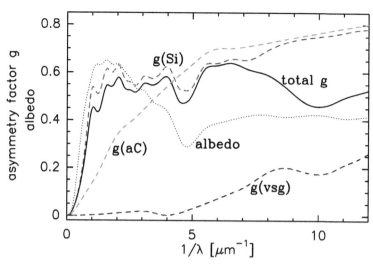

FIGURE 10.5 The total albedo of our dust model (dotted line) and the asymmetry factor $\bar{g}$ (solid). The average g's of the components are shown separately (three broken lines).

10.2.2.2 Albedo, asymmetry factor and phase function

The albedo of the dust mixture is plotted explicitly in figure 10.5, although one could extract it from figure 10.4. The albedo is close to zero in the infrared, peaks at optical wavelengths, dips at $4.6\mu m^{-1}$ and is constant and around 0.4 in the far UV.

We also determine for the dust mixture the phase function and the asymmetry factor. They specify the direction of light scattering (see 2.7). Our model contains 12 grain types differing from one another in their size or chemistry. Let those of type i have a total projected geometrical surface area A_i, a scattering efficiency $Q_{i\lambda}^{sca}$ and a phase function $f_{i\lambda}(\cos\theta)$. The average phase function, $\bar{f}_\lambda$, is then defined by

$$\bar{f}_\lambda(\cos\theta) \sum_i A_i Q_{i\lambda}^{sca} = \sum_i A_i Q_{i\lambda}^{sca} f_{i\lambda}(\cos\theta)$$

(see section 2.4.2 how to compute $f_\lambda(\cos\theta)$ from $S_{11}(\cos\theta)$). The average asymmetry factor, $\bar{g}_\lambda$, follows from $\bar{f}_\lambda$ via equation (2.7) and is also plotted in figure 10.5. Its behavior with wavelength is similar to that of the albedo; for $\lambda < 1\mu m$, it is around one half.

Figure 10.6 depicts the average phase function, $\bar{f}_\lambda(\cos\theta)$ at $2.2\mu m$, $0.55\mu m$ and 1050Å. It describes the angular dependence of scattering and is normalized to one in the forward direction. The corresponding asymmetry factors are 0.095, 0.543 and 0.470, respectively. At $2.2\mu m$, scattering is almost isotropic. Because the plots are normalized, one has the impression as if scattering were more forward at 1050Å than at V, however, $\bar{g}_V > \bar{g}_{1050}$.

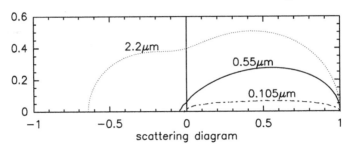

FIGURE 10.6 Selected phase functions of our dust model.

10.2.3 Extinction and absorption mass coefficients
10.2.3.1 The diffuse medium

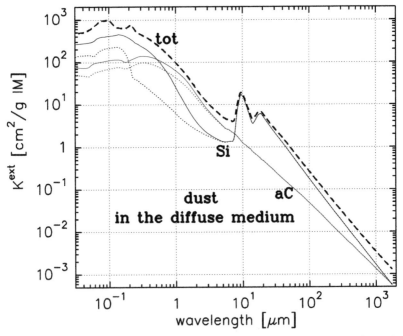

FIGURE 10.7 The total extinction coefficient (MRN+vsg+PAHs, dashed) for the model dust of the diffuse medium. It refers to $1\,\mathrm{g}$ of interstellar matter adopting a dust-to-gas ratio $R_\mathrm{d} = 1 : 140$. Also shown are the extinction (solid) and absorption (dots) coefficients for big silicates (Si, $a_- = 300\mathrm{\AA}$, $a_+ = 2400\mathrm{\AA}$) and big amorphous carbon grains (aC, $a_- = 150\mathrm{\AA}$, $a_+ = 1200\mathrm{\AA}$). One can read off the amount of scattering as the difference between extinction and absorption. The bumps at 2200 and 700Å are due to very small graphite particles ($a_- = 10\mathrm{\AA}$, $a_+ = 40\mathrm{\AA}$; see figure 5.13).

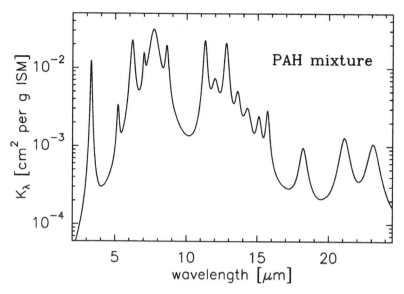

FIGURE 10.8 Mass absorption coefficient, K_λ, of a mixture of two kinds of PAHs (see text), each with an abundance $Y_C^{PAH} = 3\%$. The parameters of the major the resonances are listed in table 9.1. K_λ refers to 1 g of interstellar matter. Below $3\mu m$, K_λ slumps, but increases again when the wavelength falls below the cutoff for electronic excitation, λ_{cut}.

10.2.3.2 Protostellar cores

We extend now the model to dust in cold protostellar clouds. The grains there are modified by coagulation and by ice mantles. Consequently, their size increases and the dust mass doubles ($R_d = 1 : 70$), possibly even triples (see discussion on icy and fluffy grains in section 2.7). The effect of all three processes (fluffiness, frosting and grain growth) on the particle cross sections is explored in section 2.7.1.3 and 2.7.2.

The mass extinction coefficient of such icy and porous grains is displayed in figure 10.9. The particles are spheres and have a size distribution $n(a) \propto a^{-3.5}$ with lower limit $a_- = 0.03\mu m = 300\text{Å}$; the upper limit a_+ is varied. All grains are fluffy and composed of four components: silicate (Si), amorphous carbon (aC), ice (optical constants from figure 5.14) and vacuum. Average optical constants are calculated from the Bruggeman mixing rule (section 2.7). In all grains, the ice mass is equal to the mass of the refractory components, Si plus aC, whereas the ice volume is 2.5 times bigger than the volume of Si plus aC. The vacuum fraction is always constant, $f^{vac} = 0.5$. Silicate has twice the volume of carbon, $f^{Si}/f^{aC} = 2$. The PAHs have disappeared and are dirty spots in the ice mantles. Without UV radiation, the vsg are irrelevant. Note the ice resonances, for instance, the one at $3.1\mu m$.

The curve marked $a_+ = 0.3\mu m$ is an educated guess for dust in protostellar clouds before it is warmed up by the star and the ices evaporate. When the grains are very big, as in stellar disks, the lines with $a_+ = 30\mu m$ or 3mm might be appropriate.

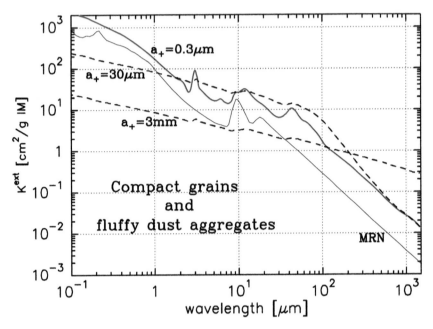

FIGURE 10.9 Mass extinction coefficient per gram of interstellar matter (IM) as a function of wavelength for composite grains which are fluffy aggregates of silicate, carbon, ice and vacuum. The lower limit a_- of the size distribution is in all curves $0.03\mu m$; the upper limit a_+ is indicated. The line labeled MRN shows the diffuse medium for reference; there the dust consists of pure (unmixed) silicate and amorphous carbon grains with $a_- \simeq 0.02\mu m$ and $a_+ \simeq 0.3\mu m$. The MRN curve is almost identical to the dashed curve of figure 10.7, except in the far UV due to slight differences in the vsg abundance.

10.2.3.3 The dust absorption coefficient at long wavelengths

For $\lambda \gtrsim 100\mu m$, the dust absorption coefficient (MRN curve in in figure 10.9) can be approximated by

$$K_\lambda = 4 \times 10^{-5} \cdot \left(\frac{\lambda}{cm}\right)^{-2} \quad cm^2 \text{ per g of interstellar matter .} \quad (10.16)$$

K_λ is the crucial parameter in determining the masses of interstellar clouds. Indeed, at long wavelengths, clouds are transparent and the observed flux S_λ

is directly proportional to the dust mass M_d,

$$S_\lambda = \frac{K_\lambda B_\lambda(T_d) M_d}{D^2} .$$ (10.17)

The distance D is usually known; the dust temperature T_d can be derived as a color temperature from flux ratios (section 6.2.4). The Planck function is not sensitive to its exact value as long as $hc/\lambda kT < 1$, therefore, the resulting mass depends all on K_λ.

In dense and cold clouds, K_λ may be five to ten times bigger than given in (10.16). One then has to use the curve labeled "$a_+ = 0.3\mu m$" in figure 10.9.

11

Radiative transport

This chapter deals with the transfer of radiation in a dusty medium. The photons that an observer picks up when looking at a source do not reach the telescope on a straight path, but have usually been scattered, absorbed and reemitted. To extract from the data as much information as possible about the source, one has to understand how the photons propagate.

We therefore summarize the elements of radiative transfer (see, for instance, the classical monograph [Mih78], or [Shu91]), discuss the general transfer equation and sketch how to derive estimates on the temperature, density and masses of a homogeneous medium. We then present a set of equations for spherical dust clouds that treats radiative cooling and heating in an energetically consistent way and outline how to solve it numerically. We append solution strategies of radiative transfer for two other astronomically relevant geometries: dusty accretion disks and dust-filled star clusters, both of which can, under minor assumptions, be approximated in one dimension. How to handle any three-dimensional configuration is shown in the section on the Monte Carlo method.

11.1 Basic transfer relations

11.1.1 Definition of intensity, mean intensity and flux

The intensity $I_\nu(\mathbf{r}, \mathbf{e})$ is the fundamental parameter describing the radiation field. We define it in the following way (figure 11.1): Consider a surface element $d\sigma$ at location $\mathbf{r}$ with normal unit vector $\mathbf{n}$. It receives within the frequency interval $\nu \ldots \nu + d\nu$ from the solid angle $d\Omega$ out of the direction $\mathbf{e}$, which is inclined by an angle θ against $\mathbf{n}$, the power

$$dW = I_\nu(\mathbf{r}, \mathbf{e}) \cos\theta \, d\sigma \, d\Omega \, d\nu . \tag{11.1}$$

The intensity $I_\nu(\mathbf{r}, \mathbf{e})$ has the dimension $\text{erg s}^{-1} \text{ cm}^{-2} \text{ Hz}^{-1} \text{ ster}^{-1}$ and is a function of
  - place $\mathbf{r}$
  - direction $\mathbf{e}$
  - frequency ν.

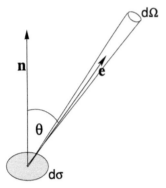

FIGURE 11.1 Defining the radiative intensity I_ν.

Imagine an extended astronomical source on the sky of solid angle Ω_S. The direction **e** towards it can be expressed by two angles, say, θ and ϕ. Let $I_\nu(\theta, \phi)$ be the intensity distribution over the source. The total flux which we receive on the Earth is given by

$$F_\nu^{\text{obs}} = \int_{\Omega_S} I_\nu(\theta, \phi)\, d\Omega ,\qquad(11.2)$$

where

$$d\Omega = \sin\theta\, d\theta\, d\phi$$

is the element of solid angle. In practice, the integration may be performed by mapping the source with high spatial resolution. If the intensity is constant over the source, $I_\nu(\theta, \phi)$ can be taken out of the integral and then

$$F_\nu^{\text{obs}} = I_\nu\, \Omega_S .\qquad(11.3)$$

Without intervening material, I_ν is independent of the distance D to the source because the solid angle Ω_S and the flux F_ν^{obs} are both proportional to D^{-2}. If we recede in a space ship from the Sun along an axis s, we receive less light, but the intensity in the direction of the Sun stays the same, therefore $dI/ds = 0$.

In (11.1), we defined the intensity from an observer's point of view: it determines the energy that one receives when looking at the sky in the direction **e**. In the discussion of radiative transfer, one lets the unit vector **e** point into the direction where the photons are flowing, and not from where they are coming, but otherwise nothing changes.

11.1.1.1 Moments of the intensity

A remark on the notation: All quantities related to light depend on the frequency ν (intensity I_ν, flux F_ν, cross section K_ν, Planck function $B_\nu(T), \ldots$). For clarity of exposition, we often drop the frequency index and write simply $I, F, B(T), \ldots$ instead of $I_\nu, F_\nu, B_\nu(T), \ldots$.

Although we also denote the frequency-integral $\int J_\nu d\nu$ by J, or $\int F_\nu d\nu$ by F, and likewise for other integrals, it should always be clear whether we mean the monochromatic or the frequency-integrated quantity.

We introduce the following moments of the intensity I:

- The *mean intensity* J is the average of I over the total solid angle,

$$J = \frac{1}{4\pi} \int_{4\pi} I(\mathbf{e})\, d\Omega \qquad (11.4)$$

or, using polar coordinates (θ, ϕ) and putting $\mu = \cos\theta$,

$$4\pi J = \int_0^\pi d\theta \int_0^{2\pi} d\phi\, I(\theta, \phi) \sin\theta = \int_{-1}^1 d\mu \int_0^{2\pi} d\phi\, I(\mu, \phi) . \qquad (11.5)$$

The result is independent of how we choose the direction $(\theta, \phi) = (0,0)$.

- The *net flux* F through a unit surface (see figure 11.1 and and equation 11.1),

$$F = \int_0^\pi d\theta \int_0^{2\pi} d\phi\, I(\theta, \phi) \cos\theta \sin\theta = \int_{-1}^1 d\mu \int_0^{2\pi} d\phi\, I(\mu, \phi)\, \mu .$$

$$(11.6)$$

The angle $(\theta, \phi) = (0,0)$ now gives the direction of the surface normal **n**. Generally speaking, the flux is a vector quantity,

$$\mathbf{F} = \int_{4\pi} \mathbf{e}\, I(\mathbf{e})\, d\Omega \qquad (11.7)$$

- The second moment of the intensity, K, is defined through[*]

$$4\pi K = \int_{-1}^1 d\mu \int_0^{2\pi} d\phi\, I(\mu, \phi)\, \mu^2 . \qquad (11.8)$$

In case of an isotropic radiation field, $I = J$ and the net flux is zero, $F = 0$. The ratio

$$f = \frac{K}{J} , \qquad (11.9)$$

which is called *Eddington factor*, equals then $\frac{1}{3}$ and $(4\pi/c)$ K gives the radiation pressure P. It is generally a symmetric tensor whose component P_{ij} gives the transfer rate of momentum by photons in the j-direction across a surface oriented perpendicular to the i-direction.

---

[*]Volume coefficients for absorption, scattering and extinction are denoted by the slightly different symbol K.

11.1.2   The general transfer equation

When a light ray propagates through a dusty medium along an axis s in the direction of the unit vector $\mathbf{e}$, its intensity I_ν changes. It is weakened by extinction and reinforced by thermal emission and by scattering of light from other directions into the direction $\mathbf{e}$. This is written in the following way

$$\frac{dI_\nu}{ds} = -K_\nu^{\text{ext}} I_\nu + \epsilon_\nu . \tag{11.10}$$

The equation is taken at a position $\mathbf{r}$, so the intensity should be fully written as $I_\nu(\mathbf{r}, \mathbf{e})$. K_ν^{ext} is the extinction coefficient per unit volume which includes absorption as well as scattering. The source term, ϵ_ν, comprises emission plus scattering. If T is the grain temperature at locus $\mathbf{r}$,

$$\epsilon_\nu = K_\nu^{\text{abs}} B_\nu(T) + \frac{K_\nu^{\text{sca}}}{4\pi} \int_{4\pi} p_\nu(\mathbf{e}', \mathbf{e}) I_\nu(\mathbf{e}') d\Omega' . \tag{11.11}$$

Note that the coefficient ϵ_ν defined in (6.1) stands for "true" emission only, equal to $K_\nu^{\text{abs}} B_\nu(T)$.

We now drop again the frequency index. The factor $p(\mathbf{e}', \mathbf{e})$ in (11.11) is the phase function for scattering from direction $\mathbf{e}'$ into $\mathbf{e}$. It equals $f(\cos\alpha)$ of (2.6) if α denotes the angle bewteen $\mathbf{e}'$ and $\mathbf{e}$. When the directions $\mathbf{e}$ and $\mathbf{e}'$ are given by the angle pairs (θ, ϕ) and (θ', ϕ'),

$$p(\mathbf{e}', \mathbf{e}) = p(\theta', \phi'; \theta, \phi) ,$$

the cosine theorem of spherical trigonometry states

$$\cos\alpha = \cos\theta \cos\theta' + \sin\theta \sin\theta' \cos(\phi - \phi')$$
$$= \mu\mu' + \sqrt{1 - \mu'^2}\sqrt{1 - \mu^2} \cos(\phi - \phi') . \tag{11.12}$$

Substituting the scalar vector product $\mathbf{e} \cdot \operatorname{grad} I(\mathbf{e})$ for dI/ds and using (11.11) for ϵ_ν, equation (11.10) may be reformulated as

$$\mathbf{e} \cdot \operatorname{grad} I(\mathbf{e}) = -K^{\text{ext}} I(\mathbf{e}) + K^{\text{abs}} B(T) + \frac{K^{\text{sca}}}{4\pi} \int p(\mathbf{e}', \mathbf{e}) I(\mathbf{e}') d\Omega' . \tag{11.13}$$

Both K^{ext} and ϵ refer to unit volume and their dimensions are cm^{-1} and erg s^{-1} cm^{-3} Hz^{-1} ster^{-1}, respectively.

11.1.2.1   Optical depth and source function

A fundamental quantity in light propagation is the *optical depth* or *optical thickness*, τ_ν. When light travels from point P_1 to point P_2, its value is given by the integral

$$\tau_\nu = \int_{P_1}^{P_2} K_\nu^{\text{ext}} ds , \tag{11.14}$$

so

$$d\tau_\nu = K_\nu^{\text{ext}} ds .$$

For a dusty medium, the volume coefficient K_ν^{ext} is insensitive to temperature and proportional to density. The integral in (11.14) is then simple. Let N_d denote the column density, i.e., the number of grains in a column of 1 cm² cross section. If all grains are identical, the extinction optical depth, τ_ν^{ext}, equals N_d times the extinction cross section of one grain,

$$\tau_\nu^{\text{ext}} = N_d C_\nu^{\text{ext}} . \tag{11.15}$$

Likewise, one defines an optical depth for absorption, $\tau_\nu^{\text{abs}} = N_d C_\nu^{\text{abs}}$, or for scattering, $\tau_\nu^{\text{sca}} = N_d C_\nu^{\text{sca}}$; by definition, $C_\nu^{\text{ext}} = C_\nu^{\text{abs}} + C_\nu^{\text{sca}}$.

An object of low optical depth ($\tau_\nu \ll 1$) is transparent and radiation reaches us from any internal point. An object of large optical depth ($\tau \gg 1$) is opaque and we can only look at its surface and do not receive photons from its interior. When we introduce the *source function*

$$S_\nu = \frac{\epsilon_\nu}{K_\nu^{\text{ext}}} , \tag{11.16}$$

we can rewrite (11.10) as

$$\frac{dI_\nu}{d\tau} = -I_\nu + S_\nu . \tag{11.17}$$

In this form, the equation tells us that the intensity increases with optical depth as long as it is smaller than the source function ($I_\nu < S_\nu$) and decreases when it is stronger. The source function is thus the limiting value of the intensity for high optical thickness.

The source function has for a dusty medium the general form

$$S_\nu = \frac{1}{K_\nu^{\text{ext}}} \left[K_\nu^{\text{abs}} B_\nu(T) + + \frac{K_\nu^{\text{sca}}}{4\pi} \int_{4\pi} p_\nu(\mathbf{e}', \mathbf{e}) I_\nu(\mathbf{e}') d\Omega' \right] . \tag{11.18}$$

If J_ν denotes the mean intensity of the radiation field, S_ν simplifies in case of isotropic scattering to

$$S_\nu = \frac{K_\nu^{\text{abs}} B_\nu(T) + K_\nu^{\text{sca}} J_\nu}{K_\nu^{\text{ext}}} . \tag{11.19}$$

Even when scattering is anisotropic, one often need not compute the full source function (11.18). For most applications it is sufficient to replace K_ν^{sca} by $(1 - g_\nu) K_\nu^{\text{sca}}$ and still use (11.19). The asymmetry factor g_ν of (2.7) is rarely negative and when it is, its absolute value is small (see figure 10.5). The substitution of K_ν^{sca} by $(1 - g_\nu) K_\nu^{\text{sca}}$ amounts to neglecting the forward scattered part and assuming that the rest is scattered isotropically; the replacement is exact for $g_\nu = 0$ and $g_\nu = 1$.

It is straightforward to generalize the optical depth τ_ν^{ext} in (11.15) or the source function S_ν in (11.19) for a mixture of grains, or to include grains with temperature fluctuations.

11.1.2.2   Zeroth moment of the transfer equation

Integration of (11.13) over all directions **e** gives

$$\text{div } \mathbf{F} \;=\; -4\pi J K^{\text{ext}} + 4\pi\,\epsilon \;=\; 4\pi K^{\text{abs}}\Big[B(T) - J\Big]. \qquad (11.20)$$

This is the zeroth moment equation. It expresses energy conservation. J is the mean radiative intensity from (11.5). The left side of (11.20) follows from (11.13) because (see B.21)

$$\text{div } \mathbf{F} = \text{div}\int \mathbf{e}\, I(\mathbf{e})\, d\Omega = \int \text{div}\,(\mathbf{e}I(\mathbf{e}))\, d\Omega = \int \mathbf{e}\cdot\text{grad}\, I(\mathbf{e})\, d\Omega\ ,$$

and its right side is correct because

$$\int d\Omega' \int d\Omega\, p(\mathbf{e}',\mathbf{e})\, I(\mathbf{e}') = \int d\Omega'\, I(\mathbf{e}') \int d\Omega\, p(\mathbf{e}',\mathbf{e}) = 4\pi J\cdot 4\pi\ .$$

Note that the scattering coefficient, K^{sca}, has canceled out in (11.20).

11.1.3   Transfer equation in spherical and slab symmetry

11.1.3.1   Spherical symmetry

In spherical geometry, the dust density $\rho(r)$, the dust temperature $T(r)$, the mean intensity $J(r)$ or the flux $F(r)$ are functions of distance r to the cloud center only. But the basic quantity, the intensity I depends also on the inclination angle θ to the radial vector, so fully written $I = I_\nu(r,\theta)$. With $\mu = \cos\theta$, we now have (see 11.5 to 11.8)

$$J = \tfrac{1}{2}\int_{-1}^{1} I(\mu)\, d\mu, \quad F = 2\pi\int_{-1}^{1} I(\mu)\,\mu\, d\mu, \quad K = \tfrac{1}{2}\int_{-1}^{1} I(\mu)\,\mu^2\, d\mu \quad (11.21)$$

The transfer equation (11.13) becomes

$$\mu\,\frac{\partial I}{\partial r} + \frac{1-\mu^2}{r}\,\frac{\partial I}{\partial \mu} = -K^{\text{ext}}I + K^{\text{abs}}B(T) + \frac{K^{\text{sca}}}{4\pi}\int_{4\pi} p(\mathbf{e}',\mathbf{e})\, I(\mathbf{e}')\, d\Omega'\ . \quad (11.22)$$

The left side is just the derivative dI/ds of (11.10) in spherical coordinates. The zeroth moment equation reads (see 11.20)

$$\frac{1}{r^2}\frac{d}{dr}\,(r^2 F) \;=\; 4\pi K^{\text{abs}}\Big[B(T) - J\Big]. \qquad (11.23)$$

It looks much simpler than (11.22) because there is no angle dependence. However, we cannot solve it for $J(r)$ because it contains the new variable $F(r)$. To find an additional equation that might help us out, we multiply

(11.22) by μ and integrate over 4π to obtain the first moment equation. All terms are straightforward, except for the double integral

$$\frac{K^{\text{sca}}}{4\pi} \int_{-1}^{1} d\mu\, \mu \int_{0}^{2\pi} d\phi \int_{1}^{1} d\mu' \int_{0}^{2\pi} d\phi'\, p(\mu',\phi';\mu,\phi)\, I(\mu',\phi')\, d\mu' \ . \quad (11.24)$$

To evaluate it, one expands the phase function $f(x)$ of (2.6) into Legendre polynomials [Cha46],

$$f(x) = \sum_{n=0}^{} a_n P_n(x) \ .$$

The normalization condition (2.6) gives $a_0 = 1$, and $a_1 = 3g$ where g is the asymmetry factor of (2.7) (see B.15). In view of the cosine theorem (11.12) and the addition theorem (B.17), we get

$$p(\mu',\phi';\mu,\phi) = \sum_{n=0}^{} a_n \left\{ P_n(\mu)P_n(\mu') + 2 \sum_{m=1}^{n} \frac{(m-n)!}{(m+n)!} P_n^m(\mu)P_n^m(\mu') \cos m(\phi-\phi') \right\}$$

Inserted into (11.24), the double integral over the second sum in (11.24) vanishes and over the first sum it yields $\frac{4\pi}{3} a_1 F$ (see B.15), therefore

$$\frac{dK}{dr} + \frac{3K-J}{r} = -\frac{F}{4\pi} \left[K^{\text{abs}} + (1-g)K^{\text{sca}} \right] \ . \quad (11.25)$$

But the *closure problem* of having one more unknown than equations remains because now K comes into play. The introduction of the next higher moment (multiplication of (11.20) by μ^2) will not remedy the situation. To be able to use formulae that do not depend explicitly on the direction angle, one needs physical guidance. For example, at low optical depth, the radiation field is streaming radially outwards and $J_\nu = K_\nu = F_\nu/4\pi$, and at high optical depth, the radiation field is isotropic and the Eddington factor $f = \frac{1}{3}$.

11.1.3.2 Slab symmetry

We repeat the discussion for a slab. The dust density and temperature, the mean intensity or the flux are then functions of the height z. The intensity I depends on z and the inclination angle θ to the vertical z-axis (see figure 11.3), so $I = I_\nu(z,\theta)$. The transfer equation (11.13) takes the form

$$\mu \frac{dI}{dz} = -K^{\text{ext}} I + K^{\text{abs}} B(T) + \frac{K^{\text{sca}}}{4\pi} \int_{4\pi} p(\mathbf{e}',\mathbf{e})\, I(\mathbf{e}')\, d\Omega' \quad (11.26)$$

where again $\mu = \cos\theta$. The zeroth and first moment equation read now (cf. 11.23 and 11.25)

$$\frac{dF}{dz} = 4\pi K^{\text{abs}} \left[B(T) - J \right] \quad (11.27)$$

$$\frac{dK}{dz} = -\frac{F}{4\pi} \left[K^{\text{abs}} + (1-g)K^{\text{sca}} \right] \ . \quad (11.28)$$

Slab symmetry is mathematically simpler than spherical symmetry, but the closure problem remains.

11.1.4   Frequency averages

One can radically simplify the radiative transfer by taking frequency averages.

- In an optically thick configuration, where the mean intensity J_ν is close to the Planck function $B_\nu(T)$, one uses the *Rosseland mean* $K_R(T)$ defined through

$$\frac{1}{K_R(T)} = \frac{\int \frac{1}{K_\nu} \frac{\partial B_\nu(T)}{\partial T} d\nu}{\int \frac{\partial B_\nu(T)}{\partial T} d\nu} . \tag{11.29}$$

As $\int B_\nu(T) \, d\nu = \sigma T^4/\pi$, where σ is the radiation constant of (A.61), the denominator in (11.29) equals $4\sigma T^3/\pi$. The Rosseland mean gives the greatest weight to the lowest values of K_ν, as in a room darkened by thick curtains: it is not their thickness, but the open slits between them that determine how much light enters from outside. If F_ν is the energy flux at frequency ν per unit area and time, the definition of K_R in equation (11.29) must be, and indeed is, although not evident at first glance, equivalent to

$$K_R \int F_\nu \, d\nu = \int K_\nu F_\nu \, d\nu .$$

K_R is used, for example, for the radiative energy transfer in stellar interiors. Assuming $J_\nu = B_\nu(T)$ and $f_\nu = \frac{1}{3}$, the total flux $F = \int F_\nu \, d\nu$ is then after (11.25) determined by the temperature gradient in the star,

$$F = -\frac{16 \sigma T^3}{3 K_R(T)} \frac{dT}{dr} . \tag{11.30}$$

The radiation pressure is $P_r = \frac{1}{3}aT^4$, with a from (A.65). Equation (11.30) corresponds to Fick's first law of diffusion, applied to a photon gas. The Rosseland mean opacity may also be acceptable in protostellar cocoons or accretion disks.

- In tenuous clouds, the *Planck mean*, K_P, is the appropriate average. It is now not a question of which frequency penetrates deepest, they all pass through; instead one takes the average over the emitted energy,

$$K_P(T) = \frac{\int K_\nu B_\nu(T) \, d\nu}{\int B_\nu(T) \, d\nu} . \tag{11.31}$$

With the approximation $K_\nu \propto \nu^m$ (see 6.9), which is fine for interstellar dust at long wavelengths, one can compute the coefficients analytically and finds per gram of interstellar matter

$$K_R \simeq K_P \simeq 3 \times 10^{-4} T^2 \quad \text{cm}^2 . \tag{11.32}$$

K_R and K_P are not much different and both have a quadratic dependence on temperature.

11.1.5 Analytical solutions to the transfer equation

We solve equation (11.10) for three scenarios:

- Pure extinction without emission ($\epsilon = 0$). An example is starlight weakened on the way to us by an intervening dust cloud. The obvious solutions to (11.10) and (11.17) are

$$I(s) = I(0) \exp\left[-\int_0^s K(s')\,ds'\right] \qquad (11.33)$$

$$I(\tau) = I(0)\,e^{-\tau} . \qquad (11.34)$$

$I(0)$ is the intensity of the background star if it were not attenuated. The observed intensity diminishes exponentially with optical depth.

- Pure emission without extinction ($K^{\text{ext}} = 0$). This situation applies quite well to submillimeter radiation from an interstellar dust cloud, now

$$I(s) = I(0) + \int_0^s \epsilon(s')\,ds' . \qquad (11.35)$$

Without a background source ($I(0) = 0$), which is the standard scenario, and for constant emission coefficient ϵ, the observed intensity grows linearly with the path length s.

- Emission plus absorption. In this most general case, we multiply (11.17) by e^τ and integrate by parts which leads to

$$I(\tau) = I(0)\,e^{-\tau} + \int_0^\tau S(\tau')\,e^{-(\tau-\tau')}\,d\tau' . \qquad (11.36)$$

Although it appears as if we had mastered the radiative transfer problem, in reality, we have not. Equation (11.36) is only a formal solution to the transfer equation (11.17). The hard part is to find the source function $S(\tau)$ in formula (11.36). If $S(\tau)$ is constant, it may be taken out from under the integral yielding

$$I(\tau) = I(0)\,e^{-\tau} + S(\tau) \cdot \left[1 - e^{-\tau}\right] . \qquad (11.37)$$

11.1.5.1 A dust cloud of uniform temperature

In a dust cloud, emission occurs only in the infrared. At these wavelengths, grain scattering is practically absent. For dust at temperature T, the source function is then equal to the Planck function at that temperature

$$S_\nu = B_\nu(T) .$$

Towards a uniform cloud of optical depth τ, one therefore observes according to (11.37) the intensity (frequency index omitted)

$$I(\tau) = I(0)\,e^{-\tau} + B(T) \cdot \left[1 - e^{-\tau}\right] . \qquad (11.38)$$

If the background is a point source of flux $F(0)$, one receives out of a solid angle Ω the flux

$$F(\tau) = F(0)\,e^{-\tau} + B(T) \cdot \left[1 - e^{-\tau}\right]\Omega\ . \qquad (11.39)$$

When the background is absent or its spectrum known and multi-frequency observations are available, equation (11.39) yields an estimate of the source temperature and its optical depth, and thus its mass. The limiting solutions to (11.38) are

$$I(\tau) = \begin{cases} B(T)\,\tau + I(0) & : \text{ for }\quad \tau \ll 1 \\[2mm] B(T) & : \text{ for }\quad \tau \gg 1\ . \end{cases} \qquad (11.40)$$

They imply that without a background source and if the cloud is translucent ($\tau \ll 1$), the intensity is directly proportional to the optical depth: doubling the column density, doubles the strength of the received signal. In this case, the intensity I_ν carries information about the amount of dust, but also about the spectral behavior of the grains via the frequency dependence of the optical depth. However, if the source is opaque ($\tau \gg 1$), all information about the grains is wiped out and the object appears as a blackbody of temperature T.

11.2 Spherical clouds

The dust emission of a cloud is observed at various wavelengths with the goal to derive its internal structure. The observational data consist either of maps, or of single point measurements towards different positions on the source obtained with different spatial resolution. When the cloud is heated from inside by a star, it is obviously inhomogeneous and estimates of the dust column density and temperature based on the assumption of a uniform cloud (see 11.39) are no longer acceptable. A more refined model is needed.

An easy-to-model source geometry is presented by a sphere (see, for instance, [Hum71], [Leu75], [Yor80]; see [Tsc77] for the treatment of shock discontinuities). The structure of the medium depends then only on one spatial coordinate, the radius, and the problem is one-dimensional. The radiative transfer may be solved in many ways and there exist fast and efficient public computer codes. In practice, however, one may encounter difficulties using a public code because it usually has to be modified to meet the demands of a specific astronomical problem. It may then be better to write a code oneself, even at the expense of less professionality. We suggest below a set of equations and sketch the numerical strategy for its solution.

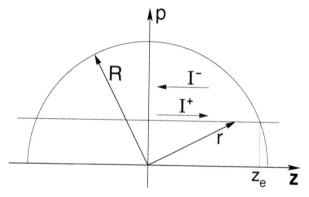

FIGURE 11.2 The coordinate system for the intensity in a spherical cloud of outer radius R. Light rays at constant impact parameter p are directed either to the left (I^-) or to the right (I^+). The ray I^+ enters the cloud at $(p, -z_e)$ and leaves at (p, z_e). The cloud is immersed into a radiation field of isotropic intensity I_0.

11.2.1 Integral equations for the intensity

As illustrated in figure 11.2, light is propagating at constant impact parameter p parallel to the z-axis either to the left (I_ν^-) or to the right (I_ν^+). The coordinates p and z, which are used instead of r and μ, are connected to the radius by

$$r^2 = p^2 + z^2 ,$$

and the intensities obey the integral equations (see 11.36)

$$I^+(\tau) = e^{-\tau} \left[I^+(0) + \int_0^\tau S(x)\, e^x\, dx \right] \tag{11.41}$$

$$I^-(t) = e^{-t} \left[I_0 + \int_0^t S(x)\, e^x\, dx \right] . \tag{11.42}$$

In equation (11.41), the optical depth τ increases from left to right, so $\tau(z=0) = 0$ and $\tau(z > 0) > 0$. In (11.42), the optical depth t equals zero at the cloud surface ($t(z_e) = 0$) and grows from right to left, so $t(z < z_e) > 0$.

If the dust particles scatter isotropically, the source function, $S(x)$, is given by (11.19). K_ν^{abs}, K_ν^{sca} and K_ν^{ext} are the volume coefficients for absorption, scattering and extinction referring to $1\,\mathrm{cm}^3$ of interstellar space. In view of the symmetry of the spherical cloud, (11.41) and (11.42) have to be solved only in the right quadrant of figure 11.2, i.e., for $z > 0$. The equations must be complemented by two boundary conditions:

- The first states that at the cloud's edge, where $z = z_e$, the incoming intensity I_ν^- is equal to the intensity of the external radiation field $I_{0\nu}$,

$$I_\nu^-(p, z_e) = I_{0\nu} \quad \text{with} \quad z_e = \sqrt{R^2 - p^2} . \tag{11.43}$$

- The formulation of the second boundary condition depends on whether there is a central star or not. Without it, or when there is a star, but the impact parameter p is larger than the stellar radius R_*, one has for reasons of symmetry

$$I_\nu^+(p, z) = I_\nu^-(p, z) \quad \text{at} \quad z = 0 . \tag{11.44}$$

With a star of stellar surface flux $F_{*\nu}$ and when $p \leq R_*$ (see A.55),

$$I_\nu^+(p, z_*) - I_\nu^-(p, z_*) = \frac{F_{*\nu}}{\pi} \quad \text{with} \quad z_* = \sqrt{R_*^2 - p^2} . \tag{11.45}$$

Solving (11.41) and (11.42) is not enough, one still has to determine the dust temperature. One arrives at a full solution to the radiative transfer problem iteratively. Suppose that after iteration cycle i we approximately know the grain temperature $T(r)$, the mean intensity of the radiation field $J_\nu(r)$, and thus after (11.19) the source function $S_\nu(r)$.

To improve the values of $T(r)$, $J_\nu(r)$ and $S_\nu(r)$, we determine for each impact parameter p the intensity I_ν^- from (11.42). With the help of the inner boundary condition, we then calculate the intensity I_ν^+ after (11.41), and from I_ν^+ and I_ν^- we form the quantities

$$u_\nu(p, z) = \tfrac{1}{2} \left[I_\nu^+(p, z) + I_\nu^-(p, z) \right] \tag{11.46}$$
$$v_\nu(p, z) = \tfrac{1}{2} \left[I_\nu^+(p, z) - I_\nu^-(p, z) \right] . \tag{11.47}$$

(The splitting of the intensity into I^- and I^+, and the introduction of u and v is due to [Fea64]). The u_ν yield an updated mean intensity

$$J_\nu(r) = \frac{1}{r} \int_0^r \frac{p\, u_\nu(p, r)}{\sqrt{r^2 - p^2}} \, dp . \tag{11.48}$$

It is not difficult to see that this agrees with the definition of J_ν from (11.5). The new $J_\nu(r)$ are inserted into equation (6.7) or into

$$\int J_\nu(r)\, K_\nu(r)\, d\nu = \int B_\nu(T(r))\, K_\nu(r)\, d\nu ,$$

which, when solved for T, gives new temperatures. Updating the source function $S_\nu(r)$ (see 11.16) finishes iteration cycle $i{+}1$.

11.2.2   Practical hints

- As starting values for the first cycle ($i = 1$), an initial mean intensity $J_\nu(r) = 0$ should be acceptable. For the temperature, one may put $T(r) = \text{const}$ if there is no central star and the outer radiation field is not too hard and strong. With a star, $T(r) \sim 20\, (L_*/r^2)^{1/6}$ should work when r and L_* are expressed in cgs-units (see 6.16).

- The iterations converge when the changes of the temperature $\Delta T(r)$ become smaller with each new cycle. As a check on the accuracy of the calculations, one can compute for each radius r the net flux

$$F_\nu(r) = \frac{4\pi}{r^2} \int_0^r p\,v_\nu(p,r)\,dp .\qquad (11.49)$$

This formula can be derived from (11.47) and the middle equation of (11.21). The integrated flux through a shell of radius r is

$$L(r) = 4\pi r^2 \int F_\nu\,d\nu .$$

If the cloud has a central star of luminosity L_*, the condition $L(r) = L_*$ should be numerically fulfilled everywhere to an accuracy of $\sim 1\%$.

- Without an embedded source, heating comes from an outer isotropic radiation field of intensity $I_{0\nu}$. The integrated flux $L(r)$ is then zero because what enters at the cloud surface must also leave. Nevertheless, the dust may be quite hot, especially in the outer envelope. Numerically, $|L(r)|$ should now be small compared to

$$4\pi R^2\,\pi \int I_{0\nu}\,d\nu ,$$

which is the total flux from the external radiation field through the cloud surface of area $4\pi R^2$.

- To select a sufficiently fine radial grid which is also economical in terms of computing time, the number of points and their spacings must be adapted during the iterations. For example, in a cloud of 50 mag of visual extinction with a central a hot star, the dust column density between neighboring grid points must be small near the star because UV radiation is absorbed very effectively. At the cloud edge, on the other hand, there are only far infrared photons and the grid may be coarse. Generally speaking, as one recedes from the star, the radiation field softens and the mean free path of the photons increases.

At each radius r, there exists therefore a cutoff frequency with the property that only photons with $\nu < \nu_{\mathrm{cut}}$ are energetically relevant, whereas those with $\nu > \nu_{\mathrm{cut}}$, whose mean free path is as a rule of thumb shorter, are unimportant. ν_{cut} may be defined by and computed from

$$\int_{\nu_{\mathrm{cut}}}^\infty J_\nu(r)\,d\nu = \alpha \int_0^\infty J_\nu(r)\,d\nu ,\qquad \alpha \sim 0.2 .$$

Choosing the radial step size in such a way that the optical depth at the cutoff frequency is of order 0.1 automatically ensures that the grid is fine only where necessary.

- Very close to the star, dust evaporates. The exact radius r_{evap} is part of the solution of the radiative transfer; it depends, of course, on the grain type. To account for dust evaporation, one may evaluate the grain temperature T_d in the local radiation field and add a flag at all radii to each grain type indicating whether T_d is above or below the evaporation threshold T_{evap}. The flag then controls whether the grain contributes in the next cycle to the extinction and emission or not. For heavily obscured sources ($A_V > 20\,\text{mag}$), the exact value of the evaporation radius has little influence on the spectrum, if at all, only in the near infrared.

- For N radial grid points $r_1 = R_*, r_2, r_3, \ldots, r_N = R$, there are $N+1$ impact parameters $p_i = r_i$ with $i = 1, \ldots, N$ and $p_0 = 0$. Because to first order the intensity over the stellar disk does not change, one needs only two impact parameters ($p=0$ and $p=R_*$) for the star. The second radial grid point r_2 should only be a tiny bit (0.1%) bigger than r_1.

- The dust temperatures T_d are computed from (6.7) which can be solved numerically as follows: Let T_j be the value of the grain temperature in iteration step j. Writing (6.7) in the form

$$\int Q_\nu^{abs} J_\nu \, d\nu = \overline{Q} \frac{\sigma}{\pi} T^4 ,$$

where σ is the radiation constant of (A.61) and $\overline{Q}$ is defined by

$$\overline{Q} = \int Q_\nu^{abs} B_\nu(T) \, d\nu \cdot \left[\int B_\nu(T) \, d\nu \right]^{-1} ,$$

one gets an improved estimate T_{j+1} for the new cycle $j+1$ from

$$T_{j+1}^4 = T_j^4 \int Q_\nu^{abs} J_\nu \, d\nu \cdot \left[\int Q_\nu^{abs} B_\nu(T_j) \, d\nu \right]^{-1} . \qquad (11.50)$$

The sequence $T_j, T_{j+1}, T_{j+2}, \ldots$ quickly converges.

11.3 Passive disks

11.3.1 Radiative transfer in a plane parallel layer

All stars are surrounded during their early evolution by an accretion disk. A disk is passive when it is only heated by the star and internal dissipational losses are negligible. Imagine a flat and geometrically thin disk with density structure $\rho(r, z)$ given in cylindrical coordinates (r, z). Let it be illuminated

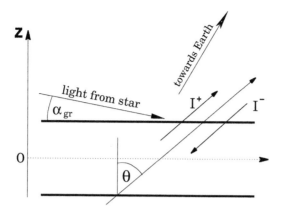

FIGURE 11.3 A small portion of a geometrically thin disk. The star is far away to the left at a distance r, its light falls on the disk under a grazing angle α_{gr}. The transfer equation (11.17) is solved under all inclination angles θ for incoming (I^-) and outgoing (I^+) radiation. The situation is symmetric with respect to the mid-plane which is indicated by the dotted line at $z = 0$.

by a central star of luminosity L_*, radius R_* and temperature T_* whose mid-point is at $(r = 0, z = 0)$.

To determine the emission from the disk, we divide it into rings of radius r and small radial width Δr. For each ring, we compute the radiative transfer in z-direction (figure 11.3). The formalism developed for spheres can be carried over to slab geometry. Starting from some first-guess vertical temperature profile $T_0(z)$, one solves the pair of equations (11.41) and (11.42) for I^+ and I^-, which are now the intensities along straight lines under various angles θ with respect to the z-axis. From I_ν^+ and I_ν^- one computes the mean intensity

$$J_\nu = \tfrac{1}{2} \int_0^1 (I_\nu^+ + I_\nu^-)\, d\nu \;,$$

(the corresponding formula for spheres is 11.48) and then an improved temperature run $T_1(z)$ using (6.6). The procedure is iterated until the sequence $T_i(z)$ $(i = 0, 1, 2, ...)$ converges.

The effective grazing angle, α_{gr}, under which starlight falls on the disk is evaluated in equation (11.54) for a flat disk and in (11.58) for an inflated disk. Viewed from the disk surface, the upper hemisphere of the star subtends a solid angle (for $r \gg R_*$)

$$\Omega_* = \frac{\pi R_*^2}{2r^2} \;.$$

In order not to break the slab symmetry, the stellar hemisphere is replaced by a luminous band of the same intensity and solid angle. This band encircles the whole sky at an elevation α_{gr}; its width is $\Omega_*/2\pi \ll \alpha_{\mathrm{gr}}$.

The first boundary condition of the transfer equations states that at the top of the disk the incoming intensity I^- is zero except towards the luminous band where it equals $B_\nu(T_*)$. The second boundary condition reads $I^+ = I^-$ at $z = 0$ and follows from symmetry requirements.

The solution to the radiative transfer at radial distance r yields the temperature structure $T(r, z)$ and the intensities $I_\nu^+(r, z, \theta)$, $I_\nu^-(r, z, \theta)$ at all frequencies ν, heights z and angles θ. To determine the flux that one will detect on Earth under a viewing angle $\theta_{\rm obs}$, one needs the intensity I_ν^+ at the top of the disk for $\theta = \theta_{\rm obs}$. Summing up the contributions from all rings and adding the direct emission from the star yields the total observed flux.

11.3.2   Disks of high optical thickness

When the opacity is high, as happens in young stellar disks where $A_{\rm V}$ in vertical direction is 10^4 mag or greater, one may encounter numerical difficulties which are exacerbated by the smallness of the grazing angle $\alpha_{\rm gr}$. We then recommend, as a slight modification of the above method, to split the disk in its vertical structure into a completely opaque mid layer sandwiched between two much thinner top layers, each of visual optical thickness $\tau_{\rm top}$ in z-direction (figure 11.4).

The optically very thick mid layer is isothermal at temperature $T_{\rm mid}$ because a temperature gradient would imply in the diffusion approximation of (11.30) a net energy flux. However, there can be none because there are no internal sources and the radiation field is the same above and below the disk.

For each ring of radius r and width Δr, we evaluate the radiative transfer for the top and mid layer separately, and connect the two by boundary conditions. In the top layer, we use the method described before but position the zero point of the z-axis at the transition between the two layers. The radiative transfer in the isothermal mid layer is trivial. The intensity that enters the top layer from below under the direction $\mu = \cos\theta$ is

$$I_\nu^+(z = 0, \mu) = B_\nu(T_{\rm mid}) \cdot \left[1 - e^{-\tau_\nu(\mu)}\right] \tag{11.51}$$

where $\tau_\nu(\mu)$ is the optical depth of the mid layer. The monochromatic flux received by the top layer from below is therefore

$$f_\nu = 2\pi B_\nu(T_{\rm mid}) \int_0^1 \left[1 - e^{-\tau_\nu(\mu)}\right] \mu \, d\mu \ .$$

The frequency integral $\int f_\nu \, d\nu$ must be equal to the downward flux at $z = 0$ because the net flux, for reasons of disk symmetry, vanishes everywhere. This condition yields $T_{\rm mid}$, of course, iteratively. In mathematical form, it reads

$$\int d\nu \, B_\nu(T_{\rm mid}) \int_0^1 d\mu \left[1 - e^{-\tau_\nu(\mu)}\right] \mu = \int d\nu \int_0^1 d\mu \, \mu \, I_\nu^-(z = 0) \ .$$

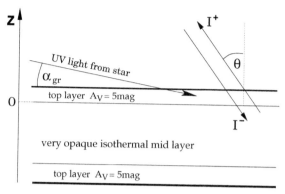

FIGURE 11.4 A small portion of a geometrically thin, but optically very thick disk. The zero point of the z-axis is different from figure 11.3. The UV light from the star on the left is completely absorbed in the top layer. The mid layer is isothermal at temperature T_{mid}. One calculates the radiative transfer for the top layer ($z \geq 0$) using the boundary condition (11.51) for I^+.

Ideally, the top layer extends far down so that the temperature at its bottom asymptotically approaches T_{mid}. But this happens only when τ_{top} is a few hundred and it would require many vertical grid points. In practice, it suffices to choose $\tau_{top} \simeq 5$. The resulting spectral energy distribution becomes then indistinguishable from one calculated with $\tau_{top} > 100$. When $\tau_{top} \simeq 5$, almost all starlight that enters the disk is absorbed or scattered in the top layer. The photons that are scattered there or reemitted (at IR wavelengths) may be scattered or absorbed again in the top layer, or enter the mid layer, or leave the disk upwards (in figure 11.4) into space.

11.3.3 The grazing angle

11.3.3.1 Flat disk

We compute the monochromatic flux f_ν that falls in z-direction on the disk and heats it. If the disk is flat and thin or if it has a constant opening angle, a surface element of unit area receives out of a solid angle $d\Omega$ pointing towards the star the flux $B_\nu(T_*)\, d\Omega\, \sin\theta \sin\phi$ (figure 11.5); $B_\nu(T_*)$ denotes the uniform stellar intensity. The whole upper hemisphere of the star illuminates the surface element by the flux

$$
\begin{aligned}
f_\nu &= B_\nu(T_*) \int_0^{\theta_{max}} \sin^2\theta\, d\theta \int_0^\pi \sin\phi\, d\phi = B_\nu(T_*) \left[\theta_{max} - \tfrac{1}{2}\sin 2\theta_{max} \right] \\
&= B_\nu(T_*) \cdot \left[\arcsin(R_*/r) - (R_*/r)\sqrt{1 - (R_*/r)^2} \right] .
\end{aligned} \tag{11.52}
$$

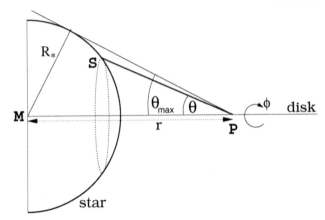

FIGURE 11.5 At position P on the disk, a unit area is illuminated by an infinitesimal surface element S of the star by the flux $B_\nu(T_*) \cdot d\Omega \cdot \sin\theta \sin\phi$. Seen from P, the surface element S subtends the solid angle $d\Omega = \sin\theta \, d\theta \, d\phi$, where θ is the angle between the lines $\overline{PM}$ and $\overline{PS}$. The factor $\sin\theta \sin\phi$ takes into account the inclination of the disk towards the incoming light. Here S is drawn near $\phi = 90°$ and the dotted line through S is a circle with $\theta = \text{const}$.

We express f_ν through an effective grazing angle, α_{gr}, defined by

$$f_\nu = \Omega_* \cdot B_\nu(T_*)\,\alpha_{gr} , \qquad (11.53)$$

where Ω_* is the solid angle of the star. At large distances ($x \equiv R_*/r \ll 1$), $\arcsin x - x\sqrt{1 - x^2} \simeq 2x^3/3$ and therefore $f_\nu = \frac{2}{3}x^3 B_\nu(T_*)$ and

$$\alpha_{gr} = \frac{4}{3\pi}\frac{R_*}{r} . \qquad (11.54)$$

To obtain the total flux F_ν falling on an infinite disk, we have to integrate (11.52) over the entire plane outside the star,

$$F_\nu = \int_{R_*}^\infty 2\pi r f_\nu \, dr = 2\pi \, B_\nu(T_*) \int_{R_*}^\infty \left[r \cdot \arcsin(R_*/r) - R_*\sqrt{1 - (R_*/r)^2} \right] dr$$
$$= \tfrac{1}{2}\pi^2 R_*^2 \, B_\nu(T_*) .$$

Therefore, each side of an infinite disk intercepts exactly one eighth of the stellar monochromatic luminosity $L_\nu = 4\pi R_*^2 \cdot \pi B_\nu(T_*)$, and thus also of the total luminosity $L_* = 4\pi R_*^2 \sigma T_*^4$, most of it close to the star (57% within $r = 2R_*$ and over 90% within $r = 9R_*$).

If the disk is a blackbody, it will have a surface temperature

$$T_{disk}^4(r) = \frac{T_*^4}{\pi}\left[\arcsin(R_*/r) - (R_*/r)\sqrt{1 - (R_*/r)^2} \right] . \qquad (11.55)$$

For $r \gg R_*$, the radiative heating rate has a $r^{-3/4}$-dependence,

$$T_{\text{disk}}(r) = T_* \left(\frac{2}{3\pi}\right)^{1/4} \left(\frac{R_*}{r}\right)^{3/4} . \qquad (11.56)$$

11.3.3.2 Inflated disk

When the gas in the disk is in hydrostatic equilibrium and gas and dust are well mixed, the disk height increases with the distance r from the star (figure 11.6); the disk is said to be inflated or flared. To derive the grazing angle α_{gr}, we define the height H by the condition that the visual optical depth A_V from there towards the star be one. If $\rho(z)$ denotes the dust density distribution in vertical direction and K_V the visual absorption coeffcient,

$$1 = \frac{K_V}{\alpha_{\text{gr}}} \int_H^\infty \rho(z)\, dz . \qquad (11.57)$$

The level $z = H$ represents the photosphere of the dust disk; it is the approximate place where stellar photons are absorbed. Because of flaring, the effective grazing angle α_{gr} of the incident radiation increases over the expression (11.54) by an amount (see figure 11.6)

$$\beta - \beta' = \arctan\frac{dH}{dr} - \arctan\frac{H}{r} \simeq \frac{dH}{dr} - \frac{H}{r} = r\frac{d}{dr}\left(\frac{H}{r}\right) ,$$

so that now we have to use in (11.53)

$$\alpha_{\text{gr}} = \frac{4}{3\pi}\frac{R_*}{r} + r\frac{d}{dr}\left(\frac{H}{r}\right) . \qquad (11.58)$$

If $H \propto r$, a cut through the disk (figure 11.6) is wedge-shaped, the illumination is very similar to the flat disk. If $dH/dr > 0$, the disk has the shape of a biconcave lens. The second term in (11.58) is then usually more important than the first. We estimate H by setting it equal to the disk scale height h assuming that the gas is isothermal in z-direction. The vertical component of the gravitational acceleration, g_z, is balanced by the pressure gradient $\partial P/\partial z$,

$$g_z = -\frac{z}{r}\frac{GM_*}{r^2} = \frac{1}{\rho}\frac{\partial P}{\partial z} .$$

Because $P = \rho kT/m$, where m is the atomic mass, one finds

$$\rho = \rho_0\, e^{-z^2/2h^2} \quad \text{with} \quad h^2 = kTr^3/GM_*m . \qquad (11.59)$$

The surface temperature of a blackbody disk, whether it is inflated or flat, has the form

$$T_{\text{disk}}^4(r) = \Omega_* \frac{\alpha_{\text{gr}}}{\pi} T_*^4 . \qquad (11.60)$$

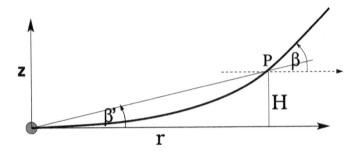

FIGURE 11.6 For a surface element P on a flared disk, the effective grazing angle α_{gr} of stellar light is enhanced with respect to the flat disk by $\beta - \beta'$. The fat line marks the top of the flared disk atmosphere; r and H are the radial distance and height above the mid-plane. The small shaded circle represents the star.

11.4   Galactic nuclei

11.4.1   Hot spots in a spherical stellar cluster

Another class of dusty objects whose infrared appearance one is eager to model and which possesses a fair degree of spherical symmetry is galactic nuclei, like the center of the Milky Way. Many galactic nuclei have large infrared luminosities as a result of internal activity. Its origin may be either rapid star formation or accretion onto a massive black hole. In either case, the hard primary radiation, from the star or the black hole, is converted into infrared photons by dust.

When the luminosity of the black hole dominates, one may, to first order, approximate it by a spherical central source with a power law emission spectrum and use the formalism outlined in section 11.2 for single stars. However, when the luminosity comes from a star cluster, the configuration is intrinsically three-dimensional because the dust temperature varies not only on a large scale with galactic radius r, which is the distance to the center of the stellar cluster, but also locally with the separation to the nearest star.

Loosely speaking, the temperature of a grain in the immediate vicinity of a star is determined by the distance to and the properties of that star, whereas a grain at the same galactic distance, but not very close to any star, absorbs only the mean interstellar radiation field $J_\nu^{\mathrm{ISRF}}(r)$ and will thus be cooler. The spherical symmetry is broken on a small scale because the immediate surroundings of a star constitute a hot spot (abbreviated HS). Nevertheless, an approximate one-dimensional treatment is possible [Krü78]. The presence of hot spots significantly changes the spectral appearance of a galactic nucleus.

Let us assume that the hot spots are spheres, each with a star of monochromatic luminosity L_ν at the center. It turns out that their total volume fraction is small, so they are distributed over the galactic nucleus like raisins in a cake. The radius R^{HS} of a hot spot is determined by the condition that outside, a dust grain is mainly heated by the interstellar radiation field (ISRF), whereas inside, heating by the star dominates. If τ_ν is the optical depth from the star to the boundary of the hot spot, R^{HS} follows from

$$\int Q_\nu J_\nu^{\mathrm{ISRF}}\, d\nu \;=\; \frac{1}{(4\pi R^{\mathrm{HS}})^2} \int Q_\nu L_\nu\, e^{-\tau_\nu}\, d\nu \;. \tag{11.61}$$

The radius R^{HS} of a hot spot is a function of galactic radius r, but it depends also on the monochromatic stellar luminosity L_ν, the dust distribution within the hot spot, and on the grain type via the absorption efficiency Q_ν.

11.4.2 Low and high luminosity stars

It is useful to divide the stars in a galactic nucleus into two categories, each containing identical objects. Stars of the first class have small or moderate luminosity, a space density $n^*(r)$ and a monochromatic and bolometric luminosity L_ν^* and L^*, respectively, where

$$L^* \;=\; \int L_\nu^*\, d\nu \;.$$

They represent the population of old stars. Stars of the second class are of spectral type O and B. They are young and luminous and were formed in a starburst. The corresponding values are denoted $n^{\mathrm{OB}}(r)$, L_ν^{OB} and L^{OB}.

- The low luminosity stars are very numerous, altogether typically 10^9, and the contribution of their hot spots to the overall spectrum may be neglected. To see why consider a nucleus of integrated luminosiy L_{nuc} containing N identical stars of luminosity L^*. When we fix $L_{\mathrm{nuc}} = NL^*$, but increase N and thus lower L^*, the intensity of the interstellar radiation field in the galactic nucleus is to first order independent of N. In view of the definition for R^{HS} in (11.61), one finds that $R^{\mathrm{HS}} \propto \sqrt{L^*} \propto N^{-1/2}$. Therefore, the total volume $N\,(R^{\mathrm{HS}})^3$ of all hot spots decreases as $N^{-1/2}$.

  A large population of low luminosity stars may be smeared out smoothly over the galactic nucleus and the structure of their hot spots need not be evaluated. To account for the radiation of these stars, one only has to introduce in the numerator of the source function (see 11.62) the volume emission coefficient

  $$\Gamma_\nu^*(r) \;=\; n^*(r)\, L_\nu^*$$

  If the stellar atmospheres are black bodies of temperature T^*, one may put $L_\nu^* \propto B_\nu(T^*)$. Note that L_ν^* need not be specified, only the product $\Gamma_\nu^*(r) = n^*(r)\, L_\nu^*$.

- The OB stars, on the other hand, are very bright and not so numerous. There are rarely more than 10^6 in a galactic nucleus and their space density $n^{OB}(r)$ is moderate, typically one star per pc^3 in the starburst region. The emission of their hot spots is not negligible and has to be evaluated explicitly.

Before solving the radiative transfer on a large scale in the galactic nucleus, one therefore has to determine for each galactic radial grid point the luminosity $L_\nu^{HS}(r)$ emerging from a hot spot. To this end, one calculates the radiative transfer of a spherical cloud centrally heated by an OB star. The cloud radius R^{HS} follows from (11.61). The hot spot is illuminated at its edge by the interstellar radiation field (ISRF). This fixes the outer boundary condition (11.43). At the inner boundary, we use (11.45) with the flux from the surface of the OB star. The volume emission coefficient due to the hot spots is

$$\Gamma_\nu^{HS}(r) = n^{OB}(r) \cdot L_\nu^{HS} \;.$$

The frequency integral $\int L_\nu^{HS} d\nu$ is, of course, equal to the luminosity $L^{OB} = \int L_\nu^{OB} d\nu$ of a single OB star, but L_ν^{OB} and L_ν^{HS} are different because much of the hard stellar UV flux is converted by dust into infrared radiation.

The source function in a spherical galactic nucleus with isotropic dust scattering is therefore (cf. 11.19)

$$S_\nu = \frac{\Gamma_\nu^{HS}(r) + \Gamma_\nu^*(r) + K_\nu^{abs} B_\nu(T_d(r)) + K_\nu^{sca} J_\nu^{ISRF}(r)}{K_\nu^{ext}} \;. \tag{11.62}$$

With this S_ν, one solves the integral equations (11.41) and (11.42) for the spherical nucleus as a whole along lines of constant impact parameter (see figure 11.2). The inner boundary condition is now $I^+ = I^-$ everywhere, and at the outer edge $I^- = 0$ (see 11.43 and 11.44). The resulting $I_\nu^+(r)$ and $I_\nu^-(r)$ yield the mean intensity $J_\nu^{ISRF}(r)$. The whole procedure has to be iterated to determine *a)* the dust temperatures T_d as outlined at the end of section 11.2.1; *b)* the size of the hot spots (see 11.61) and *c)* their internal structure.

As the galactic nucleus in this model consists of two phases, a dusty medium interspersed with hot spots, equations (11.41) and (11.42) are not strictly correct. They give, however, a good approximation as long as the volume fraction γ of the hot spots,

$$\gamma(r) = \frac{4\pi}{3} n^{OB}(r) \cdot \left[R^{HS}(r)\right]^3$$

is small; $\gamma(r)$ shrinks when $n^{OB}(r)$ becomes large. For typical space densities of OB stars in starburst nuclei, γ is of order 10^{-3}.

11.5 The pursuit of random photons

An alternative way to compute the radiative transfer is to follow the zig-zag path of many individual, but random photons, how they are scattered, destroyed through absorption and reborn by emission. This method has the advantage that it allows to handle complicated geometries. The technique is known by the name of *Monte Carlo*, after the notorious gambling place in Monaco. Although widely used, it has not lured astronomers to rashness in their financial affairs.

11.5.1 The strategy

In the description of the Monte Carlo technique we follow the ideas of [Luc99] and [Bjo01].

- Photons are binned in packets of equal energy ε, those of one packet having the same frequency ν_j. The subscript j runs from 1 to m_γ, so altogether there are m_γ different frequencies.

- Let the source (for example, a star) have a bolometric luminosity L. At each frequency point j, the star releases (per second) n photon packages. The total number of emitted packages is therefore $N = nm_\gamma$, and

$$L = N\varepsilon .$$

If the star has a spectral luminosity L_ν, the frequencies ν_j at which it emits are determined, for $j = 1, \ldots, m_\gamma$, from

$$\frac{j - \frac{1}{2}}{m_\gamma} = L^{-1} \int_0^{\nu_j} L_\nu \, d\nu , \qquad L = \int_0^\infty L_\nu \, d\nu .$$

The width of the frequency bins is $d\nu_j = \frac{1}{2}(\nu_{j+1} - \nu_{j-1})$ for $j = 2, \ldots, m_\gamma - 1$, $d\nu_1 = \frac{1}{2}\nu_2$ and $d\nu_{m_\gamma} = \nu_{m_\gamma} - \nu_{m_\gamma - 1}$.

- The cloud is partitioned into N_v small cells of volume v_i, $i = 1, 2, \ldots, N_v$, which absorb, scatter and emit radiation always in quanta of energy ε.

- When not interacting with the dust, a package of frequency ν_j traverses a cell on a straight line. Let $d\tau_j$ be the extinction optical depth between the entry and exit point of the cell (both on the cell surface) with respect to frequency ν_j. The packet is absorbed or scattered within the cell when

$$d\tau_j \geq -\log(z) , \tag{11.63}$$

otherwise it crosses the cell undisturbed. Here and in the following, z denotes a random number from the interval $[0,1]$.

- When a packet interacts in a cell, the probability for scattering is equal to the albedo, $A = K_\nu^{sca}/K_\nu^{ext}$, and the probability for absorption is $1-A$. In case of scattering, the packet retains its frequency, but changes direction. The new direction is in a probabilistic way determined by the phase function (see section 2.1.2 and 11.5.3). Upon absorption, the packet (usually) switches frequency; the emission is isotropic.

  In the numerical simulation, one follows the individual fate of all N packets as they diffuse through the cloud. When they finally cross its outer boundary, they are picked up by an observer. Because no packet gets lost, energy is exactly conserved, locally and globally.

- Initially, all cloud cells are at zero temperature. With each absorption, the temperature of a cell rises and after k such events it equals T_k which is determined from

$$\int K_\nu B_\nu(T_k)\, d\nu = \frac{k\varepsilon}{4\pi v} . \tag{11.64}$$

  K_ν denotes the volume absorption coefficient (we omit the superscript "abs") and v the cell volume.

- After the k-th absorption event, reemission of the cell occurs at the shortest frequency ν_j for which

$$\int_0^{\nu_j} K_\nu \frac{dB_\nu}{dT}\, d\nu \geq z \int_0^\infty K_\nu \frac{dB_\nu}{dT}\, d\nu \tag{11.65}$$

  where the Planck function is evaluated at $T = T_k$.

11.5.1.1 Iteration is not necessary

To understand (11.65) and its consequences, suppose the Monte Carlo method has worked and led to a *thermally relaxed* cloud for a star of luminosity L that has emitted N packages, so $L = N\varepsilon$. The emission of a cloud cell that has absorbed k packages and is at temperature T_k, according to (11.64), follows then the spectral distribution $K_\nu B_\nu(T_k)$ if the chance for releasing a packet at frequency ν_j is given by

$$w_{kj} = \frac{K_{\nu_j} B_{\nu_j}(T_k)\, d\nu_j}{\epsilon_k} , \qquad \epsilon_k = \int K_\nu B_\nu(T_k)\, d\nu . \tag{11.66}$$

The sequence of photons will be stochastic if the emission frequency ν_j is determined by the condition that ν_j be the smallest value for which

$$\int_0^{\nu_j} K_\nu B_\nu(T_k)\, d\nu \geq z\,\epsilon_k . \tag{11.67}$$

Let the star fire one more (extra) package, which raises its luminosity to $L = (N+1)\varepsilon$. When it is absorbed by a cell (it may be absorbed in many cells), the temperature there rises from the old equilibrium value T_k to

$$T_{k+1} = T_k + dT .$$

At T_{k+1}, the likelihood for radiation at frequency ν_j is

$$w_{k+1,j} = \frac{K_{\nu_j} B_{\nu_j}(T_{k+1})\, d\nu_j}{\epsilon_{k+1}}, \qquad \epsilon_{k+1} = \int K_\nu B_\nu(T_{k+1})\, d\nu .$$

However, as the MC method was assumed to be correct up to $L = \varepsilon N$, all previous k packets were emitted with the probability in (11.66), which is wrong for temperature T_{k+1}. One corrects for it by emitting the extra packet with the probability

$$p_{k+1,j} = K_{\nu_j} \frac{dB_{\nu_j}}{dT}\, d\nu_j \left[\int_0^\infty K_\nu \frac{dB_\nu}{dT}\, d\nu \right]^{-1} , \qquad (11.68)$$

where the derivative dB_ν/dT is taken at $T = T_k$. Such a probability implies that the random emission frequencies are given by (11.65). Putting $B_\nu' = dB_\nu/dT$ and $\epsilon_k' = d\epsilon_k/dT$, one gets, in view of $B_\nu(T_{k+1}) \simeq B_\nu(T_k) + B_\nu'(T_k)dT$,

$$\frac{K_{\nu_j} B_{\nu_j}(T_{k+1})}{\epsilon_{k+1}} \simeq \frac{K_{\nu_j} B_{\nu_j}(T_k)}{\epsilon_k} \frac{\epsilon_k}{\epsilon_{k+1}} + \frac{K_{\nu_j} B_{\nu_j}'(T_k)}{\epsilon_k'} \frac{\epsilon_k'\, dT}{\epsilon_{k+1}} .$$

The $(k+1)$ photon packets, taken together, have therefore the right chance of being radiated at frequency ν_j by a cell of temperatur T_{k+1}: the weight of the first k packets is $\epsilon_k/\epsilon_{k+1} = k/(k+1)1$, and the weight of the last one is $(\epsilon_{k+1} - \epsilon_k)/\epsilon_{k+1} = 1/(k+1)$.

In this way, the cloud shifts in the course of the computation from one equilibrium configuration, appropriate for the luminosity $L = N\varepsilon$, after emission of the extra package, to a new one valid for the luminosity $L' = (N+1)\varepsilon$, and so it carries on as N increases. The calculation may start from any thermal equilibrium configuration, for example, from $L = T = 0$. Because the cloud is always relaxed during the numerical simulation, iteration is not necessary, unlike the procedures described in section 11.2. Of course, the photon package energy ε should always be small compared to the thermal energy content of the cells.

Besides not needing iteration, the method has two other advantages:

- One may stop the Monte Carlo calculation at any moment (provided $N \gg N_v$). When by that time N packets have been launched, one has the correct spectrum and temperature structure for a source of luminosity $L = N\varepsilon$, and one gets for free the spectra and temperature distributions at all smaller luminosities $L' = N'\varepsilon$ with $N' < N$.

- One may place, after the stop, another star elsewhere in the cloud and continue the computation. If it emits N' packages of energy ε', the end result is that for a cloud with two sources of luminosity $N\varepsilon$ and $N'\varepsilon'$, respectively. The method is "commutative"; it does not matter with which source one begins. The number of extra sources is not limited.

11.5.2   Grains with temperature fluctuations

Dust particles with temperature fluctuations, like PAHs, may be included in the Monte Carlo scheme. The PAHs in a cell do not have a uniform temperature, like big grains, but a distribution $P(T)$ and their emissivity ϵ_ν at frequency ν is (see 6.46)

$$\epsilon_\nu = K_\nu \int_0^\infty P(T) B_\nu(T)\, dT\ .$$

We sketch two ways of how to treat PAH emission, the first is "exact", the second approximate.

- For grains at constant temperature, we wrote the emissivity after absorption of k packets as

$$\epsilon_\nu(T_k) \simeq \epsilon_\nu(T_{k-1}) + \frac{d\epsilon_\nu}{dT}\, dT$$

where $dT = T_k - T_{k-1}$ and $d\epsilon_\nu/dT = K_\nu\, dB_\nu/dT$. In case of T-fluctuations, we expand ϵ with respect to k,

$$\epsilon_\nu(k) \simeq \epsilon_\nu(k-1) + \frac{d\epsilon_\nu}{dk}\, dk\ .$$

Let $P_k(T)$ be the corresponding temperature distribution. It is numerically represented by the probabilities P_{kl} for finding an arbitrary PAH after k absorptions at temperature T_l. The sum $\sum_l P_{kl} = 1$, the index l runs from 1 to n, which is the number of temperature bins. When a packet of frequency ν_k is swallowed in a cell with r (identical) PAHs, they make altogether $\epsilon/h\nu_k$ upward (heating) transitions. Those in the temperature bin T_m will be afterwards in the bin T_l, where l follows from $\nu_l = \nu_m + \nu_k$. If $A_{lm}^{(k-1)}$ is the transition matrix after $(k-1)$ packages with frequencies $\nu_1, \ldots, \nu_{k-1}$ have been absorbed, then

$$A_{lm}^{(k)} = A_{lm}^{(k-1)} + \begin{cases} \dfrac{\varepsilon}{r h \nu_k} & : \quad \text{if}\quad \nu_l = \nu_m + \nu_k \\[2mm] 0 & : \quad \text{else}\ . \end{cases} \tag{11.69}$$

Applying the recipes of section 6.5.2, the P_{kl} are computed from the matrix equation (6.47),

$$\sum_{l=1}^{n} A_{ml}^{(k)} P_{kl} = 0 \quad \text{for}\quad m = 1, \ldots, n \tag{11.70}$$

with the definition $A_{mm}^{(k)} = -\sum_{l \neq m} A_{lm}^{(k)}$. For cooling, which is assumed to proceed only from level $l \to l - 1$ (see 6.51), the matrix elements are

$$A_{l-1,l}^{(k)} = \int_0^\infty C_\nu B_\nu(T_l) \, d\nu \cdot \left[h(\nu_l - \nu_{l-1}) \right]^{-1} .$$

They do not depend on k. C_ν is the cross section of a single PAH.

The temperature T_k and the emission frequencies of *non-fluctuating* grains are computed from (11.64) and (11.67), which is easy as the integrals can be tabulated and $dB_\nu(T)/dT$ is an analytical expression. For PAHs, the computations are more laborious. The emission frequency ν_j follows now, in analogy to (11.67), from the condition that ν_j be the smallest value for which

$$\int_0^{\nu_j} d\nu K_\nu \int_0^\infty B_\nu(T) \frac{dP}{dk} \, dT \geq z \int_0^\infty d\nu K_\nu \int B_\nu(T) \frac{dP}{dk} \, dT ,$$

where $dP/dk \simeq P_k(T) - P_{k-1}(T)$. So one has to remember for each cell the previous probability density $P_{k-1}(T)$ and matrix $A_{ml}^{(k-1)}$, then determine the new $A_{ml}^{(k)}$ from (11.69) and the new $P_k(T)$ from (11.70). As the matrix $A_{ml}^{(k)}$ should consist of not less than 50×50 elements, the demands on the storage capacity of the computer are very heavy.

- To avoid storage problems and shorten the computing time, we let the PAHs be excited by stellar photons only. Such a restriction is acceptable because temperature excursions of PAHs result from the impingement of hard photons whereas the radiation by dust is soft.

  Suppose that under this assumption N_{PAH} stellar packets are absorbed by PAHs and the remaining $N' = N - N_{\text{PAH}}$ are not. The flight paths of the latter are calculated as usual. When one of the N_{PAH} stellar packets is captured by PAHs, we remember its frequency and the cell where it happened, but do not follow the flight path to its end. Instead, we continue with the emission of the next stellar packet.

  When the star has then released N packages, one arrives at an equilibrium configuration without PAH emission that corresponds to a source of luminosity $\varepsilon N'$. In a second step, one regards each cell in which $k > 0$ packets were absorbed by PAHs as an independent source besides the star. Their total luminosity is, of course, $N_{\text{PAH}}\varepsilon$. A cell with PAH absorption has a luminosity $k\varepsilon$ and emits k packets at frequency ν_j given by

  $$\int_0^{\nu_j} d\nu K_\nu \int_0^\infty B_\nu(T) \, P_k(T) \, dT \geq z \int_0^\infty d\nu K_\nu \int B_\nu(T) \, P_k(T) \, dT .$$

  Their flight paths are followed until they leave the cloud. This simple procedure is possible *a)* because the final temperature profile of the cloud and its spectral appearance are independent of the order in which photons are emitted and *b)* because there is no restriction to the location of sources.

11.5.3 Anisotropic scattering

It is straightforward to incorporate anisotropic scattering in the Monte Carlo method. Let the unit vector

$$\mathbf{e} = (\sin\psi\cos\phi,\ \sin\psi\sin\phi,\ \cos\psi)$$

denote the direction of propagation of the photon before scattering. The direction $\mathbf{e_s}$ after scattering will form with $\mathbf{e}$ an angle β. The vector $\mathbf{u}$ in figure 11.7, which lies in the same plane as the z-axis and $\mathbf{e}$, is one such particular example of $\mathbf{e_s}$. All possible vectors $\mathbf{e_s}$ are found by rotating $\mathbf{u}$ around $\mathbf{e}$ by an angle α. After three rotations of the Cartesian coordinate system one finds

$$\mathbf{e_s} = \begin{pmatrix} \cos\phi[\sin\psi\cos\beta - \cos\psi\cos\alpha\sin\beta] - \sin\phi\sin\alpha\sin\beta \\ \sin\phi[\sin\psi\cos\beta - \cos\psi\cos\alpha\sin\beta] + \cos\phi\sin\alpha\sin\beta \\ \cos\psi\cos\beta + \sin\psi\cos\alpha\sin\beta \end{pmatrix}$$

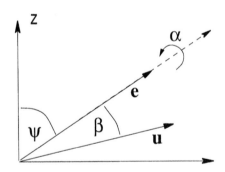

FIGURE 11.7 Light travels along the unit vector $\mathbf{e}$ and is scattered by an angle β. The scattering directions lie on a cone around $\mathbf{e}$ with an opening angle β.

Of course, $|\mathbf{e_s}|^2 = 1$. The scattering pattern is determined by the phase function $f(x)$ of (2.6) with $x = \cos\beta$. The new photon direction after scattering is given by the angles β and α which follow from

$$z = \tfrac{1}{2}\int_{-1}^{\cos\beta} f(x)\,dx$$

and $\alpha = 2\pi z$, where $z \in [0,1]$ is a random number.

11.5.4 Practical considerations

11.5.4.1 The choice of the grid

Before formulating the radiative transfer, one must choose an appropriate coordinate system. When dealing with an object that possesses strong symmetries, like a protostar fed by a disk and emitting a jet perpendicular to the disk, two dimensions are sufficient. But when there are several sources or when the medium is clumpy, the geometry is instrinsically three-dimensional. Using then 3d-Cartesian coordinates, has the advantage that the path element is simple,

$$ds = \sqrt{dx^2 + dy^2 + dz^2} \, ,$$

and that the orientation of the local with respect to the global coordinate system does not change.

The cloud has the shape of a rectangular parallelepiped and is made up of identical cubes (figure 11.8). Where higher resolution is required, the cubes are divided into subcubes and the latter, if needed, into still smaller subcubes, and so forth. The subcubes are the cells we spoke of above, and within them, the dust density and temperature are constant. The number of cells can be reduced by a factor of eight by assuming mirror-symmetry with respect to the planes $x = 0$, $y = 0$ and $z = 0$.

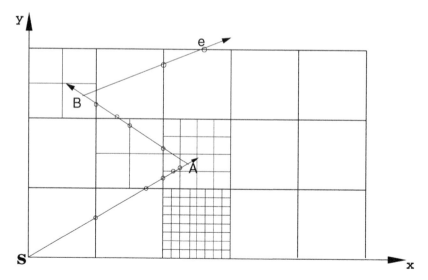

FIGURE 11.8 Photon track in Cartesian coordinates (plotted for two dimensions). Space is partitioned into cubes which may be further divided into subcubes of various degrees of fineness. The star at **s** emits a photon packet that is absorbed in subcube A and reemitted in a new direction. In subcube B, it is scattered and exits the cloud at e. The path is followed piece-wise between the intersection points of the ray with the cube or subcube surfaces.

- As the sources are usually much hotter than the cloud, one must use different frequency grids for the star(s) and the dust; the energy quantum, ε, is, of course, the same for both. For example, with $m_\gamma = 10^4$ frequency points, the photon packages of a star with $T_* = 10^4$K have wavelengths between 0.08 and 20μm, which is not sufficient to desccribe the dust emission.

- The packets escaping the cloud are binned into frequency intervals appropriate for comparison with observations. If one wants to map the (model) source or study how its appearance changes with viewing angle, further binning of the outgoing photons is needed with respect to their direction and the position on the surface from where they are leaving (see figure 11.8).

- If there are $g > 1$ different kinds of grains with extinction coefficients K_l^{ext} ($l = 1, 2, \ldots, g$), the probabilty that a photon packet interacts in a cell with the specific grain type s is $K_s^{\text{ext}} / \sum_l K_l^{\text{ext}}$.

- Because luminosity is conserved, the computed spectrum is never totally wrong, although the code may contain severe bugs. The program must therefore be thoroughly checked in benchmark tests.

12

Spectral energy distribution of dusty objects

We have learnt how cosmic grains absorb and scatter light and how they emit. In this final chapter, we apply our faculties to diverse astronomical objects. We compute and discuss the emission from globules, protostars, accretion disks, reflection nebulae, extragalactic starburst nuclei and giant stars with dusty winds. The last section is devoted to dust extinction in a clumpy medium.

Cosmic dust is now not the object of our study, but the agent by which we wish to disclose the properties and structure of a (dust related) source. Its spectrum, which is usually obtained at low resolution ($\lambda/d\lambda \lesssim 50$) and refers to the continuum, not to lines, is also called SED (= spectral energy distribution). The SED yields not only, in an obvious fashion, the infrared luminosity or an estimate of the dust temperature. Analyzing the SED by means of radiative transfer calculations often allows one to get an idea of the three-dimensional distribution of the dust with respect to the embedded star(s). For this purpose, one constructs models that are supposed to resemble the source and varies the model parameters until agreement between theory and observation is achieved. Fitting an extensive observational data set is laborious and requires occasional intuition. To solve such a task, a senior scientist needs a good student (the reverse is not true). Most of the spectra below are merely sketches, for the purpose of illustrating the magnitude of the fluxes one expects to receive.

12.1 Early stages of star formation

12.1.1 Globules

Usually, star formation is a contagious process. When a star is born in a molecular cloud, it disturbs the interstellar medium in its vicinity through its wind, jet or radiation pressure, and then neighboring clumps are pushed beyond the brink of gravitational stability, too. The isolated birth of a star is a rare event, but it can happen in globules. These are cloudlets that have been ripped off from larger complexes. They have masses of $\sim 10\,M_\odot$ and appear as dark spots on photographic plates. Globules are bathed in a radiation

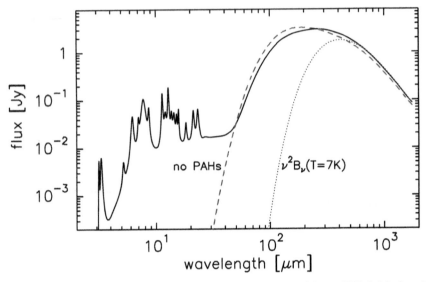

FIGURE 12.1 Dust emission of a globule irradiated by a UV field that is four times stronger than the interstellar radiation field (ISRF) of figure 6.1. The spectra show the difference in the signal towards the source and in the off-beam where one still sees the external radiation field. The dashed line gives the flux when PAHs and vsg are absent, the dotted curve the emission from dust at 7K.

field which heats them from outside and we want to evaluate their infrared emission in the absence of an internal source.

Let us consider a spherical globule of $2.6\,M_\odot$ at a distance of 200 pc with a diameter $2r = 1.2 \times 10^{17}$ cm. The hydrogen density n_H equals $4 \times 10^6\,\text{cm}^{-3}$ in the inner part ($r \le 3 \times 10^{16}$ cm) and declines linearly to $2 \times 10^6\,\text{cm}^{-3}$ at the surface. Using the standard dust model described in section 10.2, the overall visual optical thickness through the cloud center amounts to 280 mag, so the globule is very opaque.

We compute the radiative transfer for an external radiation field J_ν^{ext} that is, when integrated over frequency, four times stronger than the ISRF of figure 6.1. Such excursions in the radiative intensity can produce remarkable effects and occur when the globule is near a star (for example, at a distance of 1.5 pc from a B3V star). We assume that J_ν^{ext} consists of two components: a hard one described by a diluted blackbody of $T = 18\,000$ K that is meant to represent the UV flux from star(s), and a very soft one of spectral shape $\nu^2 B_\nu(T = 20\text{K})$ that mimics the ubiquitous far infrared radiation. The grain temperatures lie between 17 K and 27 K at the surface of the globule where UV photons are responsible for heating, and around 7 K in the cloud center.

The globule in figure 12.1 appears observationally first of all as a very cold submillimeter source. Its emission has a broad maximum around 250μm. The millimeter spectrum comes from the coldest dust; it is fit by a modified Planck curve (subsection A.4.3) of only 7 K. There is a skin layer of thickness $A_V \simeq 1$mag where the big grains are much warmer ($\sim$20K) and these grains account for the flux between 50 and 200μm.

The presence of very small grains (vsg) and PAHs in the skin layer has the remarkable effect that the source becomes visible in the mid infrared. As discussed in section 6.5, such tiny grains with heat capacities of merely a few eV make a big upward leap every time they absorb an ultraviolet photon and this produces an emission spectrum entirely different from big grains. Although the tiny grains account for only a few percent of the total dust mass, they absorb some 20% to 30% of the UV light and their excitation therefore radically changes the infrared appearance of the source. Most noticeable are the PAH features (section 9.1). Without vsg and PAHs, the globule would only be detectable in the far infrared and beyond.

The presence of the PAHs has the further interesting consequence that it leads to important heating of the interior by mid infrared photons. Without them, the core would be extremely cold ($T_d \simeq 5$ K). A rise in dust temperature from 5 K to 7 K implies an increase in the heating rate by a factor $(7/5)^6 \simeq 8$ (see 6.15).

12.1.2 Isothermal gravitationally-bound clumps

A homogenous spherical cloud of mass M and temperature T, consisting of atoms of mass m, has at the brink of stability, when thermal and gravitational energy balance, a radius

$$R \simeq \frac{GMm}{5kT} .$$

This is known as *Jeans criterion*. A typical optical thickness of the clump is

$$\tau_\lambda \simeq 2\rho K_\lambda R ,$$

where ρ is the gas density and K_λ the absorption coefficient at wavelength λ per gram of interstellar matter; it is due to dust. The optical depth becomes

$$\tau_\lambda = \frac{3K_\lambda}{2\pi} \left(\frac{5k}{Gm}\right)^2 \frac{T^2}{M} \simeq 10^{26} \frac{T^2}{M\lambda^2} .$$

To derive the factor 10^{26} (for cgs-units), we used the far infrared dust absorption coefficient (10.16), cgs units for the clump mass and the wavelength, and assumed a mean molecular mass $m = 2.34m_H$ corresponding to a normal mixture of H_2 molecules and helium atoms. At a distance D, the clump subtends a solid angle

$$\Omega = \frac{\pi R^2}{D^2}$$

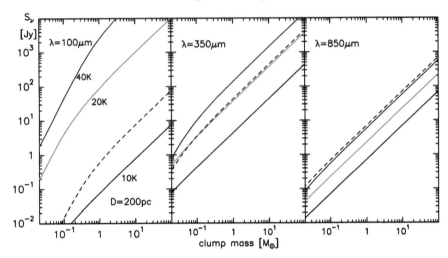

FIGURE 12.2   The flux S_ν after (12.1) emitted by dust from a gravitationally-bound, homogeneous and isothermal clump at 100μm, 350μm and 850μm. S_ν is shown as a function of cloud mass for three dust temperatures (solid curves). The dashed curve in each box refers to a clump at $10\,$K with tenfold the absorption coefficient as a result of grain coagulation and accretion of ice mantles (see figure 10.9). The source is at a distance of 200pc.

and an observer receives the flux

$$S_\lambda \;=\; \Omega \cdot B_\lambda(T)\left[1 - e^{-\tau_\lambda}\right] . \tag{12.1}$$

- If the clump is transparent $(\tau_\lambda \to 0)$: $S_\lambda \;=\; \dfrac{M K_\nu B_\lambda(T)}{D^2}$,

- if the clump is opaque $(\tau_\lambda \to \infty)$: $S_\lambda \;=\; \dfrac{\pi R^2 \, B_\lambda(T)}{D^2}$.

Cold clumps are best detected by their submillimeter dust emission. The flux S_λ of (12.1) is plotted in figure 12.2 for various temperatures and wavelengths as a function of clump mass M assuming a source distance $D = 200\,$pc.

When the optical depth is below one, which is always true at 850μm or longer wavelengths, the flux is simply proportional to the cloud mass. When τ_λ is large, which happens at $\lambda = 100\mu$m for small masses, $S_\lambda \propto M^2$. The break in the slope of the flux S_λ where the dependence on mass M changes from linear to quadratic occurs at $\tau_\lambda \sim 1$.

12.1.3   The density structure of a protostar

Protostellar clumps are strongly obscured and their emission is not detectable at optical or near infrared wavelengths. Their internal structure, whose knowledge is necessary to understand the dynamics of the protostellar collapse, must be derived from dust emission at long wavelengths (mm/submm). The conversion of measured flux into column density is, however, problematic because of the influence of the dust temperature ($\epsilon_\nu = K_\nu B_\nu(T_{\mathrm{d}})$).

The extinction of background light, on the other hand, is independent of the dust temperature. Measuring in the visual or near IR the attenuation of background stars by a foreground cloud can yield its extinction optical depth, τ^{ext}. For a dense clump, the method may still work at the cloud rim, but towards its core, one will only get a lower limit to τ^{ext}. Occasionally, however, there is strong background radiation in the *mid infrared*, and then one can determine the extinction over the whole clump, not only its rim, because mid infrared photons penetrate much deeper.

An example of such a situation is presented by the protostellar object HH108MMS. A mid infrared map of its surroundings is depicted in figure 12.3 [Sie00]. HH108MMS is the white (weak signal) spot at the map center. No far infrared luminosity is detected towards it; an embryo star has therefore not yet formed. HH108MMS appears at 14μm in *absorption* against the diffuse background and this offers the opportunity to derive from the variation of the surface brightness the optical depth and the density profile in the protostellar clump.

The geometrical configuration for our analysis is drawn in figure 12.4. Three regions contribute to the observed 14μm flux:

- The absorbing core (protostar) of optical depth τ_2 and temperature T_2.

- The molecular cloud (parameters τ_1 and T_1) which surrounds the core.

- The background with intensity I_0. The modeling of a globule in section 12.1.1 suggests that this radiation is due to emission by PAHs and very small grains heated in the surface layer on the *backside* of the molecular cloud.

The molecular cloud and the absorbing core are certainly cold ($T < 20\,\mathrm{K}$) and have no mid infrared emission of their own ($B_{14\mu\mathrm{m}}(T_i) \simeq 0$, $i = 1, 2$). The intensity in the direction of the absorbing core (source) can therefore be written as $I_{\mathrm{s}} = I_0\, e^{-(\tau_1 + \tau_2)}$, and towards the background, at a position next to the absorbing core, $I_{\mathrm{b}} = I_0 e^{-\tau_1}$, so one gets

$$I_{\mathrm{s}} - I_{\mathrm{b}} = I_0\, e^{-\tau_1} \cdot \left[e^{-\tau_2} - 1\right] . \tag{12.2}$$

I_{s} and I_{b} are known from observations. To find the optical depth τ_2 across the protostar, let us denote the maximum value of τ_2 by $\tau_{2,\mathrm{max}}$ and the corresponding intensity by $I_{\mathrm{s,max}}$. The 1.3mm flux towards that position yields for the central beam an average optical depth $\tau_{1.3\mathrm{mm}} \sim 10^{-2}$ using equation (6.18). The dust opacity appropriate for cold clumps is depicted by the upper solid line of figure 10.9; it gives $K_{14\mu\mathrm{m}}/K_{1.3\mathrm{mm}} \sim 1000$. The optical depth

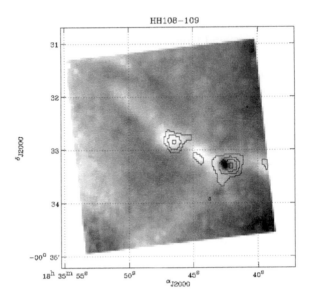

FIGURE 12.3 A 14μm image around the protostar HH108MMS. It is the white blob in the center seen in ***absorption*** against a diffuse background. The contour lines show the 1.3mm dust emission [Chi97]. HH108MMS shows up as a millimeter source. The other one, on the lower right, is a young stellar object (IRAS18331–0035) of $\sim 3\,L_\odot$ which is still heavily obscured at optical wavelengths, but ***emits*** in the mid infrared (black blob).

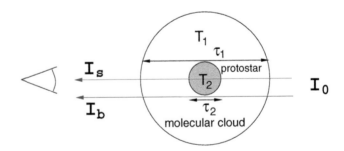

FIGURE 12.4 A sketch of how HH108MMS is embedded into a larger molecular cloud. The intensity towards the source (HH108MMS) and the background is denoted I_s and I_b, respectively. I_0 is the intensity at the edge (or behind) the molecular cloud.

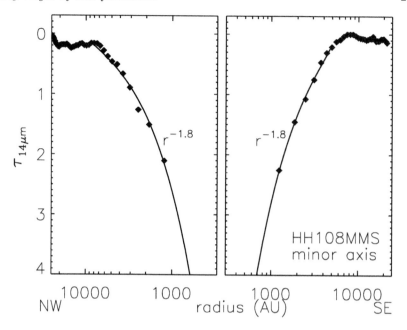

FIGURE 12.5 Optical depth at 14μm (diamonds) in HH108MMS along the minor axis. The solid lines show the variation of the optical depth in a spherical cloud with a power law density distribution $\rho \propto r^{-1.8}$.

$\tau_{2,\text{max}}$ is therefore of order 10 and well above unity. Towards the position of maximum obscuration, equation (12.2) can therefore be approximated by

$$I_0\, e^{-\tau_1} = I_\text{b} - I_{\text{s,max}} \ .$$

Inserting $I_0 e^{-\tau_1}$ into (12.2) yields the desired expression for the optical depth in the core as a function of position,

$$\tau_2 = -\ln\left[1 \ - \ (I_\text{b} - I_\text{s})/(I_\text{b} - I_{\text{s,max}})\right] \ . \tag{12.3}$$

The resulting profile is plotted in figure 12.5. If one converts it into density profile assuming that the protostellar clump is radially symmetric, one finds that the density structure is very well fit by a power-law $\rho(r) = \rho_0\, r^{-\alpha}$ with exponent $\alpha = 1.8$ (and a similar value for the major axis). According to the theory of protostellar collapse without rotation and magnetic fields, the density distribution changes from $\rho \propto r^{-2}$ at the earliest epoch towards $\rho \propto r^{-3/2}$ during the main accretion phase. The analysis of the extinction of background light by HH108MMS therefore confirms the theoretical predictions, at least, up to an optical thickness at 14μm of 4 (see figure 12.5) corresponding to a visual extinction of 200 mag.

12.2   Accretion disks

All protostars are surrounded by disks. They feed them, govern the rate at which the embryo star accretes, remove angular momentum, produce jets and determine the spectral appearance of the protostar. But, last not least, dust grains within them sometimes grow to the size of planets. So disks are important. Below we model their spectral energy distributions in various degrees of sophistication assuming that their only source of energy comes from the illumination by the star (passive disks).

12.2.1   Flat disks

12.2.1.1   The disk surface is a blackbody

Disks of young solar type stars (T Tauri stars) are during the main accretion phase optically very thick ($A_V \gtrsim 10^4$ mag). As a first approximation, we therefore assume that the disk surface is a blackbody. The flux that one observes can then be computed from

$$S_\nu = \frac{\cos i}{D^2} \int_{R_*}^{R_{out}} 2\pi r\, B_\nu(T)\, dr \; , \tag{12.4}$$

where D is the distance, $T(r)$ the temperature at radius r from (11.55) and i the viewing angle. When one looks face-on, $\cos i = 1$. In the example displayed in figure 12.6, the star has a luminosity $L_* = 2\,L_\odot$, an effective temperature $T_* = 4070$ K and the outer disk radius, R_{out}, is 20 or 100 AU. The total disk luminosity equals $\frac{1}{4}\,L_* = 0.5\,L_\odot$ (see section 11.3.3). The spectrum is dominated by the ridge between 2μm and 200μm. Its slope is constant as a result of the power law distribution of the temperature. Indeed, from $T(r) \propto r^{-q}$ follows

$$S_\nu \propto \int 2\pi r \cdot B_\nu(T)\, dr \propto \nu^{3-\frac{2}{q}} \int \frac{x^{\frac{2}{q}-1}}{e^x - 1}\, dx \; , \tag{12.5}$$

so that in our case where $q = 3/4$ (see 11.56),

$$S_\nu \propto \nu^{1/3} \; .$$

In the millimeter region, $S_\nu \propto \nu^2$. The flux has the same slope as the Planck function in the Rayleigh-Jeans part, but only because of the finite disk radius which sets a lower boundary to the temperature. If the disk were infinite, T would go to zero as $r \to \infty$ and $S_\nu \propto \nu^{1/3}$ would hold also at very low frequencies. So the radial cutoff, although irrelevant for the energy budget, determines the shape of the long wavelength spectrum.

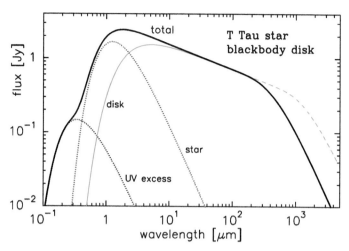

FIGURE 12.6 The SED of a T Tauri star with a flat blackbody disk viewed pole-on. Stellar parameters: $L_* = 2\,L_\odot$, $T_* = 4070\,\mathrm{K}$, $R_* = 2 \times 10^{11}$ cm; inner and outer disk radius: $R_\mathrm{in} = R_*$, $R_\mathrm{out} = 20\,\mathrm{AU}$; distance $D = 200\,\mathrm{pc}$. The heavy solid line includes the star and the UV excess (see text). If the disk is bigger and $R_\mathrm{out} = 100\,\mathrm{AU}$, the spectrum changes only beyond 200μm (dashed curve).

T Tau stars show a UV excess which is attributed to the thermalization of the kinetic energy of the accreted disk material. The excess is included in the SED of figure 12.6. The associated UV luminosity, L_UV, depends on the mass and radius of the star and the accretion rate. We assume, somewhat arbitrarily, $L_\mathrm{UV} = 0.33\,L_\odot$ in the form of $15\,000\,\mathrm{K}$ blackbody emission.

12.2.1.2 The disk is not a blackbody but isothermal in z-direction

The vertical optical thickness in the disk, $\tau_\nu^\perp$, (in z-direction, parallel to the rotation axis) is the product of the surface density, Σ, times the mass absorption coefficient per gram of circumstellar matter, K_ν,

$$\tau_\nu^\perp(r) \;=\; \Sigma(r)\, K_\nu \ .$$

For a disk of arbitrary optical depth that is isothermal in z-direction, the observed flux equals

$$S_\nu \;=\; \frac{\cos i}{D^2} \int_{R_*}^{R_\mathrm{out}} 2\pi r \, B_\nu(T(r)) \cdot \left[1 \,-\, e^{-\tau_\nu}\right] dr \qquad (12.6)$$

with $\tau_\nu = \tau_\nu^\perp / \cos i$. Compared with (12.4), there is the additional term $[1 - e^{-\tau_\nu}]$. If τ_ν is large, formula (12.4) and (12.6) are identical. At wavelengths where the emission is optically thin, the factor $\cos i$ due to the disk inclination drops out and the flux S_ν is directly proportional to the disk mass M_disk.

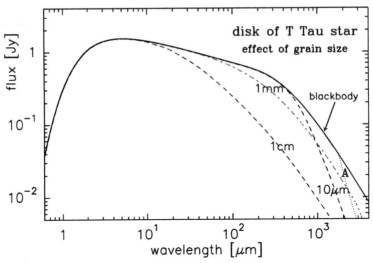

FIGURE 12.7 The spectra of a disk with a radial temperature profile $T(r)$ as in figure 12.6. The flux is computed from (12.6); the solid line shows, for comparison, the blackbody curve of figure 12.6. There is only silicate dust; the grain radius, a, is indicated. Label **A** stands for $a = 100\mu$m.

Before we can compute the flux from (12.6), we have to specify the surface density $\Sigma(r)$. Let us assume a power law distribution

$$\Sigma(r) = \Sigma_0 r^{-1}$$

which implies that the disk mass grows linearly with radius. The mass absorption coefficient, K_λ, must be calculated from Mie theory. One cannot use the approximation $K_\lambda \propto \lambda^{-2}$ of (6.9) because the grains in a disk may be bigger than the wavelength as a result of coagulation.

We apply (12.6) to a T Tau disk with parameters and radial temperature profile $T(r)$ as in figure 12.6, a surface density $\Sigma(r) = 10^{16} r^{-1}$ (cgs units) and a dust-to-gas ratio of $R_d = 1/150$. The disk mass (gas plus dust) out to 20 AU is then $\simeq 0.01\,M_\odot$.

12.2.1.3 The influence of grain size on the SED

If the disk is filled with dust particles of arbitrary radii, but all smaller than 10μm (like interstellar grains), the resulting SED is the curve in figure 12.7 labeled "10μm". It departs from the blackbody curve of figure 12.6 only longward of $\lambda_0 \simeq 500\mu$m. For $\lambda > \lambda_0$, the flux S_ν falls proportionally to ν^4, and not to ν^2, as in the Rayleigh-Jeans limit of the Planck function.

To study the influence of the particle size on the spectral energy distribution, we vary the grain radius, a, while keeping the surface density, $\Sigma(r)$, and all other parameters fixed. We now assume that all grains have equal size. Let λ_0, as above, denote the wavelength where the outer, more tenuous parts of the disk become transparent and the total disk emission falls below the blackbody case.

When a is increased from 10μm in steps of ten to 100μm, 1 mm and 1 cm, λ_0 first shifts to the right (curve tagged A where $a = 100\mu$m), and then to the left ($\lambda_0 \simeq 10\mu$m for $a = 1$ cm). The value of λ_0 is thus in this simple minded model a discriminator of particle size. We also note that for very big grains ($a = 1$ cm), the spectrum steepens very gradually as the frequency decreases.

12.2.2 Inflated disks

We refine the disk model by discarding the assumption of isothermality in z-direction. The temperature $T(z)$ is now determined from the radiative transfer outlined in section 11.3. As an example, we choose a main sequence star of type B9V ($L_* = 100\,L_\odot$, $M_* = 3\,M_\odot$, $T_* = 12\,000$ K, $R_* = 2.76\,R_\odot$).

The disk is now also inflated and the grazing angle α_{gr} taken from (11.58). The surface density $\Sigma(r) = \Sigma_0\,r^{-3/2}$ with $\Sigma_0 = 4.1 \times 10^{22}$ cgs. This implies that there are $0.01 M_\odot$ of gas within 100 AU and that at one astronomical unit from the star, the visual optical thickness of the disk due to dust is $\tau_V \simeq 2.1 \times 10^5$ ($R_d = 1/150$). The dust model is from section 10.2, except that the MRN grains have an upper size limit $a_+ = 1\mu$m. They are thus, on the average, a bit bigger because of likely coagulation in the disk.

We first look at the emission from a narrow annulus thus avoiding the complexity that arises from the radial variations in temperature and surface density. The flux from an annulus of radius $r = 20$ AU, where the grazing angle $\alpha_{gr} = 3.32°$, is displayed in figure 12.8. The temperature in the mid plane ($z = 0$) is for all sorts of grains equal, $T_{mid} = 91$ K. At the disk surface, the dust temperatures vary between 210 and 510 K. A blackbody would after (11.60) acquire a temperature of 115 K.

The solid curves in figure 12.8 result from a full radiative transfer in z-direction and include PAHs and very small grains (vsg). The viewing angle i for the upper solid curve is $90°$, for the two lower ones, $i = 78°$ and $52°$. In the far infrared, the disk annulus is opaque up to $\sim 150\mu$m. The flux is then reduced because of geometrical contraction if one looks at the disk from the side ($i < 90°$). In the submillimeter region, the disk becomes translucent and the curves begin to converge. At 1 mm (not shown), the dependence on the position of the observer has disappeared and the slope approaches $S_\nu \propto \nu^4$. Note that an increase in surface density Σ would raise the submillimeter emission without changing much in the spectrum below 50μm. For $\lambda \le 15\mu$m, the solid lines coincide. At these wavelengths, one sees only the hot dust from a thin and transparent surface layer. The dashed line, again for an inclination angle $i = 90°$, is calculated without very small particles. It merges with the

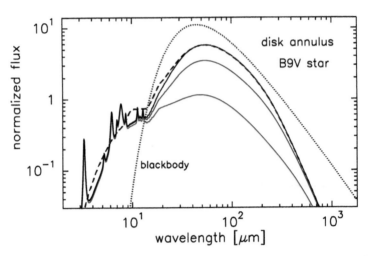

FIGURE 12.8 The flux, in arbitrary units, from a narrow annulus of a disk around a B9V star. The radius of the annulus is 3×10^{14} cm. The model and its parameters are expounded in the text and in figure 12.9. The dots show a blackbody surface. The dashed line is for dust without PAHs or vsg. The full curves refer to inclination angles $i = 90°$ (face-on), 78° and 52°, respectively.

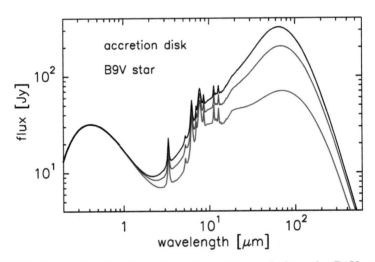

FIGURE 12.9 The flux from the whole disk, including the B9V star, at inclination angles $i = 90°$, 78° and 52°. Stellar distance $D = 100$ pc, $R_{\rm in} = 0.1$ AU, $R_{\rm out} = 100$ AU, surface density $\Sigma = \Sigma_0/r^{3/2}$ with $\Sigma_0 = 4.1 \times 10^{22}$ cgs so that at 1 AU, $\Sigma = 700$ g/cm^2 and $\tau_{\rm V} \simeq 2.1 \times 10^5$.

upper solid curve for $\lambda > 30\mu$m. The dotted line shows, for comparison, the flux from a blackbody annulus $T = 115$ K viewed face-on. The integrated flux is the same as in the other curves with $i = 90°$.

Figure 12.9 shows the emission from a full disk of outer radius 100 AU. The inner radius $R_{\rm in} = 0.1$ AU $= 9.2R_*$ is about equal to the evaporation temperature of dust. The dust-free zone between one and nine stellar radii explains the trough in the specrum at $\sim 3\mu$m with respect to the spectral energy distribution in figure 12.6. There might be even much larger holes if planetesimals or planets have formed. The rise in the far infrared in figure 12.9 is due to the large grazing angle. It is the result of inflation. Should only the gas disk be inflated, but not the dust disk because the grains have sedimented, the calculations would yield quite different numbers. Such uncertainties about the basic structure of the disk are more grave than the shortcomings of our simple one-dimensional radiative transfer code.

A Monte Carlo 2d-calculation is displayed in figure 12.10 for nearly edge-on, nearly face-on and $60°$ viewing angle. The accretion phase is at its end; the extinction through the disk plane is only $A_{\rm V} \simeq 100$ mag. The disk has a constant opening angle of $27°$ and contains pure silicate dust. The density declines with $1/r$. There is also tenuous material above and below the disk; its density is constant and in z-direction $A_{\rm V} = 0.17$ mag. To an edge-on observer, the star is heavily obscured and he sees a deep silicate absorption feature. Nevertheless, the star can be detected visually in light that has been scattered from above and below the disk; it would be polarized. Looking face-on, the star is bright and the 10μm feature is weakly in emission. The far infrared part of the SED is independent of the viewing angle.

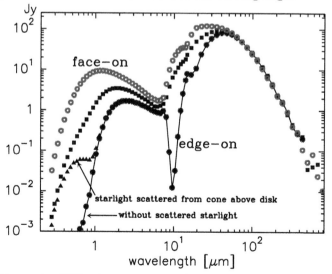

FIGURE 12.10 SED of a young star ($L = 10$ L$_\odot$, $T_* = 4000$ K) with a disk of 6 AU inner and 200 AU outer radius. See text.

12.3   Reflection nebulae

Reflection nebulae are illuminated by stars, typically of type B3V, that are
hot, but not hot enough to create an HII region. Nevertheless, the radiation
field within them is hard and the grains are in a harsh environment. This
leads to ample excitation of very small grains and PAHs. The models below
demonstrate the tremendous effect which they have on the SED.

The visual optical depth of a reflection nebula is never large ($\lesssim 1$) and we
may therefore assume that the grains are heated directly by the star, without
any foreground attenuation. Computing the radiative transfer is therefore not
really necessary. The infrared emission from the dust itself is, of course, also
optically thin. To evaluate it, we adopt the dust model outlined in section
10.2, but to make the role of PAHs more transparent, we use only one kind
with $N_C = 50$ carbon atoms, hydrogenation parameter $f_{H/C} = 0.3$ and cutoff
wavelegth $\lambda_{cut} = 0.42\mu m$. The mass fraction of the PAHs is 6%. The very
small grains (vsg), which are made of graphite and silicate, contain 10% of
the mass of the big particles and have uniform radii of 10Å. The big (MRN)
grains are exactly as described in section 10.2.

The dust emission at various positions in the nebula is shown in figure
12.11. The spectra are normalized at their maxima; absolute fluxes can be
read off from figure 12.12. Far from the star (right frame, distance $D =
5 \times 10^{18}$ cm), the big grains are cool, their mid infrared emission is negligible
and the maximum occurs beyond $100\mu m$. The spectrum below $50\mu m$ is then
entirely due to PAHs. Closer to the star, the big grains are hotter. But even
at $D = 5 \times 10^{16}$ cm, the maximum in the spectrum has shifted to $\sim 25\mu m$; the
PAHs dominate below $15\mu m$.

The *relative* strengths of the PAH features, the band ratios, are not sensitive
to the stellar distance; only the absolute flux weakens if the dust is far out. A
similar situation was encountered in figure 6.12 with respect to the very small
grains where the colors (flux ratios) are almost constant over the nebula.

To exemplify the influence of PAH size, we vary the number of carbon atoms
in a PAH, N_C, from 30 to 500 (figure 12.13). The hydrogenation parameter,
$f_{H/C}$, changes according to (10.15). The big PAHs are, on the average, cooler
than the small ones and it is therefore harder to excite their high frequency
bands. This is best seen by comparing the C–H resonances at 3.3 and $11.3\mu m$.
The emission ratio, $\epsilon_{3.3}/\epsilon_{11.3}$, is larger for small (hot) than for big (cool) PAHs.

It is also instructive to study how the temperature of the big (MRN) particles changes with distance, grain size and chemical composition.

  – Here, and in any hard radiation field, silicates are always colder than
    the amorphous carbon grains, and

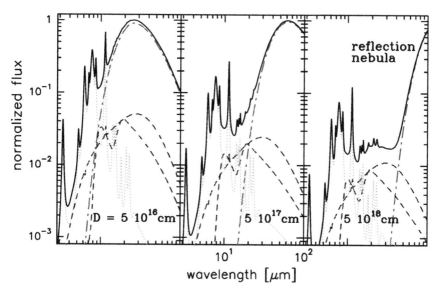

FIGURE 12.11 Normalized dust emission in a reflection nebula near a B3V star ($L = 10^3 \, \mathrm{L_\odot}$, $T_{\mathrm{eff}} = 18000 \, \mathrm{K}$) at various distances. Full curves: sum of all dust components; dash-dotted lines: big (MRN) grains; broken lines: very small graphite and silicate grains (vsg) of 10Å radius (the silicates have a hump at $10\mu\mathrm{m}$); dotted lines: PAHs.

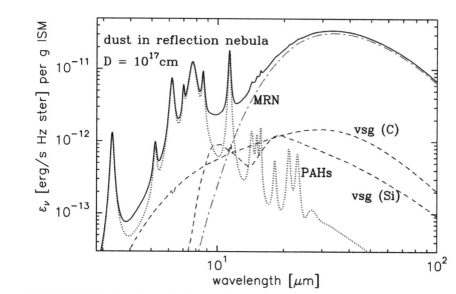

FIGURE 12.12 Very similar to figure 12.11, same kind of PAHs and vsg, but now with **absolute** fluxes per gram of interstellar matter and a distance $D = 10^{17} \, \mathrm{cm}$.

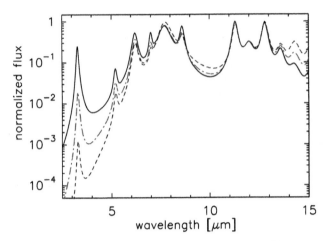

FIGURE 12.13 How PAH emission varies with PAH size. ***Big ones:*** broken curve, $N_C = 500$, $f_{H/C} = 0.13$. ***Medium sizes:*** dashes and dots, $N_C = 120$, $f_{H/C} = 0.26$. ***Small ones:*** solid curve, $N_C = 30$, $f_{H/C} = 0.5$. The grains are at a distance $D = 5 \times 10^{17}$ cm from the star, again $L = 10^3 \, L_\odot$ and $T_{\text{eff}} = 18000$ K. Arbitrary, but identical flux units for all three curves.

- big particles colder than small ones of the same chemistry (figure 6.2).

- At a distance $D = 5 \times 10^{18}$ cm, the temperature of the silicates, $T(\text{Si})$, ranges from 16 to 20 K; for carbonaceous particles, $T(\text{aC}) = 20 \ldots 25$ K.

- At $D = 5 \times 10^{17}$ cm, $T(\text{Si}) = 33 \ldots 43$ K and $T(\text{aC}) = 46 \ldots 58$ K.

- At $D = 5 \times 10^{16}$ cm, $T(\text{Si}) = 71 \ldots 93$ K and $T(\text{aC}) = 107 \ldots 137$ K.

12.4   Starburst nuclei

About 10% of the total luminosity in the local universe is generated in galactic nuclei by the rapid conversion of a large amount of gas into predominantly massive (OB) stars. Although the region where they form is relatively small (a few hundred parsec), its luminosity often exceeds that of the host galaxy. The rate at which stars form is so high ($> 10 \, M_\odot \, \text{yr}^{-1}$) that it can only be maintained over a cosmologically short spell ($< 10^8$ yr). The phenomenon is therefore called starburst. It is probably triggered by the gravitational interaction between two galaxies in which gas is funneled from the disk into the nucleus. The nucleus harbors a massive cluster of old stars (the bulge).

Starbursts are almost pure infrared objects, opaque to stellar photons because of dust. To interpret their spectral energy distribution and to arrive at a self-consistent picture for the spatial distribution of stars and interstellar matter in the nucleus, one has to compute the radiative transfer. This can only be done in an approximate way and one strategy was outlined in section 11.4. We now present results of such model calculations.

We first convince ourselves that it is indeed necessary to treat the OB stars and their surroundings separately, as hot spots. Figure 12.14 displays five spectra of basically the same nucleus, but computed under different assumptions. The discrepancies among the SEDs appear foremost in the region between 5μm and 20μm. The lowest curve (at 10μm) is calculated without hot spots. Here the luminosity of the OB stars has been smeared out. The next upper three curves (at 10μm) were computed with hot spots. The gas density in the hot spots varies from $n(\mathrm{H}) = 10^2\,\mathrm{cm}^{-3}$ (*solid*) to $10^3\,\mathrm{cm}^{-3}$ (*broken*) and $10^4\,\mathrm{cm}^{-3}$ (*dash-dots*), and the dust density increases accordingly assuming a standard dust-to-gas ratio. The immense influence of the hot spots on the mid infrared emission is evident.

To get closer to reality, one has to include PAHs and very small grains (vsg). This has been done in figure 12.15. We see that PAHs and vsg further enhance the mid infrared flux over the already high level produced by the hot spots. Strong mid infrared emission is therefore the hallmark of starburst galaxies. Interestingly, as emission by small particles dominates at these wavelengths, it is sometimes not clear at all how to derive from the silicate feature, which lies between PAH bands, a 10μm absorption optical depth.

Of not only academic interest is a point source model of the same luminosity and spectral appearance as the exciting stars, but where the stars are squeezed into a tiny volume at the center of the galactic nucleus. It is shown by the upmost dotted curve in figure 12.14 and 12.15. A point source yields a very strong mid infrared flux, too, but the model differs from those with extended emission by its small angular size. Mid infrared maps of high spatial resolution allow to discriminate against a stellar cluster.

A massive black hole with an accretion disk would also appear to an observer as a point source. But it is one with a considerably harder emission spectrum than OB stars. Usually such black holes have a power law spectrum $S_\nu \propto \nu^\alpha$ with $\alpha \simeq -0.5$. If the dust is exposed to the flux from such an object, the PAHs will evaporate in its vicinity and the ratio of mid infrared over total luminosity will thus be reduced. The radiative transfer code must obviously allow for the possibility of photo-destruction of PAHs and vsg (see section 9.1.5). The absence of PAH features in a galactic nucleus is thus an indirect indicator of a powerful non-thermal source. Spectral lines from multiple ionized atoms, which require for their excitation more energetic photons than are found in HII regions, are another (and even better) diagnostic for the presence of a massive black hole.

TABLE 12.1 Parameters for the starburst nucleus in figures 12.14 and 12.15. Numbers are typical and would apply to the archtype M82. See section 10.2.1 for nomenclature of PAHs and very small grains (vsg).

OB stars	integrated luminosity	$3 \times 10^{10} \, L_\odot$
	radius of stellar cluster	$R^{OB} = 200 \, pc$
	stellar density	$n^{OB}(r) \propto r^{-1/2}$
	surface temperature	$T^{OB} = 30\,000 \, K$
	luminosity of one OB star	$L^{OB} = 10^5 \, L_\odot$
low luminosity stars	integrated luminosity	$10^{10} \, L_\odot$
	radius of stellar cluster	$R^* = 700 \, pc$
	stellar density	$n^*(r) \propto r^{-1.8}$
	surface temperature	$T^* = 4000 \, K$
dust	radius of dust cloud	$R_{dust} = 800 \, pc$
	density $\rho_{dust}(r)$	$= const$ for $r \leq 230 \, pc$
	$\propto r^{-1/2}$ for $r > 230 \, pc$	
	A_V to cloud center	$29 \, mag$
PAHs	small ones $(N_C = 50)$	$Y_C^{PAH} = 0.05$
	big ones $(N_C = 300)$	$Y_C^{PAH} = 0.05$
very small grains	only graphites	$Y_C^{vsg} = 0.05, \; a = 10 \text{Å}$

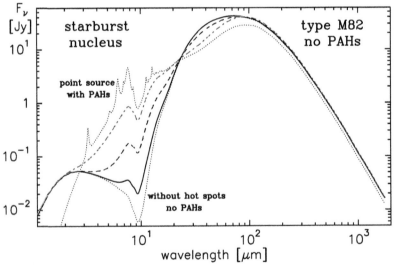

FIGURE 12.14 Radiative transfer models of a starburst nucleus, similar to M82, at a distance of 20 Mpc; further parameters in table 12.1 and text. The two dotted curves are benchmarks for comparison with figure 12.15: the lower is computed without hot spots; the upper depicts a point source and is the only curve in this figure where PAHs are included. The other three curves are models with hot spots of varying gas density (see text).

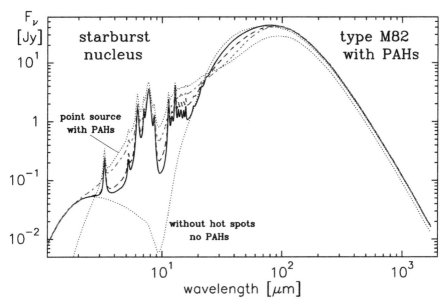

FIGURE 12.15 As figure 12.14, but now with hots spots *and* PAHs and vsg. Benchmark models as in figure 12.14 (upper and lower dotted curves).

12.5 Mass loss giants

When a star of intermediate mass ($1\,M_\odot < M \leq 6\,M_\odot$) leaves the main sequence after exhaustion of the central hydrogen reservoir, it ascends in the Hertzsprung-Russell diagram the red giant branch (RGB). Its brightness continues to grow until helium is ignited. Then the luminosity suddenly drops and the star makes a loop in the HR-diagram along the horizontal branch. When the central helium supply is consumed, leaving behind a core of carbon and oxygen, the star returns to the RGB on what is called the asymptotic giant branch (AGB). There helium burns sporadically (in thermal pulses) at the bottom of the helium shell, thus increasing the mass of the CO core. During this stage, the star develops a strong wind ($\dot{M} = 10^{-7}\ldots10^{-4}\,M_\odot\,\mathrm{yr}^{-1}$) in which most of the stellar mass is expelled: the star is on the way to its final destination as a white dwarf. Dust forms in the wind and is blown out into the interstellar medium, together with the gas, enriching it with solid particles (recommended reading [Hab96]).

12.5.1 Flow equations

In describing the interaction between gas and dust in the outflow of an AGB star of luminosity L_* and mass M_* we use the notations

- for the gas: velocity $u > 0$, density ρ, pressure $P = \rho kT/m$,
  atomic mass m, mean thermal velocity u_{th}

- for the dust: velocity v, density ρ_d.

The dust is pushed outwards by the photons and therefore ploughs through the gas at a relative velocity $v - u > 0$. The force exerted by this drift on *one* spherical grain of radius a and bulk density ρ_{gr} equals $\pi a^2 \rho |u - v| w$ according to (7.18) and (7.17), where

$$w = \begin{cases} u_{th} = \sqrt{8kT/\pi m} : & |v - u| \ll u_{th} \\ |u - v| & : & |v - u| \gg u_{th} . \end{cases}$$

At intermediate values ($|v - u| \simeq u_{th}$), one finds $w = \sqrt{u_{th}^2 + (v - u)^2}$ and this expression is also valid in the limits.

If $Q_\nu^{rp} = Q_\nu^{ext} - g_\nu Q_\nu^{sca}$ is the efficiency for radiation pressure of grains (see 2.9), we define a frequency-average by

$$\overline{Q}^{rp}(a) = \frac{\int F_\nu Q_\nu^{rp}(a) \, d\nu}{\int F_\nu \, d\nu} \tag{12.7}$$

where F_ν denotes the flux. To evaluate $\overline{Q}^{rp}$, one has to simultaneously compute the flow dynamics and the radiative transfer. Fortunately, the geometry is almost spherical. The frequency-averaged cross section of radiation pressure per gram of dust equals

$$K^{rp} = \frac{3\overline{Q}^{rp}}{4\rho_{gr}a} .$$

Because the radiation in the stellar envelope is predominantly infrared, one is in the Rayleigh limit where $Q \propto a$ (see 3.5). The acceleration of the dust by light is thus independent of the particle size.

12.5.1.1 Steady flow with constant grain size

Assuming that the flow is stationary, which is usually a very good approximation in the wind of giants, the gas outflow rate, $\dot{M}$, is constant

$$\dot{M} = 4\pi r^2 \rho u = \text{const} . \tag{12.8}$$

When the grains form abruptly at the condensation radius, r_1, with their final size, there is an equivalent equation for the dust flow,

$$\dot{M}_d = 4\pi r^2 \rho_d v = \text{const} , \tag{12.9}$$

but also the number of dust grains produced per second, $\dot{N}_d$, is constant,

$$\dot{N}_d = \frac{3\dot{M}_d}{4\pi a^3 \rho_{gr}} = \text{const} . \tag{12.10}$$

The dynamics of the two components, dust and gas, are then described by

$$v \frac{\partial v}{\partial r} + \frac{GM_*}{r^2} + \frac{3}{4a\rho_{gr}}\rho |u - v| w - \frac{K^{rp}L_*}{4\pi c r^2} = 0 . \tag{12.11}$$

$$u \frac{\partial u}{\partial r} + \frac{GM_*}{r^2} - \frac{3}{4a\rho_{gr}}\rho_d |u - v| w + \frac{1}{\rho}\frac{\partial P}{\partial r} = 0 . \tag{12.12}$$

The velocity changes with radius because of gravitational attraction, radiation pressure (dust only), gas pressure gradient (gas only) or friction between the components. The two ordinary differential equations can readily be brought into the form $u' = f_1(r, u, v)$, $v' = f_2(r, u, v)$, where a prime denotes the space derivative $\partial/\partial r$, and then be solved with standard methods (Runge-Kutta).

A few qualitative features of the two-stream flow can be found analytically by neglecting the gas pressure and assuming that the drift speed is supersonic whence, with $\beta = 3/16\pi\rho_{gr}$, (12.11) and (12.12) simplify to

$$v' = -\frac{GM_*}{r^2 v} - \beta \frac{\dot{M}}{a}\frac{(u - v)^2}{r^2 uv} + \frac{\beta}{ac} \frac{\overline{Q}^{rp}L_*}{r^2 v} . \tag{12.13}$$

$$u' = -\frac{GM_*}{r^2 u} + \beta \frac{\dot{M}_d}{a}\frac{(u - v)^2}{r^2 uv}$$

- In the outer envelope ($r \to \infty$), velocities are constant ($v' = 0$) and the drift velocity $v - u$ asymptotically approaches

$$v - u = \sqrt{\frac{\overline{Q}^{rp} Lu}{\dot{M}c}} . \tag{12.14}$$

When the mass loss rate $\dot{M}$ is high, coupling between dust and gas is strong and $u \simeq v$ (subsonic drift velocity). When it is low, friction is weak and the grains move much faster than the gas ($v \gg u$) (see figure 12.16).

- The condition that per gram of interstellar matter, which includes gas and dust, the force exerted by the radiation exceeds the gravitational attraction yields

$$\dot{M} > \frac{G^2 M_*^2 c}{\beta^2}\left(\frac{\dot{M}}{\dot{M}_d}\right)^2 \frac{a^2}{\overline{Q}^{rp}Lu} . \tag{12.15}$$

A steady flow is therefore possible only if the mass loss rate is above the minimum value given by the right side of the inequality (12.15). For a flow with parameters as in figure 12.16, the minimum mass loss rate is about 3×10^{-8} M$_\odot$/yr.

12.5.2 Solutions to the flow equations

12.5.2.1 No gas pressure and constant radiation pressure efficiency

First, we solve the dynamical equations (12.11) and (12.12) for a cold gas ($T = 0$) and assume a constant radiation pressure efficiency $\overline{Q}^{\mathrm{rp}}$ (see 12.7). Results for various mass loss rates of the gas, $\dot{M}$, a fixed ratio $\dot{M}_{\mathrm{d}}/\dot{M} = 0.01$, and representative AGB parameters are shown in figures 12.16 and 12.17. Dust forms momentarily at a distance $r_1 = 3 \times 10^{14}$ cm. At the formation radius, dust and gas have the same outward velocity of 2.5 km/s (as a result of dynamical processes in the stellar atmosphere). The dust is accelerated to more than half of its final velocity in a shell with a width of $\sim 10^{14}$ cm dragging the gas along through friction.

The escape velocities of the dust are around 20 km s^{-1}, quite independent of the mass loss rate $\dot{M}$ (figure 12.16). When $\dot{M}$ is small, the acceleration of the gas is not very efficient and it moves much slower than the dust. Their relative velocity, $v - u$, is then large. As $\dot{M}$ increases, the friction between the two components grows and $v - u$ decreases, tending to zero in the limit $\dot{M} \to \infty$. Gas velocities can be observed via the Doppler effect of molecular lines; dust velocities must be inferred indirectly.

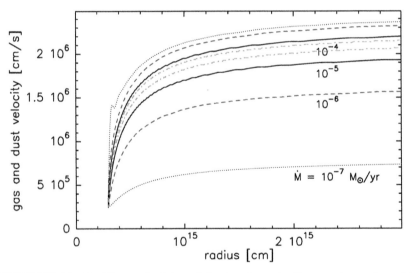

FIGURE 12.16 Flow velocities of dust and gas in the envelope of an AGB star ($L_* = 10^4$ L$_\odot$, $M_* = 1.2$ M$_\odot$, $T_* = 2800$ K). The grains have a radius $a = 6 \times 10^{-6}$ cm and a radiation pressure efficiency $\overline{Q}^{\mathrm{rp}} = 0.018$. There are four pairs of curves refering to four gas mass loss rates $\dot{M}$ from $10^{-7} \ldots 10^{-4}$ M$_\odot$ yr^{-1}. Curves belonging to one pair are drawn in the same mode (dots, broken, dash-dotted, solid); the lower one always applies to gas, the upper to dust.

Figure 12.17 displays the corresponding density profiles. In the outer envelope, the density falls off like $1/r^2$; in the acceleration zone the decline is much steeper. For better comparison, the gas densities have been reduced by a factor $\dot{M}_{\rm d}/\dot{M} = 0.01$. If the velocity of gas and dust were the same everywhere, their density profiles would, after the reduction, be identical. However, for low mass loss rates they are noticably offset. Although $\dot{M}/\dot{M}_{\rm d}$ is spatially constant, the local dust-to-gas ratio, $\rho_{\rm d}/\rho$, is proportional to u/v.

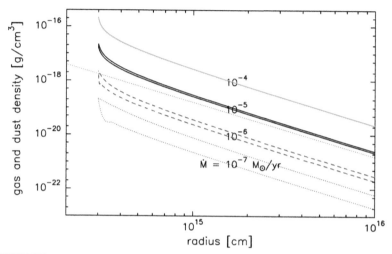

FIGURE 12.17 Density distributions of gas and dust for the models in figure 12.16. The lower curve of each pair refers to dust, the upper to gas. For better comparison, gas densities have been reduced by a factor $\dot{M}_{\rm d}/\dot{M} = 0.01$. The straight line presents a $\rho \propto r^{-2}$ density profile.

12.5.2.2 Coupling the radiative transfer and the dynamics

Next we solve the radiative transfer and the momentum equations (12.11 and 12.12) simultaneously. We include now gas pressure assuming $T_{\rm gas} = T_{\rm dust}$. The efficiency for radiaton pressure, $\overline{Q}^{\rm rp}$, depends on the spectral luminosity L_ν. The spectrum softens as the radius increases, but the integral $L_* = \int L_\nu d\nu$ stays, of course, constant.

$\overline{Q}^{\rm rp}$, together with the visual optical depth and the dust temperature, are plotted in figure 12.18 for the case $\dot{M} = 10^{-5}$ M$_\odot$ yr^{-1}. $\overline{Q}^{\rm rp}$ plummets in a thin ($\sim 3 \times 10^{13}$ cm) layer from 0.4 to under 0.1; at large distances, it is constant and around 0.03. This thin layer is of optical thickness one with respect to the stellar photons, corresponding to $A_{\rm V} \simeq 10$ mag; the dust reaches there half of its terminal velocity (figure 12.19), and the dust temperature falls from 1100 K at the formation radius to 800 K.

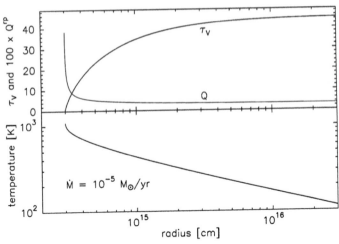

FIGURE 12.18 Variation of the dust temperature (bottom), the visual optical depth τ_V and the radiation pressure efficiency $\overline{Q}^{\mathrm{rp}}$ (top). $\overline{Q}^{\mathrm{rp}}$ is calculated from (12.7) for amorphous carbon dust. Stellar parameters are as in figure 12.16, here $\dot{M} = 10^{-5}$ M$_\odot$.

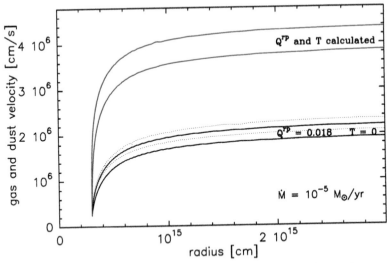

FIGURE 12.19 *Upper two curves:* Gas and dust velocity in a carbon AGB star with $\dot{M} = 10^{-5}$ M$_\odot$ yr^{-1}, stellar parameters as in figure 12.16. The two-component flow and the radiative transfer have been solved together; $\overline{Q}^{\mathrm{rp}}$ and T_{dust} are plotted in figure 12.18. *Lower two solid lines:* they are from figure 12.16 for $\dot{M} = 10^{-5}$ M$_\odot$ yr^{-1} and shown for comparison. *Lower two dotted lines:* the influence of gas pressure alone.

Figure 12.19 demonstrates how the inclusion of the radiative transfer changes the flow: the escape velocity doubles from 20 to 40 km s^{-1} and the initial acceleration becomes much stronger and more concentrated towards the dust formation radius. In comparison, the effect of the gas pressure gradient, when $\overline{Q}^{\mathrm{rp}} = 0.018$ is kept constant, is relatively mild.

A radiative transfer model is shown in figure 12.20, not for the scenario depicted in figures 12.18 and 12.19, but for the famous AGB star IRC+10216. It is a carbon star and there are only carbon grains in the wind. The spectrum peaks around 8μm with a flux of 50,000 Jy, which makes IRC+10216 one of the brightest near or mid IR sources in the sky.

The modification of the stellar spectrum through the dusty envelope becomes evident when contrasting the observed SED to the emission from the star alone. The optical and near infrared part are exponentially suppressed by extinction; the maximum brightness shifts from $\lambda_{\max} = 2\mu$m to $\lambda_{\max} = 8\mu$m, which corresponds to a dust temperature of about 300 K. It is attained at a distance of $\sim 2 \times 10^{15}$ cm. The far infrared emission is proportional to ν^4, but flattens at very long wavelengths ($\lambda > 1$mm) to ν^2 when the blackbody stellar surface dominates over dust emission.

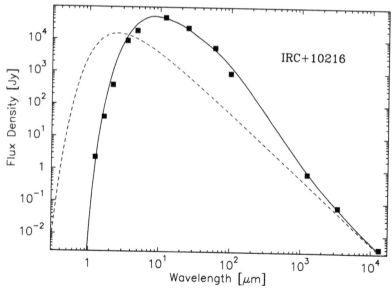

FIGURE 12.20 Spectral energy distribution of the variable carbon star IRC+10216 near maximum luminosity (from [Men06]). The squares are data points; the solid curve is the model ($L = 1.1 \times 10^4 \, L_\odot$, $T_{\mathrm{eff}} = 2000$ K, inner and outer shell radius: 3×10^{14} and 3×10^{17} cm, dust density distribution $\rho \propto r^{-2}$, $D = 110$ pc). The dashed line is the flux from the star alone, without the envelope of $A_V = 48$ mag.

12.6 The effective extinction curve

When we argued in section 10.1.2 that the interstellar extinction curve can be written as $K_\lambda^{\mathrm{ext}}/K_V^{\mathrm{ext}}$, where K_λ^{ext} is the dust extinction coefficient and V denotes the wavelength $\lambda = 0.55\mu$m, it was assumed that no scattered light falls into the telescope. This seems very reasonable with regard to the photometry of stars in the Milky Way which are not too distant and observed at high spatial resolution (a few arcsec). However, when the stars are several kiloparsec away, or even in another galaxy, they are usually within an extended dusty medium of low to moderate optical thickness and contamination of the detected flux by scattered photons is unavoidable.

We now compute the extinction curve for such a case. Because the medium is generally clumpy, we solve the radiative transfer with a Monte Carlo technique. As dust emission from a tenuous medium will only show up in the far infrared, the spectral energy distribution of a star in such an environment is governed at shorter wavelengths by extinction.

12.6.1 The effective optical thickness

Let F_λ be the flux that one receives from an object (star) obscured by dust and F_λ^{nd} the flux that one would receive if there were **no dust**. We define the effective optical thickness by

$$F_\lambda = F_\lambda^{\mathrm{nd}} e^{-\tau^{\mathrm{eff}}} \tag{12.16}$$

and call $\tau_\lambda^{\mathrm{eff}}/\tau_V^{\mathrm{eff}}$ the effective extinction curve. Observations can only yield this effective curve. The procedure by which it is obtained was described in section 10.1.2. It consists of comparing the fluxes from two identical stars, one of which is reddened, the other unreddened.

Let τ^{ext} be the extinction optical thickness along the line connecting the reddened star and the observer. One such geometrical configuration is depicted in figure 12.21. Here the star sits in the middle of an inhomogeneous plane-parallel cloud which we call the screen. If the star emits Z^* photons and the telescope subtends, as seen from the star, the (extremely small) solid angle $d\theta$, an observer would detect $Z^{\mathrm{nd}} = Z^* d\theta/4\pi$ photons if there were no dust. In the presence of the screen, he sees

$$Z^{\mathrm{nd}} e^{-\tau^{\mathrm{eff}}} = Z^* \frac{d\theta}{4\pi} e^{-\tau^{\mathrm{ext}}} + Z^{\mathrm{s}} \tag{12.17}$$

photons. Of these, $Z^* e^{-\tau^{\mathrm{ext}}} d\theta/4\pi$ come towards him on a straight line, without interaction in the cloud, and Z^{s} reach him after having been scattered, at least once. The region where they come from is determined by the field of

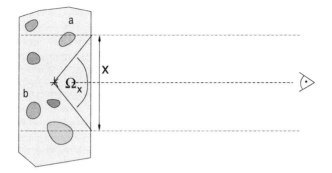

FIGURE 12.21 Observing a star in an extended clumpy medium with an aperture that corresponds to a linear size x at the distance of the medium, one detects photons coming directly from the star and scattered ones from the area indicated by the arrow. When the star is far away or the aperture large, the angle Ω_x will not be small and the scattered photons that reach the observer should not be neglected.

view of the detector which covers, at the distance of the screen, an area A_x of diameter x (figure 12.21). A photon scattered at point a can never reach the observer directly, one scattered at b can, if it has the right direction. The effective optical thickness follows from (12.17),

$$\tau^{\mathrm{eff}} \;=\; \tau^{\mathrm{ext}} \;-\; \ln\left[1 + \frac{Z^{\mathrm{s}}}{Z^{*}}\,\frac{4\pi}{d\theta}\,e^{\tau^{\mathrm{ext}}}\right]. \tag{12.18}$$

We mention the following points.

- $\tau^{\mathrm{eff}} \le \tau^{\mathrm{ext}}$.

- τ^{eff} approaches τ^{ext} as the aperture becomes small. In the limit, $\tau_\lambda^{\mathrm{eff}}/\tau_V^{\mathrm{eff}}$ coincides with the standard extinction curve $\tau_\lambda^{\mathrm{ext}}/\tau_V^{\mathrm{ext}}$.

- τ^{eff} may be negative, for example, when the medium is clumpy but the view to the star unobscured ($\tau^{\mathrm{ext}} = 0$).

- Whereas the standard extinction curve is independent of the dust column density, the effective extinction curve, $\tau_\lambda^{\mathrm{eff}}/\tau_V^{\mathrm{eff}}$, depends on how much dust there is and, furthermore, on the way it is distributed.

As a simple example that the standard dust, which is invoked to explain the standard galactic extinction, can produce various kinds of reddening, we consider a star at the center of a homogeneous and optically thin sphere. Then $Z^{\mathrm{s}} \simeq Z^{*}\tau^{\mathrm{sca}}d\theta/4\pi$ and, because $\ln(1+x) \simeq x$ for $x \ll 1$,

$$\tau^{\mathrm{eff}} \;\simeq\; \tau^{\mathrm{ext}} - \ln[1 + \tau^{\mathrm{sca}}e^{\tau^{\mathrm{ext}}}] \;\simeq\; \tau^{\mathrm{ext}} - \tau^{\mathrm{sca}} \;\simeq\; \tau^{\mathrm{abs}}.$$

The effective extinction is then not equal to $K_\lambda^{\mathrm{ext}}/K_V^{\mathrm{ext}}$, but to $K_\lambda^{\mathrm{abs}}/K_V^{\mathrm{abs}}$ and thus quite different.

12.6.2 Monte Carlo simulations

To give an impression of the variety of extinction curves that can be obtained, we present in figure 12.22 four examples of Monte Carlo computations. In the first three, we use the "standard" dust for which the ratio $K_\lambda^{ext}/K_V^{ext}$ equals exactly the mean galactic extinction curve with $R_V = 3.1$. In both boxes of the figure, this curve is shown by a solid line as a reference. In the final example, the dust is modified by lowering the graphite abundance.

(1) The case of a homogeneous medium is depicted by the filled squares. The optical thickness to the star $\tau_V^{ext} = 0.63$ and the observations are carried out with an aperture corresponding to $\Omega_x = 90°$ (see figure 12.21). Because of scattering, τ_V^{eff} defined in (12.16) equals 0.52 and is thus smaller than τ_V^{ext}.

The effective extinction curve, $\tau_\lambda^{eff}/\tau_V^{eff}$, lies in the far UV slightly above the standard reddening curve. The difference is more pronounced in the repesentation $\text{Ext}(\lambda) = (\tau_\lambda^{eff} - \tau_V^{eff})/(\tau_B^{eff} - \tau_V^{eff})$.

(2) When the medium is very clumpy, while Ω_x, the *average* column density and the mass of the screen stay constant, τ_V^{eff} becomes much smaller than τ_V^{ext}. In the particular, but nevertheless rather typical model shown by the open squares, the volume filling factor equals 0.12, τ_V^{eff} has dropped to 0.024 and the effective extinction curve obtains very large values in the far UV.

(3) When the telescope aperture is very small, Z^s in (12.18) tends to zero. We simulate this by lowering Ω_x to 10° (the computations would take too long for really small angles), without changing anything else with respect to example (2). One then obtains almost the standard reddening curve (open circles in lower box) confirming that deviations from it are due to insufficient spatial resolution.

(4) The Small Magellanic Cloud (SMC), a nearby companion of the Milky Way at a distance of 60 kpc, has a very peculiar reddening curve [Pré84]. It is very steep and the bump at $4.6\mu m^{-1}$ is missing. Its absence is ascribed to a lack of graphite. The heavy element abundance in the SMC is also much reduced and therefore also the dust-to-gas ratio, but this by itself would not change the extinction curve.

When we repeat example (2), but with dust where the graphite abundance is reduced by 75% (relative to the abundace of silicate and amorphous carbon grains), we obtain the asterisks which are not unlike the extinction curve observed towards the SMC.

Other interesting extinction scenarios well suited for Monte Carlo studies are stellar clusters. One is then not dealing with one star, as in figure 12.21, but with many. The cluster may be dust-enshrouded or have a screen in front

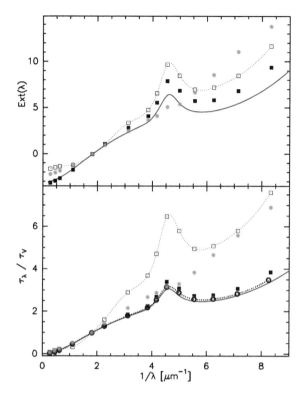

FIGURE 12.22 Effective reddening towards dust-enshrouded stars (see figure 12.21) displayed as $\tau_\lambda^{\text{eff}}/\tau_V^{\text{eff}}$ and as $\text{Ext}(\lambda) = (\tau_\lambda^{\text{eff}} - \tau_V^{\text{eff}})/(\tau_B^{\text{eff}} - \tau_V^{\text{eff}})$. **Solid lines:** mean galactic extinction. **Filled squares:** large aperture ($\Omega_x = 90°$) and homogeneous medium with $\tau^{\text{ext}} = 0.63$. **Open squares:** $\Omega_x = 90°$ and a clumpy medium. **Open circles** (only in lower box): $\Omega_x = 10°$ and a clumpy medium. **Asterisks:** $\Omega_x = 90°$, clumpy medium and a reduced graphite abundance. See text.

of it, the medium may be smooth or clumped according to some law, but irrespective of the particlar geometry, the definition (12.16) for the effective extinction stays in force: multiplying the observed flux by $e^{\tau^{\text{eff}}}$ corrects for the attenuation by dust and yields the intrinsic emission. Semi-analytical approaches to such configurations are presented in [Cal94] and [Fis05].

A

Various dust related physics

This chapter deals with purely physical topics selected for our study of dust in the interstellar medium. They concern the statistical distribution of atoms and photons, basic thermodynamic relations, blackbody radiation, transition probabilities and the quantum mechanical motion of a particle in a simple potential. We compile basic concepts and equations and sketch how they are derived. Such a biased compendium does, of course, not replace a textbook on physics, but may occasionally be convenient for quick reference.

A.1   Boltzmann statistics

For a large number of atoms in thermodynamic equilibrium, their distribution in energy space is just a function of temperature and described by the Boltzmann equation. We outline the derivation of this fundamental statistical relation and append a few loosely connected items.

A.1.1   The probability of an arbitrary energy distribution

We first compute the number of ways in which the energy of N identical particles, for example atoms, can be distributed, and then find among them the most likely distribution. To this end, we divide the energy space, which ranges from zero to some maximum $E_{\max}$, into n ordered cells with mean energy E_i so that

$$E_1 \; < \; E_2 \; < \; \ldots \; < \; E_n \; .$$

Each cell has a size ΔE_i which determines its statistical weight $g_i = \Delta E_i / E_{\max}$. The statistical weight gives the fractional size of a cell, so

$$\sum_i g_i \; = \; 1 \; .$$

Let each level E_i be populated by N_i atoms, their sum $\sum N_i = N$. If the atoms are labeled, from a_1 to a_N, a particular configuration is represented, for example, by the sequence

$$\underbrace{a_1 a_2 \ldots a_{N_1}}_{N_1} \underbrace{a_{N_1+1} \ldots a_{N_1+N_2}}_{N_2} \cdots \underbrace{a_{N-N_n+1} \ldots a_N}_{N_n} .$$

The N_i under the horizontal brackets give the number of atoms in the energy bins E_i. But there are many other ways in which the same energy distribution can be realized. According to elementary probability theory, a configuration where N particles are grouped such that N_i are in cell E_i (for $i = 1, ..., n$ and $\sum N_i = N$) can be achieved in

$$\Omega = \frac{N!}{N_1! N_2! \ldots N_n!} \tag{A.1}$$

ways. Here it is assumed that the particles are distinguishable. Ω is called thermodynamic probability and is a very large number. When the N atoms are arbitrarily distributed over the available cells, the probability that the first N_1 atoms fall into cell E_1, the following N_2 atoms into cell E_2, and so forth equals

$$g_1^{N_1} \cdot g_2^{N_2} \cdot \ldots \cdot g_n^{N_n} .$$

The probability ω (small Greek letter) for the particular configuration of the N atoms is therefore

$$\omega = \frac{N!}{N_1! N_2! \ldots N_n!} \cdot g_1^{N_1} \cdot g_2^{N_2} \cdot \ldots \cdot g_n^{N_n} . \tag{A.2}$$

Equation (A.2) is at the heart of classical statistics. ω is of course smaller than one. Summing the probabilities ω of all possible distributions gives the total probability. In view of the binomial theorem, this sum is $(g_1 + g_2 + \ldots + g_n)^N$, which equals one, as it should. When all cells are of the same size, each cell has the same chance that a particular atom will fall into it. Then $g_i = 1/n$ for all i and the probability ω in equation (A.2) becomes

$$\omega = \frac{N!}{n^N \prod N_i!} .$$

A.1.1.1 The distribution of maximum probability

As the numbers N_i in the expression for the probability ω are large, their faculties are much larger still. To handle them, one uses Stirling's formula

$$\ln N! \simeq N \cdot \ln N - N + \ln \sqrt{2\pi N} + \frac{\vartheta_N}{4N} \qquad (0 \le \vartheta_N \le 1) . \tag{A.3}$$

Note the remarkably small error estimate. Retaining only the first term, yields for (A.2)

$$\ln \omega \simeq N \ln N + \sum_{i=1}^{n} N_i \ln \frac{g_i}{N_i} .$$

In the equilibrium distribution achieved by nature, the probability has the maximum possible value. Because $N \ln N$ is constant, we are looking for an extremum of the function

$$f(N_1, N_2, \ldots, N_n) = \sum_{i=1}^{n} N_i \cdot \ln \frac{g_i}{N_i} .$$

There are two additional requirements:

- The total number of particles N is constant and equals $\sum N_i$,

$$\phi = \sum_{i=1}^{n} N_i - N = 0 .$$

- The mean particle energy $\langle E_i \rangle$ is related to the total energy E of all particles through

$$E = N \cdot \langle E_i \rangle ,$$

therefore

$$\psi = N \langle E_i \rangle - \sum_{i=1}^{n} N_i E_i = 0 . \tag{A.4}$$

The maximum of f, subject to the conditions expressed by the auxiliary functions ϕ and ψ, can be found using the method of Lagrangian multiplicators. If we denote them by α and β, we get the equation for the differentials

$$df + \alpha \, d\phi + \beta \, d\psi = 0$$

leading to

$$\sum dN_i \left(\ln \frac{g_i}{N_i} - 1 + \alpha - \beta E_i \right) = 0 .$$

In this sum, all brackets must vanish. So for every i,

$$N_i = g_i \, e^{\alpha - 1} \, e^{-\beta E_i} .$$

Exploiting $\sum N_i = N$ eliminates α and gives the fractional population

$$\frac{N_i}{N} = \frac{g_i \, e^{-\beta E_i}}{\sum_j g_j \, e^{-\beta E_j}} . \tag{A.5}$$

A.1.2 Partition function and population of energy cells

We have found N_i/N, but still have to determine β. When we define in (A.5)

$$Z = \sum_j g_j \, e^{-\beta E_j} ,$$

equations (A.1), (A.3) and (A.5) yield

$$\ln \Omega = - \sum N_i \ln \frac{N_i}{N} = - \sum N_i \left(\ln g_i - \beta E_i - \ln Z \right)$$
$$= \beta E + N \ln Z - \sum N_i \ln g_i . \tag{A.6}$$

We now plug in some thermodynamics from section A.3. According to (A.28), the entropy S of the system equals the Boltzmann constant times the logarithm of the thermodynamic probability Ω of (A.1),

$$S = k \ln \Omega ,$$

and the temperature is given by (see A.29)

$$\frac{1}{T} = \frac{\partial S}{\partial E} = k \frac{\partial \ln \Omega}{\partial E} .$$

When we evaluate this expression using (A.6) and assume that all energy bins are of equal size, which is no restriction, $\sum N_i \ln g_i = N \ln g_i = \text{const}$ and therefore

$$\frac{1}{T} = k\beta + kE \cdot \frac{\partial \beta}{\partial E} - kN \cdot \frac{\sum g_i E_i e^{-\beta E_i}}{\sum g_j e^{-\beta E_j}} \cdot \frac{\partial \beta}{\partial E} .$$

Here the second and third term on the right-hand side cancel and thus

$$\beta = \frac{1}{kT} . \tag{A.7}$$

The expression Z introduced before now becomes

$$Z(T) = \sum_j g_j e^{-E_j/kT} . \tag{A.8}$$

It is called the partition function and depends only on temperature. The fractional abundance of atoms with energy E_i is therefore

$$\frac{N_i}{N} = \frac{g_i e^{-E_i/kT}}{\sum g_j e^{-E_j/kT}} = \frac{g_i e^{-E_i/kT}}{Z} , \tag{A.9}$$

and the abundance ratio of two levels i and j,

$$\frac{N_j}{N_i} = \frac{g_j}{g_i} e^{-(E_j - E_i)/kT} . \tag{A.10}$$

Although the population numbers are statistical averages, they are extremely precise and practically important deviations from the mean never occur.

A.1.3 The mean energy of harmonic oscillators

With some reinterpretation, equations (A.8) to (A.10) are also valid in quantum mechanics. For atoms with internal energy levels E_i, the statistical weight g_i denotes then the degeneracy of the level.

Let us compute the partition function of a system of N quantized identical harmonic oscillators in equilibrium at temperature T. The energy levels are non-degenerate $(g_i = 1)$ and equidistant, $E_i = (i + \frac{1}{2})\hbar\omega$ for $i = 0, 1, 2, \ldots$ (see A.87). With $\beta^{-1} = kT$,

$$Z(T) = \sum_{i=0}^{\infty} e^{-\beta E_i} = e^{-\frac{1}{2}\beta\hbar\omega} \sum_{i=0}^{\infty} e^{-i\beta\hbar\omega} .$$

The last sum is an infinite series where each term is a factor $e^{-\beta\hbar\omega}$ smaller than the preceding one, therefore

$$Z(T) = \frac{e^{-\frac{1}{2}\beta\hbar\omega}}{1 - e^{-\beta\hbar\omega}} . \tag{A.11}$$

The mean oscillator energy $\langle E \rangle$ follows from

$$\langle E \rangle = \frac{\sum N_i E_i}{N} = \sum E_i \frac{e^{-\beta E_i}}{Z} = -\frac{1}{Z}\frac{\partial Z}{\partial \beta} = -\frac{\partial \ln Z}{\partial \beta} .$$

Because $\ln Z = -\frac{1}{2}\beta\hbar\omega - \ln(1 - e^{-\beta\hbar\omega})$, we get

$$\langle E \rangle = \frac{1}{2}\hbar\omega + \frac{\hbar\omega}{e^{\hbar\omega/kT} - 1} . \tag{A.12}$$

At high temperatures $(\hbar\omega \ll kT)$, we get the classical result that the sum of potential plus kinetic energy of an oscillator equals kT,

$$\langle E \rangle = kT .$$

At low temperatures $(\hbar\omega \gg kT)$, the average $\langle E \rangle$ is close to the zero point energy,

$$\langle E \rangle = (\tfrac{1}{2} + e^{-\hbar\omega/kT})\,\hbar\omega .$$

A.1.4 The Maxwellian velocity distribution

The probability that an arbitrary gas atom has a velocity in a certain range follows as a corollary from the equilibrium energy distribution (A.9). We substitute in this formula for E_i the kinetic energy of an atom, $p^2/2m$. All directions of motion are equally likely, and the square of the length of the momentum vector is

$$p^2 = p_x^2 + p_y^2 + p_z^2 .$$

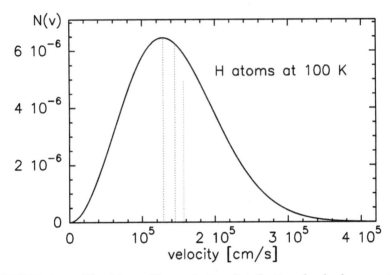

FIGURE A.1 The Maxwellian velocity distribution for hydrogen atoms at 100 K. The ordinate gives the probability density $N(v)$ after (A.14) with respect to one atom ($N=1$). The vertical dotted lines show from left to right the means v_p, $\langle v \rangle$ and $\sqrt{\langle v^2 \rangle}$ according to (A.16) to (A.18).

For the size of an energy cell, i.e., for the statistical weight, we take

$$g = 4\pi p^2 \, dp \, . \tag{A.13}$$

In the ratio N_i/N of (A.9), it is the relative, not the absolute value of the statistical weight g that matters; therefore the sum over all g need not be one. When we replace in the partition function $Z(T)$ of (A.8) the sum by an integral, we obtain the momentum distribution. Changing from momenta $p = mv$ to velocities gives the familiar Maxwellian velocity distribution (figure A.1) which states that if there are N gas atoms altogether, the number of those whose absolute velocities are in the range $v \ldots v + dv$ equals

$$N(v) \, dv = N \cdot 4\pi v^2 \left(\frac{m}{2\pi kT}\right)^{3/2} e^{-mv^2/2kT} \, dv \, , \tag{A.14}$$

with

$$\int_0^\infty N(v) \, dv = N \, .$$

In any arbitrary direction x, the number of atoms with velocities in the interval $[v_x, v_x + dv_x]$ is given, except for a normalizing constant, by the Boltzmann factor $e^{-mv_x^2/2kT}$,

$$N(v_x) \, dv_x = N \left(\frac{m}{2\pi kT}\right)^{1/2} e^{-mv_x^2/2kT} \, dv_x \, . \tag{A.15}$$

$N(v_x)$ is symmetric about $v_x = 0$ and also has its maximum there (contrary to $N(v)$ in A.14) because $N(v_x)$ includes all particles with any velocity in the y or z-direction. The total number of atoms, N, follows from (A.15) by integrating from $-\infty$ to ∞.

There are three kinds of averages of the absolute velocity v. To compute them, we exploit the integrals (B.7) to (B.9) associated with the bell-curve:

- The most probable velocity, v_p, is the one at the maximum of the curve $N(v)$,

$$v_p = \sqrt{\frac{2kT}{m}} \,. \tag{A.16}$$

- The mean velocity $\langle v \rangle$ is relevant for calculating collision rates,

$$\langle v \rangle = \sqrt{\frac{8kT}{\pi m}} \,. \tag{A.17}$$

- The mean energy of the atoms is given by the average of v^2 such that $\frac{1}{2}m\langle v^2 \rangle = \frac{3}{2}kT$ or

$$\langle v^2 \rangle = \frac{3kT}{m} \,. \tag{A.18}$$

The averages are ordered (see figure A.1),

$$v_p < \langle v \rangle < \sqrt{\langle v^2 \rangle} \,. \tag{A.19}$$

A.2 Quantum statistics

A.2.1 The unit cell h^3 of phase space

In deriving the energy distribution for an ensemble of particles after Boltzmann, we divided the energy or momentum space into arbitrarily small cells. The real coordinate space was not considered. In quantum statistics, one divides the *phase space*, which consists of coordinates and momenta, into cells. If V is the volume and p the momentum, we form cells of size

$$d\Phi = V \cdot 4\pi p^2 \, dp \,,$$

where $4\pi p^2 dp$ is the volume of a shell in momentum space with radius p and thickness dp. According to Heisenberg's uncertainty principle, position q and momentum p of a particle can only be determined to an accuracy Δp and Δq such that

$$\Delta p \cdot \Delta q \gtrsim \hbar \,.$$

This relation results from the wave character possessed by all particles, not just photons. The wavelength of a particle is related to its momentum after *de Broglie* by $p = h/\lambda$. Because of the wave character, particles in a box of volume V can have only discrete momenta; otherwise they would be destroyed by interference. The situation is formally almost identical to standing waves in a crystal, which we treat in section 6.4. We may therefore exploit equation (6.30) of that section and find that the possible number of eigenfrequencies within the momentum interval p to $p + dp$ in a box of volume V is

$$g = \frac{V \cdot 4\pi p^2 \, dp}{h^3} .\qquad\qquad (A.20)$$

The eigenfrequencies are the only allowed states in which a particle can exist and correspond to standing waves. The quantity g (named dZ in 6.30) is called statistical weight (see also A.13). It states how many *unit cells*, each of size h^3, fit into the phase space element $d\Phi$. Note that classically the number of possible (physically different) states is unlimited, which would correspond to a unit cell of size zero.

To determine the state of a particle with non-zero spin, one needs besides place and momentum the spin direction. Therefore, the number of quantum states of an electron (spin $\frac{1}{2}$) is twice the value given in (A.20). The same is true for photons (spin 1) which, traveling at the speed of light, also have two (not three) spin directions.

A.2.2   Bosons and fermions

Particles are described by their position in phase space. Any two particles of the same kind, for example, two photons or two electrons, are otherwise identical. This means when we swap their positions in phase space, nothing changes physically; the particles are indistiguishable.

In quantum mechanics, particles are represented by their wave functions. Let $\Psi(x_1, \ldots, x_n)$ denote the wave function of a system of n particles, all of one kind and not interacting. Their identity then implies that $|\Psi|^2$ stays the same when any two of the particles are interchanged. So after they have been swapped, either Ψ itself is not altered, then one speaks of a symmetric wave function, or Ψ is transferred into $-\Psi$, then the wave function is called anti-symmetric.

A.2.2.1   Fermi statistics

Particles with an anti-symmetric wave function are called fermions. They have half-integer spin, like electrons, protons or neutrons and obey Pauli's exclusion principle which states that two such particles cannot occupy the same quantum state or unit cell.

When one determines the distribution of maximum likelihood for fermions in phase space (this is similar to the procedure in section A.1 for deriving the Boltzmann distribution), one finds that the number of particles per unit cell, called the occupation number, equals

$$\mathcal{N} = \frac{1}{e^{\alpha + E/kT} + 1} . \tag{A.21}$$

The parameter α follows from the requirement that the total number N of particles in a volume V is conserved. As a phase space element $d\Phi = V\, 4\pi p^2\, dp$ has $dn = d\Phi/h^3$ unit cells,

$$\frac{4\pi V}{h^3} \int \frac{p^2\, dp}{e^{\alpha + \beta E} + 1} = N . \tag{A.22}$$

One calls α the degeneracy parameter. It may be positive as well as negative (from $-\infty$ to $+\infty$). Degeneracy is absent for large positive α, when the gas is hot and rarefied. If $e^{\alpha + E/kT} \gg 1$, one is in the realm of classical physics.

When the temperature T goes to zero, the occupation number $\mathcal{N}$ changes discontinuously. It is then 1 at energies below the Fermi limit E_F and zero for $E > E_F$ (section A.8.2).

A.2.2.2 Bose statistics

Particles with a symmetric wave function are bosons. They have integer spin, like photons, and are not subject to Pauli's exclusion principle. The occupation number is

$$\mathcal{N} = = \frac{1}{e^{\alpha + E/kT} - 1} . \tag{A.23}$$

In the denominator, it is now -1, and not $+1$. The parameter α follows again from the condition of particle conservation, but it is always positive, as otherwise the denominator under the integral might vanish. For $e^{\alpha + \beta E} \gg 1$,

$$e^{-\alpha} = \frac{N}{V} \frac{h^3}{(2\pi m k T)^{3/2}} , \tag{A.24}$$

and one recovers the Maxwell distribution (A.14),

$$dn = N \frac{4\pi p^2\, dp}{(2\pi m k T)^{3/2}} e^{-p^2/2mkT} . $$

In Bose statistics, *all* particles are in their lowest (zero momentum) state at $T = 0$.

A.2.2.3   Bose statistics for photons

The radiative equilibrium of photons is established through emission and absorption processes, in which they are created and destroyed; their number is not conserved. The parameter α in (A.23) is therefore zero and

$$\mathcal{N} = = \frac{1}{e^{h\nu/kT} - 1} \ .$$

Because the photon momentum is $p = h\nu/c$, a phase space element of size $d\Phi = (V\,4\pi h^3/c^3)\nu^2\,d\nu$ has $dn = d\Phi/h^3$ unit cells. The radiative energy density u_ν follows from $V u_\nu\,d\nu = h\mathcal{N}\,dn$,

$$u_\nu = \frac{8\pi h}{c^3}\frac{\nu^3}{e^{h\nu/kT} - 1} \ . \tag{A.25}$$

There is a factor 2 included that accounts for the two possible modes of circular polarization. Formula (A.25) describes the distribution of photons in equilibrium at temperature T, so called blackbody radiation. Although we did not explicitly prescribe the total number of photons within a unit volume, $n_{\rm phot}$, the average is fixed and follows from integrating $u_\nu/h\nu$ over frequency,

$$n_{\rm phot} = \int_0^\infty \frac{u_\nu}{h\nu}\,d\nu = 8\pi\left(\frac{kT}{hc}\right)^3 \int_0^\infty \frac{x^2}{e^x - 1}\,dx$$

(see B.2 for evaluation of the integral). A unit volume of gas in thermodynamic equilibrium, on the other hand, may be filled by an arbitrary number of atoms.

A.3   Thermodynamics

A.3.1   The ergodic hypothesis

Consider a system of atoms with f degrees of freedom, spatial coordinates $q_1, ..., q_f$ and conjugate momenta $p_1, ..., p_f$. The path of the atoms in phase space contains the full mechanical information. Usually one needs three coordinates (x, y, z) for one particle, so if there are N particles,

$$f = 3N \ .$$

Let the system be isolated so that it cannot exchange energy with its surroundings. The total energy E is therefore constant, or rather it is in a narrow range from E to $E + \delta E$, because in quantum mechanics, E cannot be defined with absolute precision in a finite time.

At any instant, the system may be represented by a point in the $2f$-dimensional phase space with the coordinates

$$q_1, \ldots, q_f, p_1, \ldots, p_f.$$

As the sytem evolves in time t, it describes a trajectory parametrized by

$$(q_1(t), \ldots, p_f(t)).$$

If one divides the phase space into small cells Z of equal size, the *microstate* of the system at a particular moment is determined by the cell in which it is found. The cells may be enumerated. All cells (states) Z_i that correspond to an energy between E and $E + \delta E$ are called *accessible*. In equilibrium, the probability to find the system in a certain accessible cell Z_r is by definition independent of time.

- The fundamental (*ergodic*) postulate asserts that when the system is in equilibrium, all accessible states are equally likely.

Let $\Omega(E)$ be the total number of accessible states (cells), i.e., those with energy in the interval $[E, E + \delta E]$. If there are many degrees of freedom ($f \gg 1$), the function $\Omega(E)$ increases *extremely* rapidly with the total energy E of the system. To see how fast $\Omega(E)$ rises, we make a rough estimate:

Suppose the system consists of identical quantum oscillators, each corresponding to one degree of freedom. Their energy levels are equally spaced in multiples of $\hbar\omega$. Each oscillator has on average an energy E/f. After equation (A.88), it is spread out in one-dimensional phase space over a region $\Delta x \Delta p = E/f\omega$ where it thus occupies $E/f\hbar\omega$ cells of size $\hbar$. It does not make sense to consider cells smaller than $\hbar$ because one cannot locate a particle more accurately. The number of states, N_s, of the total system whose energy does not exceed E is therefore

$$N_s = \left(\frac{E}{f\hbar\omega}\right)^f. \tag{A.26}$$

Hence the number of cells, $\Omega(E)$, with energies from $[E, E + \delta E]$ becomes

$$\Omega(E) = N_s(E + \delta E) - N_s(E) \simeq \frac{dN_s}{dE}\, \delta E .$$

We denote by

$$\rho(E) = \frac{dN_s}{dE}$$

the density of states around energy E. It is independent of the width of the chosen energy interval δE, contrary to $\Omega(E)$ which obviously increases with δE. As a first approximation, (A.26) tells us that $\rho(E) \simeq fE^{f-1}/(f\hbar\omega)^f$. If one counts the states more properly, but neglects powers of E with exponents $f-2$ or smaller, one finds

$$\rho(E) \simeq \frac{E^{f-1}}{(f-1)!\,(\hbar\omega)^f} . \tag{A.27}$$

If the total energy E of the system is n times greater than the natural energy unit $\hbar\omega$, the density of states becomes proportional to $n^{f-1}/(f-1)!$. Note that ρ and Ω rise with E^{f-1}. When we evaluate Ω for something from everydays life, like the gas content in an empty bottle, the number of degrees of freedom of the atoms is $\sim 10^{24}$. This by itself is a huge number, but because f appears as an exponent, Ω is so large that even its logarithm $\ln\Omega$ is of order f, or 10^{24}.

The probability that all gas atoms in a box are huddled in the upper left corner and the rest of the box is void is extremely low; one will never encounter such a configuration. After the ergodic principle, all accessible cells of phase space have an equal chance of being populated. The probability of a particular microstate which actually is encountered and where the atoms are very evenly distributed must therefore be equally low. There is no contradiction because there is just one cell corresponding to all atoms being clustered in one corner, but a multitude corresponding to a very smooth distribution of atoms over the available box volume.

A.3.2   Definition of entropy and temperature

In statistical physics, the starting point is the number of accessible states $\Omega(E)$ with energies from $[E, E+\delta E]$; it determines the entropy,

$$S = k \ln \Omega . \tag{A.28}$$

Of course, $\Omega(E)$ depends on the width of the energy interval δE, but because Ω is so large, one can easily show that δE is irrelevant in (A.28). The absolute temperature T is then defined by

$$\frac{1}{k} \cdot \frac{\partial S}{\partial E} = \frac{\partial \ln \rho}{\partial E} = \frac{1}{kT} . \tag{A.29}$$

Let us see what these two definitions mean for a bottle of warm lemonade (system A) in an icebox (system A'). The two systems are in thermal contact, i.e., heat can flow between them, but the external parameters (like volume) are fixed. The total system, $A+A'$, is isolated (we do not have an electric icebox), so the joint energy $E_{\text{tot}} = E + E'$ is constant; otherwise E and E' are arbitrary. The number of accessible states of the whole system, subject to the condition that subsystem A has an energy E, is given by the product $\Omega(E) \cdot \Omega'(E_{\text{tot}} - E)$, where Ω and Ω' refer to A and A', respectively. The probability $P(E)$ to find the whole system in a state where A has the energy E is evidently proportional to this product,

$$P(E) \propto \Omega(E) \cdot \Omega'(E_{\text{tot}} - E) .$$

When $\Omega(E)$ rises, $\Omega'(E_{\text{tot}} - E)$ must fall. Because of the extreme dependence of the number of accessible states Ω on E, or of Ω' on E', the probability

$P(E)$ must have a *very sharp* maximum at some value E_{eq}. To find system A at an energy $E \neq E_{eq}$ is totally unlikely. The probability $P(E)$ is almost a δ-function; the value where it is not zero detemines equilibrium. The energy E where P, or the entropy $S+S'$ of the whole system has its maximum follows from

$$\frac{\partial \ln P}{\partial E} = 0 .$$

This equation gives immediately

$$\left.\frac{\partial S}{\partial E}\right|_{E=E_{eq}} = \left.\frac{\partial S'}{\partial E'}\right|_{E'=E'_{eq}} ,$$

which means that in equlibrium the temperature of icebox and lemonade are equal; the lemonade is cool. Note that nothing has been said about how long it takes, starting from some arbitrary microstate, to arrive at equilibrium.

We add that Ω or the entropy S have definite values only if the cell size is clearly defined. Classically, it may be arbitrarily small and then Ω or S would be unbounded, but S can be measured (from $\delta Q = T \, dS$) and has a definitive value, so there exists a unit cell size $(\hbar^3)$.

A.3.3 The canonical distribution

Let a system A be in thermal contact with a much larger heat bath A' of temperature T. The energy of A is not fixed, only the joint energy E_{tot} of the combined system $A+A'$. Suppose now that A is in a *definite* state r of energy E_r. The heat bath has then the energy $E' = E_{tot} - E_r$ and $\Omega'(E_{tot} - E_r)$ accessible states. The probabity P_r for A being in state r is proportional to the number of states of the total system $A+A'$ under that condition, therefore, as A is fixed to state r,

$$P_r = C' \Omega'(E_{tot} - E_r) .$$

The constant C' follows from the condition $\sum_s P_s = 1$ in which the sum includes all possible states s of A, irrespective of their energy. Because $E_{tot} \gg E_r$, one can develop $\ln \Omega'$ around E_{tot},

$$\ln \Omega'(E_{tot} - E_r) = \ln \Omega'(E_{tot}) - \beta E_r ,$$

where $\beta = \partial \ln \Omega'/\partial E' = (kT)^{-1}$ (see A.29), and obtains the canonical distribution

$$P_r = \frac{e^{-\beta E_r}}{\sum_s e^{-\beta E_s}} . \tag{A.30}$$

The sum in the denominator represents the partition function $Z(T)$. The probability $P(E)$ to find system A in the energy range δE around E becomes

$$P(E) = \frac{\Omega(E) \, e^{-\beta E_r}}{\sum_s e^{-\beta E_s}} \tag{A.31}$$

where $\Omega(E)$ is the corresponding number of states accessible to A. Formula (A.31) is in agreement with the Boltzmann distribution (A.9): As the size of system A has not entered the derivation of $P(E)$, the expression is also correct when A is a microscopic system. Thermal contact then just means that A can exchange energy with A'. If A consists of just one atom with discrete energy levels E_j, there is only one state for which $E = E_j$, so $\Omega(E) = 1$ and one recovers the Boltzmann distribution (A.9).

A.3.3.1   Constraints on the system

Sometimes a macroscopic system A may have to fulfill additional constraints, besides having an energy between E and $E+\delta E$. For example, some parameter Y may have to take up a certain precise value y, or lie in the range $[y, y+\delta y]$; we then write symbolically $Y = y$. The parameter Y may be anything, for example, the energy of ten selected atoms or the integrated magnetic moment of all particles. The states for which $Y = y$ form a subclass of all accessible states, and we denote their total number by $\Omega(E; y)$. In view of the ergodic postulate, the probability P of finding the system in the desired configuration $Y = y$ is

$$P(y) = \frac{\Omega(E; y)}{\Omega(E)} . \tag{A.32}$$

Likewise the probability P can be expressed by the ratio of the densities of states,

$$P(y) = \frac{\rho(E; y)}{\rho(E)} . \tag{A.33}$$

A.3.4   Thermodynamic relations

When one transfers to a system a small amount of heat δQ and does the work δA on the system, then its internal energy U is altered by the amount

$$dU = \delta Q + \delta A \tag{A.34}$$

(first law of thermodynamics). If the infinitesimal work is due to a change of volume dV,

$$\delta A = -p\,dV .$$

So when V is decreased, δA is positive. If U depends on temperature and volume, $U = U(T, V)$, and

$$dU = \left(\frac{\partial U}{\partial T}\right)_V dT + \left(\frac{\partial U}{\partial V}\right)_T dV ,$$

and because of equation (A.34),

$$\delta Q = \left(\frac{\partial U}{\partial T}\right)_V dT + \left[\left(\frac{\partial U}{\partial V}\right)_T + p\right] dV . \tag{A.35}$$

The second law of thermodynamics brings in the entropy S. When the heat δQ is added reversibly to a system at temperature T, its entropy increases by

$$dS = \frac{\delta Q}{T} = \frac{dU - \delta A}{T} \ . \tag{A.36}$$

The entropy is a state function and the integral $\int \delta Q / T$ in the (p, V)-plane along any closed path is therefore zero. Because dS is a full differential, i.e.,

$$\frac{\partial^2 S}{\partial T \partial V} = \frac{\partial^2 S}{\partial V \partial T} \ ,$$

we get from (A.36) with $\delta A = -p \, dV$,

$$\left(\frac{\partial U}{\partial V} \right)_T = T \left(\frac{\partial p}{\partial T} \right)_V - p \ . \tag{A.37}$$

Besides U and S, there are other common thermodynamic state functions:

- The free energy or Helmholtz potential

$$F = U - TS \ . \tag{A.38}$$

If one applies in an isothermal reversible process the work δA on a system, for example, by compressing a gas, the free energy of the system increases by $dF = \delta A$. This immediately follows by forming the differential $dF = dU - TdS - SdT = \delta A - SdT = \delta A$. When a grain thermally expands, no work is done, $\delta A = 0$, and therefore

$$\left(\frac{\partial F}{\partial V} \right)_T = 0 \ . \tag{A.39}$$

We will also need the relation

$$F = -kT \ln Z \tag{A.40}$$

between the free energy and the partition function, Z. Such a relation is suggested by (A.6) when putting $g_i = 1$. A more strict derivation is given below.

- The enthalpy

$$H = U + pV \ . \tag{A.41}$$

When one presses gas through a porous plug from one vessel to another, as in the famous Joule-Thomson experiment, it is the enthalpy that stays constant.

- The free enthalpy or Gibbs potential

$$G = U + pV - TS \ . \tag{A.42}$$

The full differential can be written as

$$dG = \frac{\partial G}{\partial T}\,dT + \frac{\partial G}{\partial p}\,dp + \frac{\partial G}{\partial N_j}\,dN_j \;=\; -S\,dT + V\,dp + \sum_j \mu_j\,dN_j \;,$$

where N_j is the number of atoms of the j-th gas component and

$$\mu_j \;=\; \frac{\partial G}{\partial N_j} \tag{A.43}$$

defines the chemical potential. In a homogeneous phase, in which for any α

$$S(\alpha U, \alpha V, \alpha N_j) \;=\; \alpha\,S(U, V, N_j) \;,$$

one finds

$$G \;=\; \sum_j \mu_j\,N_j \;.$$

Therefore, the free enthalpy of a system consisting of a one-component fluid and its vapor (only one kind of molecules, i.e., $j = 1$) is

$$G \;=\; G_{\mathrm{fl}} + G_{\mathrm{gas}} \;=\; \mu_{\mathrm{fl}} N_{\mathrm{fl}} + \mu_{\mathrm{gas}} N_{\mathrm{gas}} \;. \tag{A.44}$$

The state functions have natural independent variables: $U = U(S, V, N_j)$, $S = S(U, V, N_j)$, $F = F(T, V, N_j)$, $H = H(S, p, N_j)$ and $G = G(T, p, N_j)$.

A.3.4.1 The free energy expressed by the partition function

Consider a large ensemble of "identical" systems each in contact with a heat reservoir of temperature T. The probability, P_r, that an arbitrary system is in state r of energy E_r is given by the canonical distribution ($\beta = kT$)

$$P_r \;=\; \frac{e^{-\beta E_r}}{\sum e^{-\beta E_s}} \;.$$

The sum extends over all states s irrespective of the energy; $Z = \sum e^{-\beta E_s}$ is the partition function. The mean energy of the systems is therefore

$$\overline{E} \;=\; \frac{\sum E_s\,e^{-\beta E_s}}{\sum e^{-\beta E_s}} \;=\; -\frac{\partial \ln Z}{\partial \beta} \;. \tag{A.45}$$

Let x be an external parameter of the systems (like volume). The energy E_r as well as Z depend then also on x: $E_r = E_r(x)$ and $Z = Z(\beta, x)$. Therefore, when x changes, the energy of state r changes too and the work done on the system is

$$\delta A \;=\; dE_r \;=\; \frac{\partial E_r}{\partial x}\,dx, \qquad \frac{\partial E_r}{\partial x} = \text{generalized force.}$$

To find the work representative for the statistical ensemble, one has to average over all accessible states,

$$\delta A = \frac{\sum \frac{\partial E_s}{\partial x} dx \, e^{-\beta E_s}}{Z} = -\frac{dx}{\beta Z} \frac{\partial}{\partial x} \sum e^{-\beta E_s} = -\frac{1}{\beta} \frac{\partial \ln Z}{\partial x} dx \, . \quad (A.46)$$

With

$$d \ln Z = \frac{\partial \ln Z}{\partial \beta} d\beta + \frac{\partial \ln Z}{\partial x} dx \, , \quad (A.47)$$

(A.45) and (A.46) give

$$d \ln Z = -\overline{E} \, d\beta - \beta \, \delta A = -d(\beta \overline{E}) + \beta \, d\overline{E} - \beta \, \delta A \implies$$

$$d(\ln Z + \overline{E} d\beta) = \beta(d\overline{E} - \delta A) = \beta dQ = \beta T dS = \frac{dS}{k} \implies$$

$$k \ln Z + \frac{\overline{E}}{T} = S \implies$$

$$F = \overline{E} - TS = -kT \ln Z \quad (A.48)$$

A.3.5 Equilibrium conditions of the state functions

Let us recall the conditions under which a system is in equilibrium. For a *mechanical system*, all forces F_i acting on the particles must vanish. The forces derive from a potential ϕ, so $F_i = -\partial\phi/\partial x_i$. In equilibrium, the potential ϕ is at its minimum and the differential $\Delta\phi = 0$.

For an isolated *thermodynamic system* A of constant energy U and volume V, the entropy S_A attains its maximum. Therefore, the differential ΔS_A is zero under small variations of the variables (U, V, N_j). If A is not isolated, but in contact with a heat bath A', the equilibrium condition for the total system A+A' reads

$$\Delta S_{tot} = \Delta(S_A + S_{A'}) = 0 \, .$$

System A receives from the environment A' the heat $\delta Q = -T\Delta S_{A'}$ which increases its internal energy by ΔU and does the work δA on the system (see A.34),

$$\delta Q = T \Delta S_A = \Delta U - \delta A \, .$$

Because all variables here refer to system A, we may drop the index A in the entropy and get

$$\Delta U - T \Delta S - \delta A = 0 \, .$$

In case of pure compressional work, $\delta A = -p\Delta V$ and

$$\Delta U - T \Delta S + p\Delta V = 0 \, . \quad (A.49)$$

Equation (A.49) leads in a straightforward way to the following equilibrium conditions:

(1) In a system A of constant temperature and volume ($\Delta T = \Delta V = 0$) and
    in contact with a heat reservoir A', the differential of the free energy
    vanishes under small changes of the variables,

$$\Delta F = \Delta(U - TS) = 0 .$$

F has then its minimum because S_{tot} of the total system A+A' is at its
maximum. Note that all variables without an index refer to system A.

It is reassuring that also the condition $\Delta F = 0$ is in line with the
Boltzmann equation (A.10): Suppose the N particles of the system
possess only two levels, 1 and 2. In an equilibrium state a, the lower
level 1 is populated by N_1 and the upper level 2 by N_2 atoms with
$N_1 + N_2 = N$. In a nearby, almost-equilibrium state b, the corresponding
populations are $N_1 - 1$ and $N_2 + 1$. The thermodynamic probability Ω
for the two states is given by (A.1) and the entropy changes between
state a to b by

$$\Delta S = k \left[\ln \Omega_b - \ln \Omega_a\right] = k \ln \frac{\Omega_b}{\Omega_a} = k \ln \frac{N_1}{N_2 + 1} \simeq k \ln \frac{N_1}{N_2} .$$

At constant temperature and volume, the condition $\Delta F = \Delta U - T \Delta S =
0$, where ΔU is the excitation energy of level 2, implies $N_2/N_1 =
e^{-\Delta U/kT}$, as in (A.10).

Let system A depend on some parameter Y. The number of accessible
states where Y takes up values between y and $y + \delta y$ is (see A.28)

$$\Omega_{tot}(y) = e^{S_{tot}(y)/k}$$

and the probability $P(y)$ of such states is proportional to $\Omega_{tot}(y)$. Likewise, $P(y') \propto \Omega_{tot}(y') = e^{S_{tot}(y')/k}$ for another value y' of the parameter
Y. As Y varies from y to y' (this need not be a small step), the entropy
of the total system changes by

$$\Delta S_{tot} = \frac{T \Delta S - \Delta U + \delta A}{T}$$

where δA is the (not necessarily infinitesimally small) work done on
system A and $\Delta S = S(y) - S(y')$ and $\Delta U = U(y) - U(y')$. As the
volume is constant, there is no compressional work and $\delta A = 0$. With
$\Delta F = F(y) - F(y')$, we get for the population ratio of states where
$Y = y$ and $Y = y'$

$$\frac{P(y)}{P(y')} = e^{-\Delta F/kT} .$$

This result follows also from the canonical distribution (A.30).

(2) At constant temperature and pressure ($\Delta T = \Delta p = 0$), the differential of the free enthalpy is zero,

$$\Delta G = \Delta(U + pV - TS) = 0 ,$$

and G has its minimum. In an analogous manner as for the free energy, one derives $\Delta S_{\text{tot}} = -\Delta G/T$ and, putting $\Delta G = G(y) - G(y')$, a population ratio of states where the parameter Y has values y and y'

$$\frac{P(y)}{P(y')} = e^{-\Delta G/kT} . \tag{A.50}$$

(3) At constant pressure ($\Delta p = 0$) and constant entropy ($\Delta S = 0$, adiabatic process), $\Delta H = \Delta(U + pV) = 0$ and the enthalpy has its minimum.

A.4 Blackbody radiation

A.4.1 The Planck function

The Planck function $B_\nu(T)$ gives the intensity of radiation in an enclosure in thermal equilibrium at temperature T. It is a universal function depending only on T and frequency ν. Written per frequency unit,

$$B_\nu(T) = \frac{2h}{c^2} \cdot \frac{\nu^3}{e^{\frac{h\nu}{kT}} - 1} \tag{A.51}$$

with the dimension $\text{erg cm}^{-2}\,\text{s}^{-1}\,\text{ster}^{-1}\,\text{Hz}^{-1}$. $B_\nu(T)$ is related to the monochromatic radiative energy density u_ν in the enclosure by

$$u_\nu = \frac{4\pi}{c} B_\nu(T) = \frac{8\pi h}{c^3} \cdot \frac{\nu^3}{e^{\frac{h\nu}{kT}} - 1} . \tag{A.52}$$

Alternatively, the Planck function may be referred to wavelength and designated $B_\lambda(T)$; its unit is then $\text{erg cm}^{-2}\,\text{s}^{-1}\,\text{ster}^{-1}\,\text{cm}^{-1}$. The two forms are related through

$$B_\lambda(T)\,d\lambda = -B_\nu(T)\,d\nu . \tag{A.53}$$

Because of $d\nu = -(c/\lambda^2)\,d\lambda$,

$$B_\lambda(T) = \frac{2hc^2}{\lambda^5 \left(e^{\frac{hc}{kT\lambda}} - 1\right)} . \tag{A.54}$$

In view of the exponential factor, $B_\nu(T)$ is usually very sensitive to both temperature and frequency. The Planck function increases monotonously with temperature, i.e., at any frequency

$$B_\nu(T_2) > B_\nu(T_1) \quad \text{if} \quad T_2 > T_1 .$$

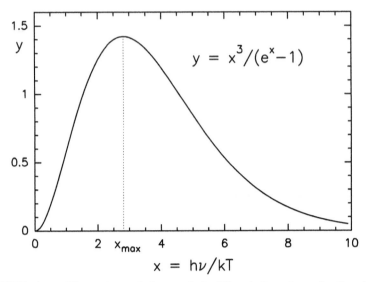

FIGURE A.2 The universal shape of the Planck function. As the abscissa x is here in the unit $h\nu/kT$, one has to multiply the ordinate y by the factor $(2k^3/h^2c^2)T^3$ to get $B_\nu(T)$.

Figure A.2 depicts the graph of the curve

$$y = \frac{x^3}{e^x - 1},$$

from which one can read off the value of the Planck function for any combination (ν, T) of frequency and temperature. It starts at the origin $(0,0)$ with a slope of zero, culminates at $x_{\max} = 2.822$ and asymptotically approaches zero for large x.

An object that emits at all frequencies with an intensity $B_\nu(T)$ is called a blackbody. The emergent flux F_ν from a unit area of its surface into all directions of the half-sphere is

$$F_\nu = \pi B_\nu(T) . \tag{A.55}$$

This expression, for example, is approximately applicable to stellar atmospheres when T is the effective surface temperature.

A.4.2   Low and high frequency limit

There are two asymptotic approximations to the Planck function depending on the ratio x of photon energy $h\nu$ over thermal energy kT,

$$x = \frac{h\nu}{kT} .$$

- In the *Wien limit*, $x \gg 1$ and

$$B_\nu(T) \ \rightarrow \ \frac{2h\nu^3}{c^2} \, e^{-\frac{h\nu}{kT}} \ . \tag{A.56}$$

With respect to dust emission, where wavelengths are typically between $1\mu m$ and $1\,mm$ and temperatures from 10 to $2000\,K$, Wien's limit is never appropriate.

- In the *Rayleigh-Jeans limit*, $x \ll 1$ and

$$B_\nu(T) \ \rightarrow \ \frac{2\nu^2}{c^2} \, kT \ . \tag{A.57}$$

As the photon energy $h\nu$ is much smaller than the thermal energy kT of an oscillator, one can expand the Planck function into powers of x and enters the realm of classical physics where the Planck constant h vanishes. The dependence of $B_\nu(T)$ on frequency and temperature is then no longer exponential. The Rayleigh-Jeans approximation is often applicable to dust emission at long wavelengths ($\lambda \gtrsim 0.3\,mm$), however, one should check whether

$$x \ = \ \frac{1.44}{\lambda T} \qquad (\lambda \ \text{in cm})$$

is really small compared to 1.

A.4.3 The laws of Wien and Stefan-Boltzmann

In the wavelength scale, the Planck function $B_\lambda(T)$ reaches its maximum at $\lambda_{\max}$ given by $\partial B_\lambda/\partial \lambda = 0$, therefore

$$\lambda_{\max} T \ = \ 0.289 \quad \text{cm K} \ . \tag{A.58}$$

In the frequency scale, maximum emission is determined by $\partial B_\nu/\partial \nu = 0$ and occurs at $\nu_{\max}$ for which

$$\frac{T}{\nu_{\max}} \ = \ 1.70 \times 10^{-11} \quad \text{Hz}^{-1} \ \text{K} \ . \tag{A.59}$$

Note that $\lambda_{\max}$ from (A.58), which refers to the wavelength scale, is a factor 1.76 smaller than the corresponding wavelength $c/\nu_{\max}$ from (A.59). The wavelength where the flux from a blackbody peaks depends thus on whether one measures the flux per Hz (F_ν) or per cm (F_λ). The total energy per s over a certain spectral interval is, of course, the same for F_λ and F_ν. If one wants to detect a blackbody with an instrument that has a sensitivity curve S_ν, one usually tries to maximize $\int S_\nu B_\nu(T)d\nu$.

When $\lambda_{\max}$ is known, the displacement law determines the temperature of a blackbody. Interstellar grains are certainly not blackbodies, but the shape of the spectral energy distribution from a dusty region may at far infrared wavelengths be approximated by $\nu^m B_\nu(T)$ (see (6.1) for the correct expression) if the emission is optically thin and the absorption coefficient has a power law dependence $K_\nu \propto \nu^m$. The term

$$\nu^m B_\nu(T)$$

is sometimes called modified Planck function. Maximum emission follows now from $\partial(\nu^m B_\nu)/\partial\nu = 0$. For $K_\nu \propto \nu^2$; one finds $\lambda_{max} T = 0.206 \, cm \, K$; so λ_{max} of radiating dust is shifted to shorter wavelengths with respect to a blackbody emitter of the same temperature.

Integrating the Planck function over frequency, we obtain (see B.2)

$$B(T) = \int_0^\infty B_\nu(T) \, d\nu = \frac{\sigma}{\pi} T^4 \tag{A.60}$$

where

$$\sigma = \frac{2\pi^5 k^4}{15c^2 h^3} = 5.67 \times 10^{-5} \, erg \, cm^{-2} \, s^{-1} \, K^{-4} \tag{A.61}$$

is called radiation or Stefan-Boltzmann constant. The total emergent flux F from a unit area of a blackbody surface into all directions of the half-sphere is given by the Stefan-Boltzmann law (see A.55)

$$F = \int F_\nu \, d\nu = \sigma T^4 . \tag{A.62}$$

Applying it to a star of radius R_* and effective temperature T_*, we find for its bolometric luminosity

$$L_* = 4\pi\sigma R_*^2 T_*^4 . \tag{A.63}$$

For the total radiative energy density u, we get

$$u = \int u_\nu \, d\nu = aT^4 \tag{A.64}$$

with constant

$$a = \frac{4\sigma}{c} = 7.56 \times 10^{-15} \, erg \, cm^{-3} \, K^{-4} . \tag{A.65}$$

A.5   The classical Hamiltonian

Consider a system of particles with generalized coordinates q_i and velocities $\dot{q}_i$. If velocities are small (non-relativistic), one finds the motion of the particles from the *Lagrange function*

$$L = T - V ,$$

where T is the kinetic energy of the particles and V the potential, by integrating the second order Lagrange equations

$$\frac{d}{dt}\left(\frac{\partial L}{\partial \dot{q}_i}\right) = \frac{\partial L}{\partial q_i} \, . \qquad (A.66)$$

Alternatively, using the conjugate momenta

$$p_i = \frac{\partial L}{\partial \dot{q}_i} \, , \qquad (A.67)$$

one constructs the *Hamiltonian*,

$$H = \sum \dot{q}_i \, p_i \; - \; L \, , \qquad (A.68)$$

and integrates the first order equations

$$\dot{q}_i = \frac{\partial H}{\partial p_i}, \qquad -\dot{p}_i = \frac{\partial H}{\partial q_i} \, . \qquad (A.69)$$

If H does not explicitly depend on time, it is a constant of motion. Moreover, in our applications, H equals the total energy E of the system,

$$E = T + V = H(q_i, p_i). \qquad (A.70)$$

A.5.1   Normal coordinates

To determine the eigenfrequencies (resonances) of a grain, one replaces its atoms by mass points connected through springs and treats them as an ensemble of harmonic oscillators. If there are N mass points in the grain, we describe their positions by *one* vector in Cartesian coordinates,

$$\mathbf{x} = (x_1, \ldots, x_n) \, , \qquad n = 3N \, .$$

At the equilibrium position, designated $\mathbf{x}_0$, all forces exactly balance, so the derivatives of the potential $V(\mathbf{x})$ vanish,

$$\left(\frac{\partial V}{\partial x_i}\right)_{\mathbf{x}_0} = 0 \, .$$

We introduce coordinates relative to equilibrium,

$$\eta_i = x_i - x_{i0} \, ,$$

and put $V(\mathbf{x}_0) = 0$, which is always possible. If the oscillations are small, one can approximate the potential by a Taylor expansion. Because V and its first derivatives are zero, the first non-vanishing term is

$$V(\mathbf{x}) = \tfrac{1}{2}\sum_{ij}\left(\frac{\partial^2 V}{\partial x_i \partial x_j}\right)_0 \eta_i \eta_j = \tfrac{1}{2}\sum_{ij} V_{ij}\eta_i\eta_j \, . \qquad (A.71)$$

Because the kinetic energy T is quadratic in $\dot{\eta}_i$,

$$T = \tfrac{1}{2} \sum_i m_i \dot{\eta}_i^2 , \qquad (A.72)$$

one derives from the Lagrange function $L = T - V$ of (A.66) the equations of motion

$$m_i \ddot{\eta}_i + \sum_j V_{ji} \eta_j = 0 \qquad i = 1, \ldots, n . \qquad (A.73)$$

Trying as usual a solution of the kind

$$\eta_i = a_i \, e^{-i\omega t} ,$$

yields a set of n linear equations for the amplitudes a_i,

$$\sum_j (V_{ji} - \delta_{ji} \, m_i \omega^2) \, a_i = 0 .$$

For non-trivial values of a_i, the determinant must vanish,

$$|V_{ji} - \delta_{ji} \, m_i \omega^2| = 0 . \qquad (A.74)$$

This is an algebraic equation of n-th order in ω^2. It has n solutions and one thus finds n frequencies. In fact, six solutions will be zero if there are more than 2 (non-linear) atoms, but the remaining frequencies refer only to internal atomic oscillations. The motion in any coordinate x_i will be a superposition of n harmonic oscillations of different frequencies ω_i and amplitudes a_i.

It is a standard procedure to obtain from the relative coordinates, η_i, through a linear transformation new so-called *normal coordinates* y_i, with the property that each normal coordinate corresponds to a harmonic motion of just one frequency, and that the kinetic and potential energy are quadratic in y_i and $\dot{y}_i$, respectively,

$$T = \tfrac{1}{2} \sum_i \dot{y}_i^2 \qquad V = \tfrac{1}{2} \sum_i \omega_i^2 \, y_i^2 .$$

So a grain of N atoms may be substituted by $f = 3N - 6$ independent oscillators, each with its personal frequency.

A.6 The Hamiltonian in quantum mechanics

A.6.1 The time-dependent Schrödinger equation

In quantum mechanics, the formulae (A.69) governing the motion of the particles are replaced by the Schrödinger equation

$$i\hbar \frac{\partial \Psi}{\partial t} = \hat{H} \, \Psi . \qquad (A.75)$$

This equation follows from $E = H(x_j, p_j)$ of (A.70) by turning the energy E and the Hamiltonian H into operators using the standard prescription for the conversion of energy, conjugate momentum and coordinates,

$$E \rightarrow i\hbar\frac{\partial}{\partial t} \qquad p_j \rightarrow \hat{p}_j = -i\hbar\frac{\partial}{\partial x_j} \qquad x_j \rightarrow \hat{x}_j = x_j \ . \tag{A.76}$$

Operators are marked by a hat on top of the letter. One has to keep in mind the commutation rules, in particular

$$[\hat{x}_j, \hat{p}_k] \ = \ \hat{x}_j\hat{p}_k - \hat{p}_k\hat{x}_j \ = \ i\hbar\,\delta_{jk} \ , \tag{A.77}$$

and

$$[\hat{x}_j, \hat{H}] \ = \ \hat{x}_j\hat{H} \ - \ \hat{H}\hat{x}_j \ = \ i\frac{\hbar}{m}\hat{p}_j \ . \tag{A.78}$$

A.6.2 Stationary solutions

The wave function Ψ in (A.75) depends generally on space and time,

$$\Psi \ = \ \Psi(\mathbf{r}, t) \ .$$

Stationary solutions correspond to fixed energy eigenvalues E. In this case, the wave function can be written as

$$\Psi(\mathbf{r}, t) \ = \ \psi(\mathbf{r}) \cdot e^{-iEt/\hbar} \ . \tag{A.79}$$

$\Psi(x, t)$ contains the time merely through the factor $e^{-iEt/\hbar}$ which cancels out on forming the probability density $|\Psi|^2$. The eigenfunction $\psi(x)$ depends only on coordinate x. The time-independent Schrödinger equation of a particle of energy E in a potential $V(x)$ follows from (A.75),

$$\frac{\hbar^2}{2m}\Delta\psi \ + \ \Big[E \ - \ U(\mathbf{r})\Big]\psi \ = \ 0 \ . \tag{A.80}$$

The eigenfunctions ψ_n, each to an eigenvalue E_n, form a complete set so that an arbitrary wave function $\Psi(x, t)$ can be expanded into a sum of ψ_n,

$$\Psi(\mathbf{r}, t) \ = \ \sum_n a_n \psi_n(\mathbf{r}) \cdot e^{-iE_n t/\hbar} \ , \tag{A.81}$$

with expansion coefficients a_n.

A.6.3 The dipole moment of a transition

The Hamiltonian operator $\hat{H}$ is Hermitian, which means (restricting the discussion to one space coordinate)

$$\int \psi^*(\hat{H}\phi)\, dx \ = \ \int (\hat{H}\psi)^*\phi\, dx$$

for any ψ and ϕ and guarantees real expectation values. The asterisk denotes the complex conjugate. If ψ_k, ψ_k are the eigenfunctions to the eigenvalues E_j, E_k, respectively, it follows with the help of equation (A.78) that

$$\frac{e}{m} \int \psi_j^* \hat{p} \, \psi_k \, dx \;=\; i \frac{e}{\hbar} (E_j - E_k) \int \psi_j^* \hat{x} \, \psi_k \, dx \;.$$

So one can substitute in the left integral for the momentum operator the coordinate operator (but for a constant). If we put $\hbar \omega_{jk} = E_j - E_k$ and define

$$\mu_{jk} \;=\; e \int \psi_j^*(x) \, x \, \psi_k(x) \, dx \;, \qquad (A.82)$$

we get

$$\frac{e}{m} \int \psi_j^* \hat{p} \, \psi_k \, dx \;=\; i \, \omega_{jk} \, \mu_{jk} \;. \qquad (A.83)$$

For $j = k$, the expression for μ_{jk} in (A.82) is the quantum mechanical analog to the classical dipole moment of equation (1.2). When two different states are involved with eigenfunctions $\psi_j(x) \neq \psi_k(x)$, one calls μ_{jk} the dipole moment of the transition $j \to k$.

A.6.4   The quantized harmonic oscillator

We review the solution of the time-independent Schrödinger equation for the undamped free linear harmonic oscillator. The oscillator obeys the equation of motion

$$\ddot{x} + \omega^2 x \;=\; 0 \;,$$

where $\omega^2 = \kappa/m$ is the square of the frequency of oscillation, m the particle mass, and $-\kappa x$ the restoring force (see section 1.2). The total energy of the system equals

$$E \;=\; T + V \;=\; \frac{p^2}{2m} + \tfrac{1}{2} m \omega^2 x^2$$

and the time-independent Schrödinger equation is therefore

$$\frac{\hbar^2}{2m} \Delta \psi_n + \left[E_n - \tfrac{1}{2} m \omega^2 x^2 \right] \psi_n \;=\; 0 \;. \qquad (A.84)$$

with $\Delta = \partial^2/\partial x^2$. This second order differential equation looks much better in the form

$$u_n'' + (\lambda_n - y^2) \, u_n \;=\; 0 \;,$$

where we introduced the function $u_n(\alpha x) = \psi_n(x)$ with

$$\alpha^2 \;=\; \frac{m\omega}{\hbar} \;, \qquad (A.85)$$

and put

$$y \;=\; \alpha x \;, \qquad \lambda_n \;=\; \frac{2E_n}{\hbar \omega} \;.$$

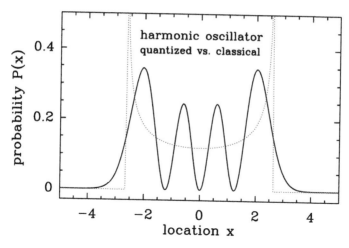

FIGURE A.3 The undulating quantum mechanical probability $P(x) = |\psi(x)|^2$ for finding the particle at locus x when the oscillator is in energy level $n = 3$. The corresponding probability of a classical oscillator of the same energy is shown by dots; here the particle is strictly confined to the allowed region. For high n, the two curves converge.

An ansatz for $u_n(y)$ of the form $e^{-y^2/2} H_n(y)$ yields the normalized eigenfunctions of the harmonic oscillator

$$\psi_n(x) = N_n \, e^{-\alpha^2 x^2/2} \, H_n(\alpha x) \qquad \text{with} \qquad N_n = \sqrt{\frac{\alpha}{2^n n! \sqrt{\pi}}} \, . \qquad (A.86)$$

H_n are Hermite polynomials (see Appendix B.2). The energy levels E_n of the harmonic oscillator in (A.84) are equidistant,

$$E_n = \left(n + \tfrac{1}{2}\right) \hbar\omega, \qquad n = 0, 1, 2, \ldots \qquad (A.87)$$

and the lowest level

$$E_0 = \tfrac{1}{2} \hbar\omega$$

is *above* zero. Figure A.3 displays as an example the square of the eigenfunction $|\psi_n(x)|^2$ for $n = 3$ and $\alpha^2 = 1$.

With the help of (B.20) we find that the mean position and momentum of an oscillator, $\langle x \rangle$ and $\langle p \rangle$, always disappear for any energy E_n. So

$$\frac{d\langle x \rangle}{dt} = \frac{\langle p \rangle}{m}$$

is fulfilled in a trivial way, but $\langle x^2 \rangle$ and $\langle p^2 \rangle$ do not vanish. For the product of the uncertainties we have

$$\Delta x \, \Delta p = \sqrt{\langle x^2 \rangle - \langle x \rangle^2} \, \sqrt{\langle p^2 \rangle - \langle p \rangle^2} = \frac{E_n}{\omega} = (n + \tfrac{1}{2})\hbar \, . \qquad \text{(A.88)}$$

The probability for a downward transition $v' \rightarrow v$ of an harmonic oscillator is calculated according to the general formulae (A.93) and (A.94). With the eigenfunctions from (A.86), we are led to the integral

$$\int H_v(y) \, y \, H_{v'}(y) \, e^{-y^2} \, dy$$

which we solve using (B.19) and (B.20). We find that it does not vanish only when $\Delta v = v' - v = 1$ and then

$$A_{v,v-1} = v \, A_{1,0} \qquad (v \geq 1) \, . \qquad \text{(A.89)}$$

The Einstein coefficient is proportional to the quantum number v; all one has to know is $A_{1,0}$, the coefficient for the ground transition.

A.7   The Einstein coefficients A and B

A.7.1   Induced and spontaneous transitions

A.7.1.1   How A and B are defined

We consider radiative transitions in atoms between an upper level j and a lower one i of energies E_j and E_i such that

$$E_j - E_i = h\nu \, .$$

Let the atoms have number densities N_j and N_i per cm^3, respectively, and be bathed in a radiation field of energy density u_ν. As atomic energy levels and lines are not infinitely sharp, but have a finite width, we define the profile function $\Phi(\nu)$ by demanding that the number of atoms in state i that can absorb radiation in the frequency interval $[\nu, \nu + d\nu]$ is equal to

$$N_i(\nu) \, d\nu = N_i \, \Phi(\nu) \, d\nu \, ;$$

of course,

$$\int_{\text{line}} \Phi(\nu)\, d\nu \; = \; 1 \;.$$

Radiative transitions between level j and i may be either spontaneous or induced. Setting

$$u \; = \; \int_{\text{line}} u_\nu \, \Phi(\nu)\, d\nu \;, \tag{A.90}$$

the Einstein coefficients A_{ji} and B_{ji} are defined such that per cm^3

- $N_j \, A_{ji}$ $\;=\;$ rate of *spontaneous downward* transitions

- $N_j \, u B_{ji}$ $\;=\;$ rate of *induced downward* transitions

- $N_i \, u B_{ij}$ $\;=\;$ rate of *induced upward* transitions.

The A's and B's are related through

$$g_j \, B_{ji} \; = \; g_i \, B_{ij} \;, \tag{A.91}$$

(g_i, g_j are the statistical weights of level i and j) and

$$A_{ji} \; = \; \frac{8\pi h \nu^3}{c^3} \, B_{ji} \;. \tag{A.92}$$

The quantum mechanical expression for the Einstein coefficient A_{ji} for spontaneous emission from upper state j to lower state i is

$$A_{ji} \; = \; \frac{64\pi^4 \nu^3}{3hc^3} \, |\boldsymbol{\mu}_{ji}|^2 \;, \tag{A.93}$$

where

$$\boldsymbol{\mu}_{ji} \; = \; e \int \psi_i^*(\mathbf{r})\, \mathbf{r}\, \psi_j(\mathbf{r})\, dV \tag{A.94}$$

is the dipole moment corresponding to the transition (see A.82) and ψ_i and ψ_j are the eigenfunctions of the states.

A.7.1.2 A classical analogy

The essence of formula (A.93) can already be grasped using classical arguments by equating the emission rate $A_{ji}\, h\nu$ to the average power radiated by a harmonic oscillator (see 1.92). If $x = x_0\, e^{-i\omega t}$ is the time-variable coordinate of the electron and $p_0 = e x_0$ its dipole moment, then

$$A_{ji}\, h\nu \; = \; \frac{p_0^2 \,\omega^4}{3c^3} \;.$$

This yields exactly (A.93) if one puts $\mu = \frac{1}{2} p_0$ and $\omega = 2\pi\nu$. Note that A_{ji} increases with the square of the dipole moment and with the third power of the frequency. We see from (A.93) that, *cum granu salis*, A_{ji} is high for optical transitions and low at radio wavelengths. To get a feeling for the numbers, we apply the formula:

(1) to an electronic transition in the hydrogen atom. For an atomic radius $x_0 \simeq 0.5 \text{Å}$, the dipole moment is $\mu = 2.4 \times 10^{-18}$ cgs $\simeq 2.4$ Debye. At an optical frequency $\nu = 6 \times 10^{14}$ Hz, corresponding to $\lambda = 5000 \text{ Å}$, the Einstein coefficient becomes $A_{ji} \sim 10^7 \, \text{s}^{-1}$.

(2) To the lowest rotational transition of the CO molecule. Now the dipole moment is weak ($\mu = 0.11$ Debye), the frequency low ($\nu \simeq 1.15 \times 10^{11}$ Hz) and therefore $A_{ji} \sim 10^{-7} \, \text{s}^{-1}$ is 14 powers of ten smaller!

A.8 Potential wells and tunneling

A.8.1 Wave function of a particle in a constant potential

Consider in one dimension a free particle of mass m and energy E in a potential $V(x)$. To facilitate writing, in this section we often put

$$U(x) = \frac{2m}{\hbar^2} V(x) , \qquad \varepsilon = \frac{2m}{\hbar^2} E . \tag{A.95}$$

In this notation, the wave function of a stationary state satisfies the equation

$$\psi'' = \left[U(x) - \varepsilon \right] \psi . \tag{A.96}$$

If U is constant, the general solution reads

$$\psi(x) = A_1 e^{i\alpha x} + A_2 e^{-i\alpha x}, \qquad \alpha = \sqrt{\varepsilon - U} \tag{A.97}$$

with complex A_i. There are two cases:

- $\varepsilon - U > 0$. Then $\alpha > 0$ and the wave has an oscillatory behavior. It could be represented by a combination of a sine and a cosine.

- $\varepsilon - U < 0$. Then $i\alpha$ is real and the two solutions correspond to an exponentially growing and declining function. The former is only allowed in finite intervals because the integral $\int \psi^* \psi \, dx$ must be bounded.

If the potential $U(x)$ is a step function, i.e., if it is constant over intervals $[x_i, x_{i+1}]$, but makes jumps at the connecting points x_i, with $x_1 < x_2 < x_3 < \dots$, the wave function ψ and its derivative ψ' must be continuous at these points; otherwise the Schrödinger equation would not be fulfilled. The boundary conditions of continuity for ψ and ψ' at all x_i fully determine the wave function. To find it is straightforward, but often tedious.

A.8.2 Potential walls and Fermi energy

If the particle is trapped in an infinitely deep well of length L, as depicted on the left of figure A.4, the wave function must vanish at the boundaries. It cannot penetrate beyond the walls. The situation is reminiscent of a vibrating string fixed at its ends. If $\lambda = h/p$ is the de Broglie wavelength and p the momentum, the condition $n\lambda/2 = L$ leads to the discrete energies

$$\varepsilon_n = \frac{2m}{\hbar^2} E_n = \frac{n^2\pi^2}{L^2} , \qquad n = 1, 2, \dots \qquad (A.98)$$

and sine-like wave functions

$$\psi_n(x) = A \sin\left(\frac{n\pi}{L}x\right) .$$

There is a zero-point energy ($\varepsilon > 0$ for $n=1$) because the particle is spatially localized in the interval from $-L/2$ to $L/2$, and so its momentum cannot vanish. The spacing between energy levels increases quadratically with quantum number n. This is very different from an harmonic oscillator which has a parabolic potential and equally spaced energy levels (see figure A.4).

If a system with N electrons is in its lowest state, all levels $n \le N/2$ are filled. The threshold or Fermi energy of the topmost filled level is

$$E_F = \frac{\hbar^2\pi^2 N^2}{8m_e L^2} \qquad (A.99)$$

and the velocity of those electrons is $v_F = \sqrt{2E_F/m_e}$.

If the walls of the potential well are finite and of height U and the particle energy $\varepsilon < U$, there is some chance, exponentially decreasing with distance, to find it outside the well. The energies of the eigenstates depend then in a more complicated manner on n.

In three dimensions, for a cube of length L and infinite potential barrier, the eigenfunctions are

$$\psi(\mathbf{r}) = A \sin\left(\frac{n_x\pi}{L}x\right) \cdot \sin\left(\frac{n_y\pi}{L}y\right) \cdot \sin\left(\frac{n_z\pi}{L}z\right) ,$$

with positive integers n_x, n_y, n_z, and the Fermi energy is

$$E_F = \frac{\hbar^2}{2m_e}\left(\frac{3\pi^2 N}{L^3}\right)^{2/3} . \qquad (A.100)$$

For example, the conduction electrons in a metal have Fermi temperatures, defined by

$$T_F = \frac{E_F}{k} ,$$

between 10^4 and 10^5 K. So at $T \sim 100$ K, the energy distribution of the conduction electrons is highly degenerate ($\alpha \ll -1$ in A.21); only few of them are thermalized. For graphite particles of interstellar dust grains, the Fermi temperature is much lower, of order 10^3 K.

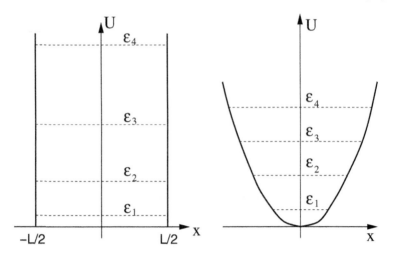

FIGURE A.4 *Left:* An electron in a square potential with walls of infinite height. The stationary states correspond to energies $E_n \propto n^2$ (see A.98). *Right:* For comparison, the parabolic potential and the equidistant energy levels of an harmonic oscillator.

A.8.3   Rectangular potential barriers

A.8.3.1   A single barrier

Let a particle travel from left to right and encounter a potential wall, as in figure A.5. If its energy ε is greater than U, it can of course overcome the barrier, but contrary to classical physics, part of the wave is reflected. Most relevant is the case when $\varepsilon < U$. Then the particle can tunnel through the barrier, which is classically forbidden. What happens, as the particle approaches the barrier, depends obviously on time and therefore stationary solutions may not seem possible in this scenario. However, one may interpret the infinite wave $e^{\frac{i}{\hbar}(px - Et)}$ of a particle with definite energy E as a stationary particle beam, and then the time drops out.

The wave function is constructed piece-wise,

$$
\psi(x) = \begin{cases}
A\,e^{i\alpha x} + B\,e^{-i\alpha x} : & x < 0 \\
C\,e^{\beta x} + D\,e^{-\beta x} : & 0 < x < L \\
F\,e^{i\alpha x} + G\,e^{-i\alpha x} : & x > L .
\end{cases}
$$

The coefficient $G = 0$, as there is no wave coming from the right, and

$$
\alpha = \sqrt{\varepsilon} > 0 \qquad \beta = \sqrt{U - \varepsilon} > 0 .
$$

Exploiting the boundary conditions and keeping in mind that $(e^x + e^{-x})^2/4 = \cosh^2 x = 1 + \sinh^2 x$, the transmission coefficient T, which is the fraction of

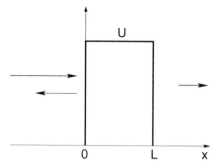

FIGURE A.5 Rectangular potential barrier for a particle coming from the left. Part of the beam is reflected and part goes through the barrier.

the particle beam that penetrates the barrier, becomes

$$T = \frac{4\varepsilon(U-\varepsilon)}{4\varepsilon(U-\varepsilon) + U^2 \sinh^2(L\sqrt{U-\varepsilon})} \, . \tag{A.101}$$

When the barrier is high compared to the kinetic energy of the particle $(U \gg \varepsilon)$ and sufficiently broad $(LU^{1/2} > 1)$, the tunneling probability can be approximated by

$$T = \frac{16\,\varepsilon}{U} e^{-2L\sqrt{U}} \, . \tag{A.102}$$

A.8.3.2 Periodic potential

Regularly arranged atoms in a solid (crystal) produce a periodic potential. We discuss this important case in one dimension. Let the potential be periodic with period a, as depicted in figure A.6. The wave function ψ of a particle with energy $\varepsilon < U$ is then given by

$$\psi(x) = \begin{cases} A\,e^{i\alpha x} + B\,e^{-i\alpha x} : & 0 < x < b \\ C\,e^{\beta x} + D\,e^{-\beta x} : & b < x < a \end{cases}$$

with

$$\alpha = \sqrt{\varepsilon} \qquad \beta = \sqrt{U-\varepsilon} \, .$$

Two equations for determining A, B, C, D are found from the condition that ψ and ψ' must be continuous at $x = 0$. Another two follow from a result detailed in section 5.2.2: In a periodic potential, any eigenfunction ψ_k, with respect to wavenumber k and energy $\varepsilon = k^2$, must have the form

$$\psi_\mathrm{k} = u_\mathrm{k}\,e^{ikx}$$

where u_k is periodic such that

$$u_\mathrm{k}(x) = u_\mathrm{k}(x+a) \, .$$

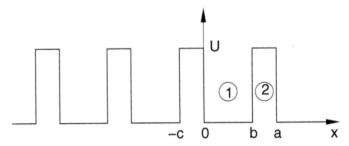

FIGURE A.6 A periodic square potential of a one-dimensional lattice. The length of the period is a; the length of one barrier is $c = a - b$. In region 1, where $0 \leq x \leq b$, the potential is zero, in region 2, where $b \leq x \leq a$, it has the value U.

Therefore, $u_k(b) = u_k(-c)$ and $u'_k(b) = u'_k(-c)$, where $u_k(x) = \psi_k(x)\,e^{-ikx}$. Altogether, this gives

$$A + B = C + D$$
$$i\alpha\,(A - B) = \beta\,(C - D)$$
$$e^{-ikb}\left[Ae^{i\alpha b} + Be^{-i\alpha b}\right] = e^{ikc}\left[Ce^{-\beta c} + De^{\beta c}\right]$$
$$ie^{-ikb}\left[Ae^{i\alpha b}(\alpha - k) - Be^{-i\alpha b}(\alpha + k)\right] = e^{ikc}\left[Ce^{-\beta c}(\beta - ik) - De^{\beta c}(\beta + ik)\right]\;.$$

Non-trivial solutions of this linear system of equations, those for which *not* $A = B = C = D = 0$, have a vanishing determinant. The condition Det $= 0$ leads after some algebra to

$$\cos ka \;=\; \frac{\beta^2 - \alpha^2}{2\alpha\beta}\,\sinh\beta c \cdot \sin\alpha b \;+\; \cosh\beta c \cdot \cos\alpha b\;. \tag{A.103}$$

For fixed values a, b, U of the potential in figure A.6, the right side of (A.103) is a function of particle energy ε only; we denote it by $f(\varepsilon)$. Because of the cosine term on the left side of (A.103), only energies for which $f(\varepsilon) \in [-1, 1]$ are permitted and those with with $|f(\varepsilon)| > 1$ are forbidden. The function $f(\varepsilon)$ is plotted in figure A.7 for a specific set of values U, a, c.

In the potential well on the left of figure A.4, which is not periodic and where the walls are infinite, the energy ε rises quadratically with the wavenumber, $\varepsilon = k^2$. If the length L of the well in figure A.4 is large, there will be a continuous spectrum as depicted by the continuous line in figure A.8. In a periodic potential, on the other hand, the function $\varepsilon(k)$ is only aproximately quadratic and there are now jumps at integral values of $|ka/\pi|$.

Although figure A.8 is derived for a one-dimensional potential, it nevertheless gives the basic explanation why electrons in crystals, which have a periodic (but of course three-dimensional) potential, are arranged in energy bands (they correspond to the fat stripes in figure A.8) with energy gaps between them.

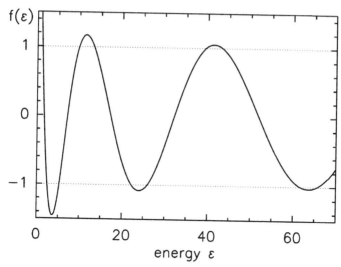

FIGURE A.7 The quasi-sinusoidal variation of the right side of equation (A.103) as a function of particle energy ε. It is calculated for the periodic potential of figure A.6 with $U = 100$, $a = 2$, $c = 0.04$. Energies for which $|f(\varepsilon)| > 1$ are forbidden. This is, for instance, the case when $\varepsilon \simeq 4, 11$ or 24.

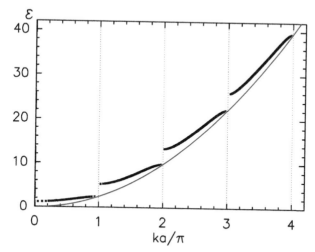

FIGURE A.8 The energy spectrum $\varepsilon(k)$ of particles in a periodic potential as in figure A.6 calculated for $U = 100$, $a = 2$ and $c = 0.04$. The function $\varepsilon(k)$ has discontinuities at integer values of ka/π. The curve without discontinuities is explained in the text. See also figure A.7.

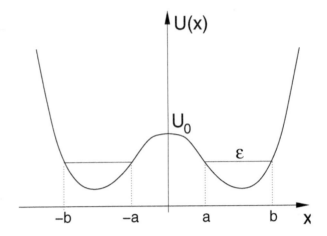

FIGURE A.9 A particle in a double potential well with a barrier U_0 between the minima. Because of tunneling, an energy level ε splits into two with a separation $\Delta\varepsilon$ given by (A.106).

A.8.4 The double potential well

A.8.4.1 Splitting of energy levels

Next we consider an atom of mass m and energy ε in a potential $U(x)$ consisting of two symmetric adjacent wells with a barrier U_0 between them as depicted in figure A.9. Classical motion is allowed in the intervals $[-b, -a]$ and $[a, b]$, where $U(x) \le \varepsilon$. When the barrier U_0 is infinitely large, the particle can only be in the left or right well. The corresponding wave functions are denoted by ψ_- and ψ_+, respectively, and satisfy the Schrödinger equation (see A.96)

$$\psi_+'' + \left[\varepsilon - U(x)\right]\psi_+ = 0 \qquad \psi_-'' + \left[\varepsilon - U(x)\right]\psi_- = 0 . \qquad (A.104)$$

If the barrier U_0 is finite, there is a certain chance of tunneling from one well to the other. The particle is now not localized in either of them and its wave function is a symmetric and anti-symmetric combination of ψ_+ and ψ_-,

$$\psi_1 = \frac{\psi_+}{\sqrt{2}} + \frac{\psi_-}{\sqrt{2}} \qquad \psi_2 = \frac{\psi_+}{\sqrt{2}} - \frac{\psi_-}{\sqrt{2}} .$$

The probability of finding the particle either in the left or right well is unity and the integral of $|\psi_i|^2$ over the whole x-axis equals one. The wave functions ψ_1 and ψ_2 also obey the Schrödinger equations

$$\psi_1'' + \left[\varepsilon_1 - U(x)\right]\psi_1 = 0 \qquad \psi_2'' + \left[\varepsilon_2 - U(x)\right]\psi_2 = 0 , \qquad (A.105)$$

but with slightly different energies ε_1 and ε_2. To obtain the difference $\varepsilon_2 - \varepsilon_1$, we multiply the left equation in (A.104) by ψ_1 and the left equation in (A.105) by ψ_+, subtract the two products, and integrate from $x = 0$ to $x = \infty$. Exploiting $\psi_1(0) = \psi_+(0)$, $\psi_1'(0) = 0$ and assuming that the tunneling probability is small such that $\psi_1 \simeq \psi_+$ for $x > 0$ gives $\varepsilon_1 - \varepsilon = -2^{\frac{1}{2}} \psi_+'(0) \psi_+(0)$. Likewise we find $\varepsilon_2 - \varepsilon$, and finally get

$$\varepsilon_2 - \varepsilon_1 = \psi_+(0) \cdot \psi_+'(0) .$$

So the energy difference is determined by the wave function ψ_+ in the classically forbidden region at $x = 0$. If $p = \hbar\sqrt{\varepsilon - U(x)}$ denotes the momentum and p_0 its value at $x = 0$, one finds

$$\psi_+(0) = \frac{\hbar}{p_0}\psi_+'(0) = \sqrt{\frac{\nu_0 m}{p_0}} \exp\left[-\frac{1}{\hbar}\int_0^a |p|\, dx\right]$$

where ν_0 is the classical oscillation frequency and ν_0^{-1} the time for the particle to go from a to b and back, therefore

$$\varepsilon_2 - \varepsilon_1 = \frac{4m\nu_0}{\hbar} \exp\left[-\frac{1}{\hbar}\int_{-a}^a |p|\, dx\right] \sim \frac{4m\nu_0}{\hbar}\exp\left[-\tfrac{2}{3}a\sqrt{U_0 - \varepsilon)}\right] .$$
$$(A.106)$$

In the forbidden region $[0, a]$, p is imaginary. The value of the integral depends on how $(U_0 - \varepsilon)$ behaves there, but (A.106) is a reasonable guess.

The possibility that the particle can tunnel to the adjacent well leads to two wavefunctions, ψ_1 and ψ_2. Their space probabilities $|\psi_i|^2$ are different, although only in the forbidden region, but there they couple differently to the potential $U(x)$ resulting in the splitting of the energy ε into two levels, ε_1 and ε_2.

A.8.4.2 Tunneling time

To find the approximate time t_{tun} that the particle needs to tunnel from the left to the right well or back, we turn to the time-dependent wave function

$$\Psi(t) = \psi_1 e^{-\frac{i}{\hbar}E_1 t} + \psi_2 e^{-\frac{i}{\hbar}E_2 t} .$$

Here we used again $E = (\hbar^2/2m)\varepsilon$ and $V = (\hbar^2/2m)U$ (see A.95). At time $t = 0$, the wave function has the value $\Psi(0) = \sqrt{2}\psi_+$, so the particle starts in the right well. It reaches the left well when $\Psi(t) = \psi_-$, which happens after a tunneling time

$$t_{\text{tun}} = \frac{\pi\hbar}{|E_1 - E_2|} . \qquad (A.107)$$

This equation expresses the uncertainty principle $\Delta E \cdot \Delta t \simeq \hbar$. A precisely defined energy means that the particle is fixed in one well and does not tunnel.

A broad energy band ΔE, on the other hand, implies a high mobility; the particle is then not localized anywhere.

From (A.106) and (A.107) we see that with increasing potential barrier V_0, the energy splitting $E_2 - E_1$ becomes smaller and the tunneling time longer. Inserting the energy difference of (A.106) into (A.107), we find

$$t_{\text{tun}} \simeq \nu_0^{-1} \exp\left[\frac{2a\sqrt{2mV_0}}{\hbar}\right] \tag{A.108}$$

B

Miscellaneous

B.1   Mathematical notations

- Vectors are written boldface: torque $\boldsymbol{\tau}$, velocity $\mathbf{v}$, electric field $\mathbf{E}$ etc.

- The absolute values of the above vectors are τ, v and E, respectively. As an exception, the position vector in Cartesian coordinates, (x, y, z), is sometimes denoted by $\mathbf{x}$, and then, of course, $|\mathbf{x}| = \sqrt{x^2 + y^2 + z^2}$ and not $|\mathbf{x}| = x$.

- A dot over a letter means time derivative. When a function ϕ depends on the space coordinates (x, y, z) and on time t, its full differential is

$$d\phi = \frac{\partial \phi}{\partial t}\, dt + \frac{\partial \phi}{\partial x}\, dx + \frac{\partial \phi}{\partial y}\, dy + \frac{\partial \phi}{\partial z}\, dz\ .$$

  In our notation, $\dot{\phi} = \partial \phi / \partial t$, which is, of course, different from $d\phi / dt$.

  A prime denotes space derivative: $f'(x) = df/dx$ where x is a space coordinate.

- For the vector operators, we usually use the notations div, rot and grad. Other common symbols are

$$\text{rot} = \text{curl} = \nabla \times = \nabla \wedge$$

$$\text{div} = \nabla \cdot \qquad \text{grad} = \nabla \qquad \Delta = \nabla^2 = \text{div grad}\ .$$

  The *Nabla* or *Hamilton* operator ∇ is mathematically more convenient, whereas div, grad and rot are physically more suggestive. Δ is the *Laplace* operator.

- The symbol $\simeq$ indicates that two quantities are quite similar; $\sim$ means that the agreement is only rough.

B.2   Mathematical formulae

B.2.1   Sums and integrals

$$\sum_{i \geq 0} y^i = \frac{1}{1-y} \qquad \Longrightarrow \qquad \sum_{i \geq 1} i y^{i-1} = \frac{d}{dy} \sum_{i \geq 0} y^i = \frac{1}{(1-y)^2} \qquad (\text{B.1})$$

$$I(s) = \int_0^\infty \frac{x^{s-1}}{e^x - 1} \, dx = \Gamma(s) \cdot \zeta(s) = (s-1)! \cdot \sum_{n \geq 1} \frac{1}{n^s} \, . \qquad (\text{B.2})$$

The last integral is expressed as the product of the Gamma-function, $\Gamma(s)$, and Riemann's Zeta-function, $\zeta(s)$. In particular,

$$\zeta(4) = \frac{\pi^4}{90} \, , \qquad I(4) = \frac{\pi^4}{15} \, .$$

$$\int \frac{\sqrt{x^2 - a^2}}{x} \, dx = \sqrt{x^2 - a^2} - a \cdot \arccos \frac{a}{x} \qquad (\text{B.3})$$

$$\int \frac{\arcsin(ax)}{x^3} \, dx = -\frac{\arcsin(ax)}{2x^2} - a \frac{\sqrt{1 - a^2 x^2}}{2x} \qquad (\text{B.4})$$

B.2.2   The bell curve

For $a > 0$,

$$\int_0^\infty e^{-ax^2} \, dx = \frac{\sqrt{\pi}}{2\sqrt{a}} \qquad (\text{B.5})$$

$$\int_0^\infty x^3 e^{-ax^2} \, dx = \frac{1}{2a^2} \qquad (\text{B.8})$$

$$\int_0^\infty x e^{-ax^2} \, dx = \frac{1}{2a} \qquad (\text{B.6})$$

$$\int_0^\infty x^4 e^{-ax^2} \, dx = \frac{3}{8} \sqrt{\frac{\pi}{a^5}} \qquad (\text{B.9})$$

$$\int_0^\infty x^2 e^{-ax^2} \, dx = \frac{\sqrt{\pi}}{4\sqrt{a^3}} \qquad (\text{B.7})$$

$$\int x e^{-ax^2} \, dx = -\frac{1}{2a} e^{-ax^2} \qquad (\text{B.10})$$

$$\int x^2 e^{-ax^2} \, dx = -\frac{xe^{-ax^2}}{2a^2} + \frac{\sqrt{\pi}}{4a^3} \, \text{erf}(ax) \qquad (\text{B.11})$$

$$\int x^3 e^{-ax^2} \, dx = -\frac{ax^2 + 1}{2a^2} e^{-ax^2} \qquad (\text{B.12})$$

B.2.3 Polynomials

B.2.3.1 Legendre polynomials

The differential equation

$$(1 - x^2)\, u'' \; - \; 2xu' \; + \; n(n+1)\, u \; = \; 0 \, , \tag{B.13}$$

where $n = 0, 1, 2, \ldots$, is satisfied on the interval $[-1, 1]$ by the Legendre polynomials $P_n(x)$. The first four read

$$P_0(x) = 1 \qquad\qquad P_2(x) = \tfrac{1}{2}\,(3x^2 - 1)$$
$$P_1(x) = x \qquad\qquad P_3(x) = \tfrac{1}{2}\,(5x^3 - 3x) \, .$$

Orders higher than $n = 1$ follow from the recursion formula

$$P_{n+1}(x) \; = \; \frac{2n+1}{n+1}\, x\, P_n(x) \; - \; \frac{n}{n+1}\, P_{n-1}(x) \, . \tag{B.14}$$

The functions $P_n(x)$ are normalized by $P_n(1) = 1$. They are orthogonal (with respect to the weight function $w(x) = 1$),

$$\int_{-1}^{1} P_j(x)\, P_n(x)\, dx \; = \; \delta_{jn}\, \frac{2}{2n+1} \, . \tag{B.15}$$

δ_{jn} is the Kronecker delta. The differential equation

$$(1 - x^2)\, u'' \; - \; 2xu' \; + \; \left[n(n+1) \; - \; \frac{m^2}{1 - x^2} \right] u \; = \; 0 \, , \tag{B.16}$$

where again $n = 0, 1, 2, \ldots$ and $m = 1, 2, 3, \ldots$, is satisfied by the associated Legendre polynomials $P_n^m(x)$.

The addition theorem reads

$$P_n\!\left[\mu\mu' + \sqrt{1 - \mu^2}\sqrt{1 - \mu'^2}\, \cos\phi \right] =$$
$$P_n(\mu)P_n(\mu') + 2 \sum_{m=1}^{n} \frac{(m-n)!}{(m+n)!}\, P_n^m(\mu)P_n^m(\mu') \cos m\phi \, . \tag{B.17}$$

B.2.3.2 Hermite polynomials

The Hermite polynomials are defined through

$$H_n(y) \; = \; (-1)^n\, e^{y^2}\, \frac{d^n}{dy^n}\, e^{-y^2} \qquad (n = 0, 1, 2, \ldots) \, . \tag{B.18}$$

They may be generated from the recurrence relation $(n \geq 1)$

$$H_{n+1}(y) \; - \; 2y\, H_n(y) \; + \; 2n\, H_{n-1}(y) \; = \; 0 \, . \tag{B.19}$$

The Hermite polynomials $H_n(y)$ are orthogonal with respect to the weight function e^{-y^2}, in particular

$$\int_{-\infty}^{\infty} H_n^2(y) \, e^{-y^2} \, dy \;=\; 2^n n! \sqrt{\pi} \; .$$

Exploiting (B.19) gives

$$\int_{-\infty}^{\infty} H_i(y) \, y \, H_j(y) \, e^{-y^2} \, dy = \frac{1}{2} \int_{-\infty}^{\infty} H_i \cdot \Big[H_{j+1} + 2j \, H_{j-1} \Big] e^{-y^2} \, dy$$

$$= 2^{i-1} \, i! \, \sqrt{\pi} \, \Big\{ \delta_{i,j+1} + 2j \, \delta_{i,j-1} \Big\} \; . \qquad (B.20)$$

The first four Hermite polynomials are

$$H_0(y) = 1 \qquad\qquad\qquad H_2(y) = 4y^2 - 2$$
$$H_1(y) = 2y \qquad\qquad\qquad H_3(y) = 8y^3 - 12y \; .$$

B.2.4 Vector analysis

Let $f(x, y, z)$ be a scalar function and $\mathbf{A}(x, y, z)$, $\mathbf{B}(x, y, z)$ vector functions. Then

$$\text{div} \, (f\mathbf{A}) = \mathbf{A} \cdot \text{grad} \, f \; + \; f \, \text{div} \, \mathbf{A} \qquad\qquad (B.21)$$

$$\text{div} \, (\mathbf{A} \times \mathbf{B}) = \mathbf{B} \cdot \text{rot} \, \mathbf{A} \; - \; \mathbf{A} \cdot \text{rot} \, \mathbf{B} \qquad\qquad (B.22)$$

$$\text{rot rot} \, \mathbf{A} = \text{grad div} \, \mathbf{A} \; - \; \Delta\mathbf{A} \qquad\qquad (B.23)$$

$$\text{grad} \, (\mathbf{A} \times \mathbf{B}) = (\mathbf{B} \, \text{grad}) \, \mathbf{A} \; + \; (\mathbf{A} \, \text{grad})\mathbf{B} \; + \; \mathbf{B} \times \text{rot}\mathbf{A} \; + \; \mathbf{A} \times \text{rot} \, \mathbf{B} \quad (B.24)$$

$$\text{rot}(\mathbf{A} \times \mathbf{B}) = (\mathbf{B} \, \text{grad}) \, \mathbf{A} \; - \; (\mathbf{A} \, \text{grad}) \, \mathbf{B} \; + \; \mathbf{A} \, \text{div} \, \mathbf{B} \; - \; \mathbf{B} \, \text{div} \, \mathbf{A} \qquad (B.25)$$

$$\text{rot grad} \, f = \mathbf{0} \qquad\qquad (B.26)$$

$$\text{div rot} \, \mathbf{A} = 0 \qquad\qquad (B.27)$$

$$(\mathbf{A} \cdot \text{grad}) \, \mathbf{A} = \tfrac{1}{2} \, \text{grad} \, \mathbf{A}^2 \; - \; \mathbf{A} \times \text{rot} \, \mathbf{A} \qquad\qquad (B.28)$$

$(\mathbf{A} \, \text{grad}) \, \mathbf{B}$ is short writing. For example, the x-component of this vector is

$$\Big[(\mathbf{A} \cdot \text{grad}) \, \mathbf{B} \Big]_x = A_x \frac{\partial B_x}{\partial x} + A_y \frac{\partial B_x}{\partial y} + A_z \frac{\partial B_x}{\partial z}$$

- Let $\mathbf{A}$ be a vector function over some region G with surface S. Gauss' theorem connects the volume and the surface integral,

$$\int_G \text{div} \, \mathbf{A} \, dV = \oint_S \mathbf{A} \cdot \mathbf{n} \, d\sigma \; . \qquad\qquad (B.29)$$

The unit vector $\mathbf{n}$ on the surface element $d\sigma$ on S is directed outwards.

- Let $\mathbf{A}$ be a vector function over some surface S with boundary B. Stokes' theorem connects the surface integral to a line integral,

$$\int_S \text{rot } \mathbf{A} \cdot \mathbf{n} \, d\sigma = \oint_B \mathbf{A} \cdot d\mathbf{s} \tag{B.30}$$

B.2.5 The time average of an harmonically varying field

Let there be two *real* physical fields with a phase difference φ between them,

$$A_{\text{R}} = A_{0\text{R}} \cos \omega t, \qquad B_{\text{R}} = B_{0\text{R}} \cos(\omega t - \varphi),$$

and real amplitudes $A_{0\text{R}}, B_{0\text{R}}$. When they are rapidly oscillating, one is mostly interested in mean values over many cycles. For the product

$$S = A_{\text{R}} \cdot B_{\text{R}},$$

we get the time average

$$\langle S \rangle = \tfrac{1}{2} A_{0\text{R}} B_{0\text{R}} \cos \varphi,$$

and therefore

$$\langle A_{\text{R}}^2 \rangle = \tfrac{1}{2} A_{0\text{R}}^2.$$

Next let us write the fields in *complex* form,

$$A = A_0 e^{-i\omega t}, \qquad B = B_0 e^{-i\omega t},$$

which is more convenient when taking derivatives, and allow the amplitudes A_0, B_0 to be complex too, so that A and B need not be in phase. Of course, only the real components have physical meaning. We now obtain for the product

$$S = \text{Re}\{A\} \cdot \text{Re}\{B\} = \tfrac{1}{4}(A + A^*)(B + B^*) = \tfrac{1}{4}(AB^* + A^*B + AB + A^*B^*).$$

An asterisk denotes the complex conjugate. Taking the mean of S, the third and fourth term in the last bracket vanish because they contain the factor $e^{\pm 2i\omega t}$. In the remaining expression, the time cancels out. So finally

$$\langle S \rangle = \tfrac{1}{2} \text{Re}\{A_0 B_0^*\}, \tag{B.31}$$

and the time average of the square of a complex quantity A is

$$\langle A^2 \rangle = \tfrac{1}{2} |A_0|^2. \tag{B.32}$$

B.3   Cosmic constants

Bohr radius $\hbar^2/m_e e^2$	a_0	$= 5.2918 \times 10^{-8}$	cm
Bohr magneton $e\hbar/2m_e c$	μ_b	$= 9.2741 \times 10^{-21}$	esu cm
Boltzmann constant	k	$= 1.3806 \times 10^{-16}$	erg K^{-1}
electron rest mass	m_e	$= 9.1096 \times 10^{-28}$	g
elementary charge	e	$= 4.8033 \times 10^{-10}$	esu
fine structure constant $e^2/\hbar c$	α	$= 1 : 137.04$	
gravitational constant	G	$= 6.673 \times 10^{-8}$	cm^3 g^{-1} s^{-2}
Planck's constant	h	$= 6.6262 \times 10^{-27}$	erg s
proton rest mass	m_p	$= 1.6726 \times 10^{-24}$	g
Rydberg constant	R_∞	$= 1.097373 \times 10^5$	cm^{-1}
velocity of light	c	$= 2.9979 \times 10^{10}$	cm s^{-1}

astronomical unit	1 AU	$= 1.496 \times 10^{13}$	cm
parsec	1 pc	$= 3.086 \times 10^{18}$	cm

Sun:

absolute visual magnitude	M_V	$= 4.87$	mag
effective temperature	T_{eff}	$= 5780$	K
mass	$M_\odot$	$= 1.989 \times 10^{33}$	g
luminosity	$L_\odot$	$= 3.846 \times 10^{33}$	erg s^{-1}
radius	$R_\odot$	$= 6.960 \times 10^{10}$	cm
surface gravity	$g_\odot$	$= 2.736 \times 10^4$	cm s^{-2}

Earth:

mass	$M_\oplus$	$= 5.973 \times 10^{27}$	g
equatorial radius	$R_\oplus$	$= 6.378 \times 10^8$	cm
equatorial surface gravity	$g_\oplus$	$= 9.781 \times 10^2$	cm s^{-2}
tropical year	1 yr	$= 3.1557 \times 10^7$	s

B.4 Problem set

The main purpose of the exercises is to familiarize readers with the two dozen or so basic formulae, figures and tables of the book. They already contain, or it is easy to extract from them, the answer to many of the practical questions one encounters in the study of astronomical objects: How much dust is there? How well does it shield from star light? How warm is it? What is its luminosity? What kind of spectral energy distribution does one expect?

1. **Temperature dependence of the absorption coefficient.** In equation (1.105), we separated the permittivity, $\varepsilon(\omega)$, into a dielectric and a conducting part. How does the temperature influence $\varepsilon(\omega)$ in case of metals, for example, iron grains?

2. **Phase function for scattering.**

   a) Show that the phase function $f(\cos\theta)$ defined in (2.8) satifies the conditions (2.6) and (2.7). The exercise is purely mathematical, but one is always well advised to check a formula before using it.

   b) Which fraction of the scattered light is scattered to the rear if the asymmetry factor $g = \frac{1}{2}$?

3. **Computing Q^{ext} and Q^{sca} for spheres.**

   It is often useful to have a program that computes the extinction and scattering efficiency, Q^{ext} and Q^{sca}, exactly from Mie theory. You may copy the one listed in Appendix A of [Boh83] (one only needs the subroutine BHMIE) or download one from the Internet, for instance, from www.astro.spbu.ru/staff/ilin2/DOP/6-SOFT/ours.html.

   First, check the program. For example, for a size parameter $x = 1$ and optical constants $n = 1.3$ and $k = 0, 2$, it should yield $Q^{\text{ext}} = 0.606446$ and $Q^{\text{sca}} = 0.0979956$. Then play with it, for example, by reproducing points in figure 5.10.

4. **Extinction and scattering efficiency of small spheres.**

   a) Write down the expressions for the extinction and scattering efficiency of the Mie formulae (2.37) and (2.38) retaining only the terms with $n = 1$ and $n = 2$. Then insert the expansion coefficients a_1, b_1 from (3.3) and (3.4). The next higher coefficients, including terms up to x^6, read (see [Boh83])

$$a_2 = -\frac{ix^5}{15}\frac{m^2 - 1}{2m^2 + 3} + O(x^7)$$
$$b_2 = O(x^7)$$

b) Confirm the Rayleigh limit approximation (3.6) for Q^{sca}. Which of the three addends on the right side of (3.3) is responsible?

c) Confirm likewise equation (3.5) for Q^{ext}. Which of the terms in a_1 and b_1 is now responsible?

d) Extend formula (3.5) for Q^{ext} to include terms up to x^3.

e) See how the value of Q^{ext} improves with the new term by comparing it with the exact value from Mie theory (exercise 3). For example, amorphous carbon has at 10μm an optical constant $n \simeq 4$ and $k \simeq 1.5$ (see figure 5.11). Assume a grain of $a = 0.1\mu$m radius, the size parameter is then $x \simeq 0.063$.

f) Confirm equation (3.8) for the magnetic dipole absorption efficiency. The sphere is small, metallic and nonmagnetic ($\mu = 1$).

5. **Dust absorption coefficients.**

a) What is the absorption coefficient, C^{abs}, of a spherical grain at $\lambda = 0.44\mu$m if the grain has a radius of 1000Å and is made of amorphous carbon (aC) or of silicate?

b) What is $Q^{\text{abs}}(\text{aC})$ at $\lambda = 0.44\mu$m if the radius is only 100Å? Compare the number with the one resulting from formula (3.5). Is the particle in the Rayleigh limit?

c) Determine the *mass* absorption coefficient, K^{abs}, of 1 g of dust (60% silicate and 40% amorphous carbon) at $\lambda = 1$ mm. Compare with formula (3.5). Does the result depend on the grain size?

d) Graphite spheres are optically anisotropic. Why does one form from the efficiencies $Q_{\parallel}$ and $Q_{\perp}$ in figure 5.13 the average $\frac{1}{3}(Q_{\parallel} + 2Q_{\perp})$? Which assumptions still enter?

e) Why is for astronomical applications the accuracy with which one can read off numbers from the figures 5.10 to 5.13 more than sufficient?

6. **Dust scattering coefficients.** Repeat the previous excercise, but now with respect to scattering. Find the cross section of one grain, C^{sca}, and of 1 g of dust, K^{sca}. Which figures do you have to use now? Why is the accuracy slightly lower than before? How do you find C^{sca} when the values for C^{abs} and C^{ext} lie close together?

7. **Dust absorption cross sections at X-rays.** What is the absorption efficiency of a silicate sphere of $a = 200$Å radius with respect to $2\,$keV radiation? Use an analytical approximation. When does it break down?

8. **Dust heating by X-rays.**

   a) Make a plot of the penetration depth in dust material of energetic electrons.

   b) A $5\,\mathrm{keV}$ photon is absorbed in the center of a spherical grain of $0.1\mu\mathrm{m}$ radius. The photon ejects an electron. Which fraction of the photon energy goes into heating the grain?

9. **The Serkowski curve of linear polarization.** The light from a moderately reddened star ($E_{B-V} \simeq 1\,\mathrm{mag}$) turns out to be linearly polarized. The percentage of polarization amounts to

   > 2.96% at $\lambda = 0.36\mu\mathrm{m}$
   > 3.64% at $\lambda = 0.44\mu\mathrm{m}$
   > 4.06% at $\lambda = 0.55\mu\mathrm{m}$
   > 3.34% at $\lambda = 0.90\mu\mathrm{m}$

   The observational errors are small ($< 5\%$). Determine the three parameters p_{max}, λ_{max} and k of the Serkowski curve (4.1). Which degree of polarization does one expect at $0.3\mu\mathrm{m}$ and $2.2\mu\mathrm{m}$?

10. **Injection of dust into the interstellar medium.**

    a) Which are the major dust suppliers in the Milky Way? What are their approximate annual input rates?

    b) What is the ratio, R_{inj}, of dust injection to gas injection by these components? How does R_{inj} compare to the dust-to-gas mass ratio, $R_d = M_{dust}/M_{gas}$, of the interstellar medium? What does this imply for the mean lifetime of a gas atom and a grain: t_{gas} and t_{dust}? How is gas removed from the ISM?

    c) An AGB star has a typical mass loss rate of $\dot{M}_{RG} = 3 \times 10^{-6}\,\mathrm{M}_\odot$ yr^{-1}, about 1% of which is dust. How many AGB stars are there in the Milky Way if they provide all dust? In which volume, V_{AGB}, does one find, on average, one AGB star?

    d) What is the mass of solid iron in the interstellar medium?

    e) Suppose all iron in the Milky Way comes from SN of type I. Derive the rate at which they explode assuming an iron input per event of $m_{Fe} = 0.5 M_\odot$ and a mean lifetime of dust grains $t_{dust} = 7.5 \times 10^8$ yr (section 7.4.1). Compare the number with the supernova rate in table 7.2?

    f) To first order, the dust-to-gas mass ratio, R_d, seems to be uniform in the Milky Way. Why? The heavy element abundance, Z, however, decreases with galactic radius, by about a factor of two from the Galactic Center towards the Sun, and continues to fall farther out. What does this imply for R_d? What for the star formation rate?

11. How many dust grains are there in the universe?

a) What is the number, N_{1g}, of grains in 1 g of interstellar dust? Assume a power law size distribution $n(a) = \beta a^{-3.5}$ (see 5.9) with lower and upper limit $a_- = 100\text{Å}$ and $a_+ = 3000\text{Å}$, and a grain material density $\rho_{gr} = 2.5$ g cm^{-3}.

b) How many PAHs are in 1 g of dust? Assume that PAHs account for 5% of all solid carbon and that a typical PAH consists of 200 C atoms.

c) What is the surface area, A, of 1 g of dust? Which fraction is due to PAHs?

d) How many grains are in the Milky Way, in the universe? If they were pressed into one block, how big would it be?

e) Which mass fraction, f, of the baryonic matter in the universe is in the form of dust?

f) Determine the number, N_{red}, of grains involved in reddening a star. Suppose the cloud lies half-way between us and the star and has a color excess $E_{B-V} = 1$ mag. The dust also radiates. How many grains (N_{em}) are involved when we observe dust emission of the cloud?

12. How many grains are identical?

a) What is the number, N_9, of grains in 1 g of dust with exactly, say, 10^9 atoms? Consider for these order of magnitude estimates carbon grains. How many of such grains are there in the universe?

b) How likely is it that two of them are identical as quantum–mechanical systems, like two ammonia molecules (of course, they may be in different excitation states)?

Interstellar grains are certainly not perfectly regular crystals. Assume that 1% of the $N = 10^9$ atomic positions in the grains with 10^9 atoms have defects: they are void or filled with atoms other than C.

13. Millimeter emission from planets.

a) Mars, a very big dust grain, is at a mean distance $D = 1.51$ AU and has an albedo $A_\sigma = 0.16$. Compute its surface temperature. What is the flux, F_{1mm}, that one receives at 1 mm when the planet is in conjunction ($R_\sigma = 3397$ km, eccentricity of the orbit $e_\sigma = 0.093$)? By how much will the surface temperature be lower in aphelion than in perihelion?

b) Repeat the calculation for Jupiter ($D_{2\!\!\!\!\;_l} = 5.2$ AU, $R_{2\!\!\!\!\;_l} = 71400$ km, $A_{2\!\!\!\!\;_l} = 0.41$).

14. **Grain temperatures, cooling and heating rate.**

a) Derive an approximate formula for the total radiation, ϵ, of a spherical grain of radius a and at temperature T assuming for the absorption efficiency $Q = Q_0 \nu^\beta$ (see 6.9). What is the dimension of ϵ? Check the formula for ϵ by comparing it with the accurate curves in figure 6.5.

b) What is the total radiation of $1\,g$ of such grains? What is now the dimension of ϵ? How does ϵ depend on grain size?

c) How warm is $1\,g$ dust heated at a rate ϵ?

d) What is the infrared luminosity of a cloud of dust mass M and dust temperature T?

15. **Color temperature of dust emission.** One observes towards a star forming region at wavelengths $\lambda_1 = 1300\mu m$ and $\lambda_2 = 100\mu m$ the fluxes $F_1 = 0.1\,Jy$ and $F_2 = 140\,Jy$.

a) Which quantity is being measured in Jy? How does it differ from intensity, power and integrated flux? Convert Jy to cgs-units.

b) What is the spectral index α defined in (6.21)? Why is it dust emission that one observes and not synchrotron or free-free radiation?

c) What is the ratio of the mass absorption coeficients at λ_1 and λ_2? Consult figure 10.7 and formula (10.16) and compare the results.

d) Determine the color temperature, T_c, between λ_1 and λ_2 from (6.20) by solving the equation numerically.

16. **Temperature jump after UV absorption.** By how much will the temperature of a silicate grain of 10Å radius increase after absorption of a $10\,eV$ photon? The grain is initially at $10\,K$ ($1000\,K$).

17. **Dust heating by collisions with gas atoms.** A grain of radius a is heated in a radiation field J_ν to a temperature T_0. What is the dust temperature T_d if the grain is additionally heated or cooled by gas collisions? Suppose the gas consists of molecular hydrogen, has a temperature $T_g > T_0$ and a number density n. What is the critical density for thermal coupling between gas and dust?

18. **The flux from a star.**

a) Which flux, F_ν, does one receives from a star of luminosity L at a distance D? The star emits like a blackbody of temperature T_* and there is no intervening dust. How much energy does a grain of cross section C_ν^{abs} at the position of the observer absorb? How is the flux related to the mean radiative intensity J_ν?

b) Suppose now the star is in a dusty medium. How much *stellar* radiation does a grain of cross section C_ν^{abs} absorb if the optical depth towards the star equals τ_ν? How much radiation does the (spherical) grain absorb altogether? How are flux and mean intensity related now?

c) The human eye can detect wavelengths between about 0.4 and 0.7μm. Which fraction, f, of the (blackbody) solar radiation falls into this range?

19. **Solving the energy equation of radiative equilibrium.**

a) Write a computer program to solve numerically equation (6.7) for the dust temperature T, when the absorption coefficient, Q_ν^{abs}, and the mean intensity of the radiation field, J_ν, are given. One solution strategy is described in section 11.2.2.

b) How can you check that your program works?

c) Derive the recursion formula (11.50).

d) Approximate the absorption efficiency of amorphous carbon grains with $a = 0.01\mu$m radius by power laws for $\lambda < 1\mu$m and $\lambda \geq 1\mu$m (see figure 5.11) and determine then with the help of the program the minimum distance of such grains from the Sun before they evaporate.

20. **The lowest resonance frequency in a grain.** What is roughly the lowest resonance frequency, $\nu_{\min}$, of a grain with $N = 10^5$ atoms and a Debye temperature of $\theta = 800$ K? How can one explain grain emission at still longer wavelengths and what is its spectral shape?

21. **How cold can dust be?** Estimate the dust temperature in an opaque cloud ($A_V > 10\,\mathrm{mag}$) at the periphery of the Milky Way, at galactic radius $R = 15$ kpc. The cloud is devoid of stars or other internal heating sources. Does the microwave background play a role in dust heating?

22. **Counting photons.**

a) When does one need the number of photons, and not the energy, that a grain absorbs?

b) How many photons, N_{phot}, does a star of luminosity L and effective surface temperature T_* emit? How many photons does the Sun emit? What is their average energy $h\bar{\nu}$ and average wavelength $\bar{\lambda}$?

c) How many photons with energy above 10 eV pass at a distance of $D = 0.1$ pc in one second through 1 cm^2 if the star has a luminosity $L = 10^4\,L_\odot$ and a surface temperature $T_* = 22\,000$ K?

d) How many times is a PAH that consists of $N_C = 200$ carbon atoms excited by these photons in one day? Assume that the cross section per carbon atom is $C_\nu = 7 \times 10^{-18}$ cm^2 (section 10.2.1).

e) How many photons per second does our eye receive from the weakest star we can detect (6.5 visual magnitude)? Assume a bandwidth $\Delta\nu = 2 \times 10^{14}$ Hz^{-1}. The pupil diameter is 7 mm. A space telescope with a 2 m mirror can detect stars or distant galaxies of 30-th visual magnitude. Why is it so much more sensitive? At which rate do photons there arrive?

23. **Field emission.** The electric field at the surface of a highly charged and small grain can become so large that ions or electrons are ejected.

    a) What is the critical grain charge for ions? How does it depend on grain radius? How does the positive charge of a grain produced by the photoelectric effect depend on the parameters of the environment (temperature, electron density, radiation field)?

    b) What about field emission of electrons? They can be emitted when the grain is negatively charged.

24. **Betelgeuse in Orion.** What are the apparent magnitudes of the red supergiant Betelgeuse in the Orion constellation at V (0.55μm) and K (2.2μm) assuming a distance $D = 130$ pc, a bolometric luminosity $L = 5 \times 10^4$ L$_\odot$, and a blackbody surface with $T = 3000$ K? What is its color V-K?

25. **Total infrared emission of the Milky Way.**

    a) Estimate the fraction, f, of stellar light that is converted in the Milky Way via dust into the infrared. Consider a star in the mid plane of the galactic disk. Let τ be the visual optical thickness for absorption perpendicular to the disk from the star to infinity. Derive the function $f(\tau)$.

    b) At the locus of the Sun, $\tau \simeq 0.1$. What is $f(0.1)$? Observations indicate that one third of the galactic stellar light is absorbed, $f_{obs} \simeq \frac{1}{3}$. How does it compare?

    c) The interpretation of the infrared observations and their conversion into f_{obs} is, however, problematic because we are *in* the galaxy. Discuss the difficulties. Why is it easier to determine the luminosity of external galaxies?

    d) How does scattering of starlight and star formation affect $f(\tau)$?

26. **Anisotropic emission of galaxies.** When we have measured the spectral energy distribution (SED) of a galaxy at distance D, we think we know its total luminosity L. If F_{obs} is the integrated observed flux and $L_{obs} = 4\pi D^2 F_{obs}$ the observed luminosity, we put $L = L_{obs}$. However, this can be wrong because of dust obscuration.

a) Estimate by how much one underrates the true luminosity L for a disk galaxy observed exactly edge-on. Assume that all stars are in the mid-plane and that the dust disk envelops the stellar disk.

b) If one underrates L viewing the object edge-on, does one get the right luminosity looking face-on?

27. **Anisotropic emission of accretion disks.**

a) Why can the estimates for L in accretion disks around stars go wrong by even larger factors than for disk galaxies (exercise 26)?

b) What is the ratio of observed luminosity L_{obs} over stellar luminosity L_* for a star with a geometrically thin accretion disk viewed face-on? L_{obs} is defined as $4\pi D^2$ times the observed flux F_{obs}; the disk is not spatially resolved.

c) If the stellar disk is geometrically very thin, one should not notice it looking exactly from the side. In this case, $L_{\mathrm{obs}} = L_*$. Looking face-on, $L_{\mathrm{obs}} > L_*$. Why is the average of L_{obs} over 4π still equal to L_* (as it should)?

28. **Dereddening a star in a two-color-diagram.** A star in a young cluster has a visual apparent magnitude V = 13.2 mag and colors B–V = 1.4 mag and U–B = 0.4 mag. What is its spectral type? How much is the star reddened? Use figure 10.2. Find the absolute visual magnitude, M_{V}, of the star, for example by consulting [All73]. What is its distance.

29. **A galaxy of constant dust temperature.** One receives from a galaxy at wavelengths $\lambda_1 = 1300\mu$m and $\lambda_2 = 100\mu$m the fluxes $F_1 = 80$ mJy and $F_2 = 28$ Jy. Assume that the emission comes from dust at constant temperature T_{c}.

a) Determine T_{c} assuming, as is appropriate for interstellar grains, $K_\nu \propto \nu^\beta$ with $\beta = 2$. What would be T_{c} if β were 1.7 or 1.4? Note the trend.

b) What is the integrated luminosity, L, emitted by this dust component ($\beta = 2$) if the galaxy is at a distance $D = 60$ Mpc? Does one have to worry about radiative tranfer?

30. **A galaxy with dust of different temperatures.** There are also measurements at shorter wavelengths for the galaxy in exercise 29. At $\lambda_3 = 60\mu$m, $\lambda_4 = 25\mu$m and $\lambda_5 = 25\mu$m, the fluxes are $F_3 = 17$ mJy, $F_4 = 5$ Jy and $F_5 = 0.4$ Jy.

a) Display the measurements in a log-log plot.

b) Verify that one cannot describe the SED with dust of constant temperature. Approximate the points with $\lambda \geq 60\mu$m by the sum of a cold (c) and a warm (w) component, $L_{\mathrm{c},\nu}$ and $L_{\mathrm{w},\nu}$, both with

$K_\nu \propto \nu^2$. Insert the graph of these components into the plot. What are the temperatures of the two components. You have to write a little program.

c) What are the luminosities of the cold and warm component?

d) The warm component fails at 25μm and even more so at 12μm. What does it imply?

e) Usually measurements at different wavelengths are obtained with different spatial resolution. Suppose the optical size of the galaxy is 3', the beam at 100μm was 1' and at 1300μm only 24". Both beams were directed towards the center of the galaxy. What does this imply for the color temperature T_c that one derives from the fluxes at $\lambda_1 = 1300\mu$m and $\lambda_2 = 100\mu$m?

31. How to display an SED.

a) Repeat the log-log plot of the SED in Exercise 30a, but now with λF_λ as the ordinate.

b) Fluxes may be given as F_ν (per unit frequency) or F_λ (per unit wavelength). Convert a flux of 0.1 Jy at 3μm into the frequently used unit erg cm^{-2} s^{-1} μm^{-1}.

c) What is the difference between λF_λ and νF_ν?

d) Which representation informs us better about the energetics: λF_λ vs. λ, or F_λ vs. λ? Show that for λF_λ vs. λ, spectral intervals of equal area contain equal luminosity.

32. The dust mass of a cloud.
One has measured towards an extended cloud at 850μm a flux $F_{850} = 100$ mJy. The beam width was $\theta = 18''$ (full width at half maximum or FWHM). The cloud is at $D = 150$ pc; molecular line observations indicate that it is cold ($T = 15$ K).

a) What is the solid angle Ω of the beam? What are the optical depth τ_{850} at 850μm and the visual extinction A_V of the cloud at that position? What is the hydrogen column density?

b) What are the dust and gas mass contained in the beam?

c) How much flux does one receive from the microwave background at that wavelength? How does one separate it from the dust emission?

33. Is there more cold or more warm dust in a galaxy?
The infrared emission of dusty objects is separated into components of different temperatures. In the decomposition of the SED of a galaxy, one typically finds between 25μm and 1000μm a cold component with $T_c \simeq 20$ K and a warm one with $T_w \simeq 60$ K of comparable luminosities, similar to exercise 30.

a) What is the mass fraction of the components, M_c/M_w, if $T_c = 20\,K$ and $T_w = 60\,K$? Explain why measurements at $60\mu m$ or even $100\mu m$ therefore tell us very little about the total dust (or gas) mass.

b) To acccount for the mid infrared emission, say from $8\mu m$ to $20\mu m$, one needs a third hot component. It is typically around 150 K and contributes less than 10% to the total luminosity. What is its mass fraction?

34. The mean dust temperature of the Milky Way and in M82.

a) The Milky Way has a luminosity $L \simeq 4 \times 10^{10}$ $L_\odot$, of which one third is absorbed by dust of mass $M_{dust} = 3 \times 10^7$ $M_\odot$ (table 7.2). What is the mean dust temperature in the Milky Way? At which wavelength does the infrared emission peak?

b) The nucleus of the starburst galaxy M82 is of similar brightness ($L = 3 \times 10^{10}$ $L_\odot$), but almost twice as hot. Why? How does the fact that the nucleus is very dust obscured ($A_V > 10$ mag) and compact (radius $R \sim 300$ pc) affect the *average* dust temperature?

35. The ultraluminous galaxy Arp220.
The high luminosity of this galaxy ($D = 75\,Mpc$) is the result of a galaxy merger and a subsequent starburst in the now obscured nuclear region.

Estimate the basic parameters of this galaxy (total luminosity L, total dust and gas mass M_d and M_g, average dust temperature T_d, and column density N_H or visual extinction A_V) from just the following three measurements: $F_{2.2\mu m} = 0.04\,Jy$, $F_{100\mu m} = 115\,Jy$ and $F_{800\mu m} = 0.82\,Jy$. Which reasonable assumptions enter the back-of-an-envelope calculations?

36. The night sky without dust.
Suppose all dust in the universe were removed. What would the night sky look like? In a clear moonless night, two thirds of the visual sky light is actually due to airglow [Lei87] and the integrated starlight comes primarily from stars with apparent magnitude $m_V = 11\ldots13$ mag [Roa73].

a) How would the integrated starlight increase if there were no dust?

b) How would the surface brightness of the band of the Milky Way increase? Assume a constant stellar density in the galactic plane and a constant disk height h. The average visual extinction in the plane of the Milky Way is 1.7 mag kpc^{-1}.

c) What would happen to the Galactic Center? The visual extinction towards it is ~25 mag. How many times would the stars appear brighter without dust? How many times in the K band at $2.2\mu m$ assuming $K_V^{ext}/K_{2.2\mu m}^{ext} \simeq 11$? The intrinsic luminosity of the central region $5'$ across is $\sim 2 \times 10^7$ L$_\odot$. What would it look like ($D = 8.5$ kpc)? How would the appearance of star forming regions, like the one in Orion near the sword, change?

d) Of course, without dust, other things, as important as the view of the night sky, would change, too. Name and explain a few.

37. **The mass loss rate of AGB stars.** Which observations are needed to determine the mass loss rate of gas and dust, $\dot{M}$ and $\dot{M}_\mathrm{d}$, in the shell of an AGB star?

38. **Photon scattering and the extinction curve.** Half-way towards a star at a distance $D = 100$ pc is a cloud of low optical thickness ($\tau \ll 1$). By which fraction, f_1, is the stellar flux reduced by the presence of the cloud through extinction, by which fraction, f_2, is it amplified through scattering? What is the ratio f_2/f_1 if the aperture in the observations is 5 arcsec. Assume isotropic scattering and an albedo of 0.5.

39. **The effective extinction of a dense circumstellar shell.** What is the effective optical thickness, τ^eff, defined in equation (12.16), towards a star surrounded by a dense homogeneous dust shell. Treat the photon path as a random walk.

40. **All we would like to know about dust.**

a) Make a wish list addressing the questions: What is dust? What does it do? How does it do it? What is the *curriculum vitae* of a typical dust grain? What is the relevance of dust in the universe?

b) To which items on your wish list do you think we have by now a fair answer. Where are we still ignorant? Could there be interesting effects produced by dust we are not yet aware of?

c) Suppose we had an exact description of dust grains and a most powerful computer. Would the study of interstellar dust then have come to an end?

B.5 List of symbols

A	$=$	area, albedo
a	$=$	radius of spherical grain or of circular cylinder
a_0	$=$	Bohr radius $= 5.29 \times 10^{-9}$ cm
a_-, a_+	$=$	lower and upper limit of grain radii in size distribution
A_{ij}	$=$	Einstein coefficient for spontaneous emission
A_V	$=$	visual extinction in magnitudes
$\mathbf{A}$	$=$	vector potential of magnetic field
B_{ij}	$=$	Einstein coefficient for induced emission
$B_\nu(T)$	$=$	Planck function
$\mathbf{B}$	$=$	magnetic induction
c	$=$	velocity of light
c_n	$=$	number density of droplets consisting of n molecules (n-mer)
C^{abs}	$=$	absorption cross section of a single grain, likewise $C^{\mathrm{sca}}, \ldots$
$C_{\mathrm{p}}, C_{\mathrm{v}}$	$=$	specific heat at constant pressure or volume
$\mathbf{D}$	$=$	displacement in Maxwell's equations
e	$=$	charge of a proton
	$=$	eccentricity of a spheroid
$\mathbf{e}$	$=$	unit vector
E	$=$	energy
$\mathbf{E}$	$=$	electric field
$E_{\mathrm{B-V}}$	$=$	standard color excess
f	$=$	number of degrees of freedom
f_i	$=$	volume fraction in a grain of chemical component i
F_ν	$=$	electromagnetic flux at frequency ν
F	$=$	free energy or Helmholtz potential
$\mathbf{F}$	$=$	force
g	$=$	asymmetry factor in scattering
	$=$	statistical weight
G	$=$	gravitational constant
	$=$	free enthalpy or Gibbs potential
h	$=$	Planck's constant
H	$=$	enthalpy
	$=$	Hamiltonian
$\mathbf{H}$	$=$	magnetic field
I	$=$	moment of inertia
I_ν	$=$	intensity at frequency ν
$\mathbf{J}$	$=$	current density
k	$=$	Boltzmann constant
	$=$	imaginary part of the optical constant m
k	$=$	wavenumber, in vacuo $\mathrm{k} = 2\pi/\lambda$, as a vector $\mathbf{k}$
K^{abs}	$=$	volume or mass absorption coefficient, likewise $K^{\mathrm{sca}}, \ldots$

L	$=$	Lagrange function
	$=$	luminosity
	$=$	angular momentum
	$=$	Loschmidt's number $= 6.02 \times 10^{23}$
L_a, L_b, L_c	$=$	shape factor of ellipsoid with respect to axis a, b, c
m, M	$=$	mass
$\mathbf{M}$	$=$	magnetization
$m = n + ik$	$=$	optical constant
$\mathbf{m}$	$=$	magnetic moment
m_e	$=$	mass of electron
n	$=$	real part of the optical constant m
	$=$	number density of particles
n_H	$=$	$n(\mathrm{HI}) + n(\mathrm{H_2})$, atomic plus molecular hydrogen density
N	$=$	column density, occasionally also number density
$\mathcal{N}$	$=$	occupation number (quantum mecahnics)
p	$=$	momentum of a particle
$p(\lambda)$	$=$	degree of linear polarization of light at wavelength λ
$\mathbf{p}$	$=$	electric dipole moment
$\mathbf{P}$	$=$	polarization of matter, related to $\mathbf{D}$
p, P	$=$	pressure
q	$=$	charge of a particle
Q^{abs}	$=$	absorption efficiency of a grain, likewise $Q^{\mathrm{sca}}, \ldots$
Q	$=$	heat (thermodynamics)
r	$=$	distance or radius
	$=$	reflectance
$\mathbf{r}$	$=$	position vector
R_d	$=$	dust-to-gas mass ratio
R	$=$	Rydberg constant
s	$=$	saturation parameter of vapor pressure
S	$=$	entropy
$\mathbf{S}$	$=$	Poynting vector
t	$=$	time
T	$=$	temperature
	$=$	kinetic energy
T_d	$=$	dust temperature
T_x	$=$	excitation temperature
u_ν	$=$	radiative energy density at frequency ν
U	$=$	potential energy
	$=$	internal energy (thermodynamics)
$\mathbf{v}, \mathbf{V}$	$=$	velocity
v_g, v_{ph}	$=$	group and phase velocity
v_0	$=$	volume of one molecule
V	$=$	volume
	$=$	potential energy

W	=	energy loss rate
	=	work
$\mathbf{x}$	=	position vector
x	=	size parameter of grain
x	=	x-coordinate
Z	=	grain charge in multiples of the proton charge
α	=	degeneracy parameter (quantum mechanics)
	=	fine structure constant
α_e, α_m	=	electric, magnetic polarizability
β	=	$1/kT$ (thermodynamics)
γ	=	damping constant
δA	=	infinitesimal work (thermodynamics)
ϵ_ν	=	emissivity at frequency ν
ε	=	$\varepsilon_1 + i\varepsilon_2$ = complex dielectric permeability
ζ	=	surface tension
η	=	efficiency
η_i	=	sticking probability of gas species i
κ	=	force constant of a spring
	=	ratio of specific heats
λ	=	wavelength
μ	=	$\mu_1 + i\mu_2$ = complex dielectric permeability
μ_{ij}	=	dipole moment for transition $j \rightarrow i$, see (A.94)
ν	=	frequency
ρ	=	density or charge density
$\rho(E)$	=	density of states at energy E
ρ_{gr}	=	density of grain material
σ	=	radiation constant
	=	electric conductivity
	=	surface charge
$\sigma(\omega)$	=	cross section of atom at frequency ω
σ_T	=	Thomson scattering cross section
σ_{geo}	=	geometrical cross section
τ_ν	=	optical depth at frequency ν
τ	=	time scale
ϕ	=	scalar potential (electrodynamics)
$d\Phi$	=	size of an infinitesimal cell in phase space
χ	=	$\chi_1 + i\chi_2$ = complex susceptibility
ψ	=	quantum mechanical wave function
ω	=	$2\pi\nu$ = circular frequency
	=	probability for a configuration of atoms
ω_p	=	plasma frequency
Ω	=	thermodynamic probability
$d\Omega$	=	element of solid angle

Bibliography

[1] [Abr74] Abraham F. F., 1974, *Homogeneous Nucleation Theory* (New York: Academic)

[2] [Abr70] Abramowitz M. and Stegun I. A., 1970, *Handbook of Mathematical Functions* (New York: Dover)

[3] [All73] Allen C. W., 1973, *Astrophysical Quantities* (Athlone Press)

[4] [All89] Alllamandola L. J., Tielens A. G. G. M. and Barker R. J., 1989, ApJS 71, 733

[5] [Asp90] Aspin C., Rayner J. T., McLean I. S. and Hayashi S. S., 1990, MNRAS 246, 565

[6] [Bie80] Biermann P. L. and Harwit M., 1980, ApJ 241, L107

[7] [Bjo01] Bjorkman J. E. and Wood K., 2001, ApJ 554, 615

[8] [Boh83] Bohren C. F. and Huffman D. R., 1983, *Absorption and Scattering of Light by Small Particles* (New York: Wiley)

[9] [Cal94] Calzetti D., Kinney A. and Storchi–Bergmann T., 1994, ApJ 429, 582

[10] [Car89] Cardelli J. A., Clayton G. C. and Mathis J. S., 1989, ApJ 345, 245

[11] [Cep98] Ceplecha Z., Borovicka J., Elford W. G., Revelle D. O., Hawkes R., Porubcan V. and Simek M. 1998, Space Science Reviews 84, 327

[12] [Cha46] Chandrasekhar S., 1946, ApJ 104, 191

[13] [Cha85] Chase M. W., Davies C. A., Downey J. R., Frurip D. J., McDonal R. A. and Syverud A. N., 1985, J Phys Chem Ref Data 14, Suppl. No. 1

[14] [Chi97] Chini R., Reipurth B., Sievers A. et al., 1997, A&A 325, 542

[15] [Chi00] Chiar J. E., Tielens A. G. G. M., Whittet D. C. B. et al., 2000, ApJ 537, 749

[16] [Deb29] Debye P., 1929, *Polare Molekeln* (Leipzig: Hirzel)

[17] [Dis98] van Dishoeck E. F., Helmich F. P., Schutte W. A., Ehrenfreund P., Lahuis F., Boogert A. C. A., Tielens A. G. G. M., de Graauw Th., Gerakines P. A. and Whittet D. C. B., 1998, ASP Conference Series, Vol. 132

[18] [Dor82] Dorschner J., 1982, Astrophys. Space Sci., 81, 323

[19] [Dra87] Draine B. T. and Sutin B., 1987, ApJ 320, 803

[20] [Dra88] Draine B. T., 1988, ApJ 333, 848

[21] [Dra01] Draine B. T. and Li A., 2001, ApJ 551, 807

[22] [Dul84] Duley W. W. and Williams D. A., 1984, *Interstellar Chemistry* (Academic Press)

377

[23] [Dwe92] Dwek E. and Arendt R. G. 1996,
      Ann. Rev. Astron. Astrophys. 30, 11

[24] [Dwe96] Dwek E. and Smith R. K. 1996, ApJ 459, 686

[25] [Fea64] Feautrier P., 1964, C. R. Acad. Sci. Paris, 258, 3189

[26] [Fed66] Feder J., Russell K. C., Lothe J. and Pound G. M., 1966, Advances in
      Physics, 15, 111

[27] [Fis05] Fischera J. and Dopita M., 2005, ApJ 619, 340

[28] [Gil73] Gillett F. C., Forrest W. J. and Merrill K. M., 1973, ApJ 183, 87

[29] [Gre01] Grevesse N and Sauval A J, 2001, Space Sci. Rev. 85, 161

[30] [Grü94] Grün E., Gustafson B., Mann I., Baguhl M., Morfill G. E., Staubach
      P., Tayler A. and Zook H. A., 1994, A&A 286, 915

[31] [Guh89] Guhathakurta P. and Draine B. T., 1989, ApJ 345, 230

[32] [Hab96] Habing H. J., 1996, A&A Review 7, 97

[33] [Hel70] Hellyer B., 1970, MNRAS 148, 383

[34] [Hen41] Henyey L. G. and Greenstein J. L., 1941, ApJ 93, 70

[35] [Hol70] Hollenbach D. J. and Salpeter E. E., 1970, J Chem Phys, 53, 79

[36] [Hon78] Hong S. S. and Greenberg J. M., 1978, A&A 70, 695

[37] [Hor93] Horanyi M., Morfill G. and Grün E., 1993, J Geophys. Res. 98, 21,245

[38] [Hul57] van de Hulst H. C., 1957, *Light Scattering by Small Particles*
      (New York: Wiley)

[39] [Hum71] Hummer D. G. and Rybicki G. B., 1971, MNRAS 152, 1

[40] [Jac99] Jackson J. D., 1999, *Classical Electrodynamics* (New York: Wiley)

[41] [Joc94] Jochims H. W., Rühl E., Baumgärtel H., Tobita S. and Leach S., 1994,
      ApJ 420, 307

[42] [Jur94] Jura M., 1994, ApJ 434, 713

[43] [Ker69] Kerker M., 1969, *Scattering of Light* (New York: Academic)

[44] [Koi06] Koike C., Mutschke H., Suto H. et al., 2006, A&A 449, 583

[45] [Kru53] Krumhansl J. and Brooks H., 1953, J Chem Phys 21, 1663

[46] [Krü78] Krügel E. and Tutukov A. V., 1978, A&A 63, 375

[47] [Lao93] Laor A. and Draine B. T., 1993, ApJ 402, 441

[48] [Leq05] Lequeux J., 2005, *The Interstellar Medium* (Springer)

[49] [Lei87] Leinert Ch., Bowyer S., Haikala L. K. et al., 1987, A&A Suppl 127, 1

[50] [Leu75] Leung C. M., 1975, ApJ 199, 340

[51] [Li97] Li A. and Greenberg J. M., 1997, A&A 323, 566

[52] [Li01] Li A. and Draine B. T., 2001, ApJ 554, 778

[53] [Luc99] Lucy L. B., 1999, A&A 344, 282

[54] [Lut96] Lutz D., Feuchtgruber H., Genzel R. et al., 1996, A&A 315, L272

[55] [Mat77] Mathis J. S., Rumpl W. and Nordsieck K. H., 1977, ApJ 217, 425

[56] [McK89] McKee C. F., 1989, IAU 135, 431

[57] [Men06] Menten K. M., Reid M. J., Krügel E., Claussen M. J. and Sahai R., 2006 A&A 453, 301

[58] [Mie08] Mie G., 1908, Ann. Phys. 25, 377

[59] [Mih78] Mihalas D., 1978, *Stellar Atmospheres* (W H Freeman & Co)

[60] [Mol01] Molster F., Lim T. L., Sylvester R. J., Waters L. B. F. M., Barlow M. J., Beintema D. A., Cohen M., Cox P. and Schmitt B., 2001, A&A 372, 165

[61] [Mol02] Molster F., Waters L. B. F. M. and Tielens A. G. G. M., 2002, A&A 382, 222

[62] [Mol05] Molster F. and Kemper C., 2005, *Space Science Review ISO Special Issue* (Springer)

[63] [Omo86] Omont A., 1986, A&A 164, 159

[64] [Pen02] Pendleton Y. J. and Allamandola L. J., 2002, ApJS 138, 75

[65] [Pen62] Penndorf R. B., 1962, J Optical Society of America 52, 896

[66] [Pep70] Pepper S. V., 1970, Journal of the Optical Society of America, 60, 805

[67] [Per87] Perault M., 1987, These d'Etat, Universite de Paris VII

[68] [Pré84] Prévot M. L., Lequeux J., Maurice E., Prévot L., Rocca–Volmerange B., 1984, A&A 132, 389

[69] [Pur69] Purcell E. M., 1969, ApJ 158, 433

[70] [Pur73] Purcell E. M. and Pennycacker C. R., 1973, ApJ 186, 705

[71] [Roa73] Roach F. E. and Gordon J. L., 1973, *The Light of the Night Sky* (Reidel Publ. Co.)

[72] [Sch80] Schmidt G. D., Cohen M. and Margon B., 1980, ApJ 239, L133

[73] [Sch93] Schutte W. A., Tielens A. G. G. M. and Allamandola L. J., 1993, ApJ 415, 397

[74] [Sch99] Schutte W. A., 1999, *Formation and Evolution of Solids in Space*, p. 177, eds. Greenberg J. M. and Li A. (Dordrecht: Kluwer)

[75] [Sea87] Seab C. G., 1987, *Interstellar Processes*, p. 491, eds. Hollenbach D. J. and Thronson H. A. (Dordrecht: Reidel)

[76] [Shu91] Shu F. H., 1991, *The Physics of Astrophysics, Volume I, Radiation* (Univ. Sci. Books)

[77] [Sie00] Siebenmorgen R. and Krügel E., 2000, A&A 364, 625

[78] [Som94] Somerville W. B., Allen R. G., Carnochan D. J. et al., 1994, ApJ 427, L47

[79] [Tie94] Tielens A. G. G. M., McKee C. F., Seab C. G. and Hollenbach D. J., 1994, ApJ 431, 321

[80] [Jon96] Jones A. P., Tielens A. G. G. M. and Hollenbach D. J., 1996, ApJ 469, 740

[81] [Tie05] Tielens A. G. G. M., 2005, *The Physics and Chemistry of the Interstellar Medium* (Cambridge University Press)

[82] [Tsc77] Tscharnuter W. M., 1977, A&A 57, 279

[83] [Tsc79] Tscharnuter W. M. and Winkler K.-H., 1979, Computer Physics Communications 18, 171

[84] [Voi91] Voit G. M., 1991, ApJ 379, 122

[85] [War84] Warren S. G., 1984, Appl. Optics 23, 1206

[86] [Wat72] Watson W. D., 1972, ApJ 176, 103

[87] [Wat75] Watson W. D., 1975, Les Houches, XXVI, *Atomic and Molecular Physics and the Interstellar Medium*, eds. Balian R., Encrenaz P., Lequeux J.

[88] [Wea77] Weast R. C. (ed), 1977, *Handbook of Chemistry and Physics* (Boca Rotan, FL: Chemical Rubber Company)

[89] [Wei01] Weingartner J. C. and Draine B. T., 2001, ApJ 548, 295

[90] [Whi78] Whitmer J. C., Cyvin S. J., Cyvin B. N., 1978, Zeitschrift für Naturforsch. 33a, 45

[91] [Whi63] Whitten G. Z. and Rabinovich B. S., 1963, J Chem Phys 38, 2466

[92] [Whi96] Whittet D. C. B., Schutte W. A., Tielens A. G. G. M. et al., 1996, A&A 315, L357

[93] [Whi03] Whittet D. C. B., 2003, *Dust in the Galactic Environment* (IoP)

[94] [Yam77] Yamamoto T. and Hasegawa H., 1977, Progress Theor. Physics 58, 816

[95] [Yor80] Yorke H. W., 1980, A&A 86, 286

[96] [Zub96] Zubko V. G., Mennella V., Colangeli L. and Bussoletti E., 1996, MNRAS 281, 1321

Index

Absorption coefficient
  by mass or volume 32, 90, 254, 256
  of ellipsoids 67
  grey approximation 266
  per grain 30
  physical background 13
  at submm wavelengths 256-257
Absorption efficiency 32
  calculation for spheres 37ff
  at long wavelengths 48
  of small spheres 60ff
  plots of 85-89, 123-124, 126
Accretion of gas on grains 130, 182, 185, 191, 195, 201ff, 212, 214-215, 292
Accretion disk 231, 266, 272, 296ff
  flat disks 296-299
  grazing angle 273-278, 299-301
  inflated disks 273, 277, 299-301
Acoustical phonon branch 147-148
Activation barrier 115, 213
Adsorption 202-203
AGB stars 111, 121, 130, 172, 233, 307ff
Albedo 30, 122, 253, 282
Ammonia 120, 186-187, 201, 215
Amorphous carbon 50-54, 69, 89-90, 111, 117, 121-122, 124, 139, 144-145, 250, 252, 254-256, 302, 312
Amphiboles 119
Amplitude scattering matrix 40ff
Anthracene ($C_{14}H_{10}$) 221-224
Arizona crater 181
Asymmetry factor 31, 64, 88-89, 92-93, 253, 263, 265,

Babinet's theorem 73-74, 76, 79, 85
Balmer decrement 245
Benzene 113, 115, 219, 221, 227
Betatron acceleration 184
Blackbody radiation 136, 328, 337ff
Bloch theorem 109-110
Bohr radius 83, 362
Bohr magneton 362
Boltzmann distribution 131, 132, 185, 207, 223, 235,
Bose statistics 319ff, 327, 328, 332, 336
Bosons 326, 327
Bragg law 107
Bravais lattices 102-105
de Broglie wavelength 326, 349
Bronzite $((Mg,Fe)SiO_3)$ 118, 119, 233
Brownian motion 50, 170-172
Bruggeman mixing rule 52-53

Calorie, definition 185
Canonical distribution 131-132, 331, 334, 336
Carbon bonding 112ff
Carbonaceous grains 111ff, 183, 304
Carnot cycle 185
Cauchy principle value 48
Charge balance 161ff
Chemical potential 204, 334
Chemisorption 202, 204
Cigars (spheroids) 65-69
Circular polarization 42, 43, 328
Clausius-Mossotti eq. 46, 52, 72-73
Clausius-Clapeyron equation 185
Clinopyroxenes 119
Close-packed structure 113, 116, 118-119

CNO-cycle 120

Coated particles 50, 56, 97-98, 209, 248, 256

CO_2 110, 120, 187, 203, 214-217

Collisional fragmentation 129ff

Color 231, 241-242, 245, 246, 302

Color excess E_{B-V} 241-243, 248

Color temperature 139-140, 159, 257

Column density 84, 216, 248, 263, 293, 316

Composite grains, see Fluffy grains

Commutation rules 343

Conventional cell 101-102, 106

Conduction band 108

Conductor, see Metal

Conductivity, electric 25-28, 38, 47, 115

Conjugate momenta 328, 341, 343

Coordination number 118

Coordination polyhedron 118

Copper 28, 115

Cosmic abundance of elements 54, 56, 111, 118, 120, 248

Cosmic rays 122, 209-212, 233

Covalent bonding 107, 108

Critical damping 15

Critical saturation 189-190

Cross section 29ff
    Volume coefficient 32, 90-91, 100, 261, 263, 269
    Mass coefficient 32, 90-91

Crystals 101ff
    binding 107
    classes 102-105
    space groups 105
    symmetry operations 104-105
    two-dimensional 105-106, 151

Cubic lattice 4, 45-46, 71, 102-105, 113, 116, 118, 145, 204

Current density 6, 8-9, 22, 25, 27

Cutoff size of grains 128

Cutoff wavelength 223, 225-226, 255, 271, 302

Cylindrical grains 50, 96-97, 250

C_2H_2 (acytelene) 114, 226

C_2H_4 (ethylene) 114

C_2H_6 (ethane) 112, 114

CH_4 (methane) 112-113, 186

CH_3OH (methanol) 115, 120, 187, 210, 214-215

CsF 108

CO 111, 120, 182, 187, 203, 208, 209, 213, 214

CO_2 110, 120, 187, 203, 214-217

DDA, see Discrete dipole approx.

Debye frequency 149

Debye temperature 149-151, 223

Degenerate electrons 65, 111, 121

Degeneracy parameter 327, 349

Degrees of freedom 143, 145, 170, 185, 226-227, 328-330

Density of vibrational modes 149-151

Depolarization field 70-71

Depolarization 95

Depletion 111, 120, 201-202

Diamond 114ff

DIB, see Diffuse interstellar bands

Dielectrics, definition 5

Dielectric permeability,
    definition 2-3
    of conducting medium 25 - 27

Diffraction pattern of sphere 78

Diffusion of photons 266

Dipole field 6, 22, 62, 176

Dipole absorption 60, 63

Dipole moment 2-4, 7, 8, 16, 18-19, 22, 24, 49
    of a transition 344, 347-348
    induced 110, 203

Dipole potential 71

Dipole radiation 21ff

Dipole scattering 60

Discrete dipole approximat. 45-46

Disorder time 170-171

Diffuse interstellar bands 229-230

Dispersion relation 4, 14, 18-19, 25, 48, 146-147

Distance modulus 241, 243

DNA 110

Drude profile 25, 27-28, 244
Dulong-Petit rule 150, 152
Dust-to-gas ratio 120, 182, 202, 248ff, 251, 254, 298, 311, 316

Earth, surface temperature 135
    annual influx of meteorites 181
Eddington factor 261, 265
Effective optical depth 314ff
Effective medium theories 50ff
Einstein coefficients 132-133, 224, 346ff
Electron radius, classical 24
Electric polarizability 3, 13, 16, 40, 45-48, 52, 62
Electric susceptibility, definition 3
Electromagnetic wave 2, 9-11, 25-26, 29, 33-34, 38-39, 77-78
Elementary cell 101
Emissivity 131-132, 140, 157, 160, 284
Energy bands 109ff
Energy density
    electric or magnetic field 8-9
    blackbody radiation 328, 340
Enstatite ($MgSiO_3$) 119, 233,234
Enthalpy 333, 337
Entropy 5, 47, 188, 189, 322, 330-337
ERE, see Extended red emission
Ergodic hypothesis 328-330, 332
Euler's constant 80, 81
Evaporation of grains 50, 110, 122, 129-130, 177-180, 183-191, 195, 199, 203, 208, 212, 215, 226-227, 256, 272, 305
    Typical temperatures 187
Extended red emission 229-230
Extinction coefficient, definition 30
Extinction efficiency, definition 30
    plots of 85-89, 123-124, 126

Fayalite Fe_2SiO_4 119
Fermi energy 108, 349
Fermi velocity 65
Fermi statistics 326
Fermi temperature 349
Fermions 326, 327

Ferromagnets 7
Ferrosilite 119
Fine structure constant 2
Fluffy grains 50-54, 255-256
    absorption coefficient 55-56
Forsterite Mg_2SiO_4
Fourier expansion 18, 25, 105, 109
Free charges, definition 8
Free energy, see Helmholtz potential
Free enthalpy, see Gibbs potential
Friction coefficient 14-17, 168, 309-310
Fullerenes 117

Galactic nuclei 278ff, 304-307
Gas constant R 185
Gas phase chemistry 210-211
Garnett mixing rule 51-52
Gibbs potential 191, 192, 195, 333-334, 337
Globules 289-293
Grain alignment 96
Grain charge 161-165, 183
Grain destruction 181ff
Grain emission 131-133, 143-144
Grain formation 163, 184-199
Grain motion 166ff
Grain size distribution 128-130
Grain temperature
    equilibrium 134-137, 139
    fluctuations 152-160
Graphite 28, 65, 99, 100, 110, 159, 160, 173, 303, 317
    interstellar extinction 252, 254
    evaporation temperature 187
    Fermi temperature 249
    optical constants 122, 125, 126
    specific heat 151, 152
    structure 115-117
Group velocity 146

HAC 117
Hamiltonian 341, 343, 344
Harmonic oscillator
    classical Lorentz model 13ff

absorption cross section 25
Dispersion relation 18-19
dissipation of energy 17-18
emission of 22
scattering cross section 24
quantum mechanical 323, 344ff
Helium 251, 291, 307
Heisenberg's uncertainty principle 325, 355
Helmholtz potential 333
Hertzsprung Russell diagram (HRD) 111, 307
Homogeneous nucleation 190ff
Homogeneous phase 334 (homopolar
Homopolar, see Covalent bonding
Huygen's principle 74, 77, 78
Hydrogen bridges 110
H_2 108, 172, 187, 207, 212, 248
formation 214-215
H_2S 110

Ice Mantles 56
Inertia ellipsoid 171
Insulators, see Dielectrics
Intensity of radiation
definition 259-260
moments of 260-261
Interstellar extinction, see Reddening curve
Interstellar radiation field (ISRF) 137-139, 278-280, 290
Ionic bonding 108
Io, moon of Jupiter 174-176
Ionization equilibrium 161ff

Jeans stability criterion 291
Joule-Thomson experiment 333
Jupiter 172, 174-176

Kimberlites 115
Kirchhoff's law 131ff, 153
Kirchhoff's wave propagation 78
Kramers-Kronig relations 20, 36, 47ff, 123, 135, 248

Lagrange function 143, 341, 342

Lagrangian multiplicators 321
Laplace equation 61
Larmor radius 176
Large particles 73-79
Lattice types 102ff
Lattice base 101-102, 235
Linear polarization 42-45, 64, 93-97
Local electric field 71ff
Lorentz force 1, 363
Lorentz profile 18, 25, 228
Lorentz transformation 176
Loschmidt number 185
Lyman band 214, 248
$Ly\alpha$ line 248

Magnetic dipole moment 7
Magnetic dipole absorption 61
Magnetic dissipation 97
Magnetic polarizability 7, 13
Magnetic polarization 6-7
Magnetic susceptibility 7
Magnetization, definition 7
Main sequence stars 241, 245-246, 299, 307
Major axis 42, 295
Maxwell's equations 9
Maxwellian velocity distribution
Metals, 6, 26-28, 61, 65, 100, 165
bonding 108
spheres 89, 93
Meteorites 172, 173, 178, 181
Meteoroids 176-179
Meteors 176, 179
Micrometeorites 178
Microcanonic emission 219, 222-223
Mie theory 37ff, 51, 60, 61, 64, 80, 165, 250, 298

Normal coordinates 221, 341-342
Normal incidence 75-76
Na (sodium) 108-109
NaCl 108
Nördlingen Ries 181

Occupation number 327

Ohm's law 25, 27, 61
Olivine $(Mg,Fe)_2SiO_4$ 118, 119, 233
Optical constant, definition 11
    Kramers-Kronig relation 47
    vs. dielectric permeability 11
    of dusty medium 35-36, 49
    of harmonic oscillators 19
    of metals 27
    of Rayleigh-Gans particles 80
    at X-rays 81
    of standard dust 123-127
Optical thickness, see Optical depth
Optical phonon branch 147-148
Optical depth 55, 68-69, 142, 216, 230, 241, 252
    definition 262-263
Optical theorem 32ff, 80
Origin of dust 111-112, 117, 172-173
Origin of elements 120-121
Orthopyroxenes 119

PAHs 111, 115-117, 138, 143, 151-152, 166, 219ff, 249, 251
    cross section and resonances 227-228, 255
    destruction 226-227
    emission 220, 224-225, 290, 300-307
    cutoff wavelength 225-226
    MC radiative transfer 284-286
Pancakes (spheroids) 65-68
Paramagnetic substances 7
Partition function 132, 236, 321-324, 333-334
Pauli's exclusion principle 107, 203, 226, 327
Penetration depth of light 76, 366
Phase function 41, 64, 92, 253, 254, 262, 265, 282, 286
    definition 31
    Henyey-Greenstein 31
Phase space 226, 325-330
Phase transition 115, 177, 185
Phase velocity 10, 28, 48, 60, 77,
Photoelectric effect 163ff

Photon yield 165-166, 229
Planck function 90, 135, 138, 257, 267, 337ff
Planck absorption coefficient 266
Plasma frequency 19, 27-28, 81, 100
Polarizability 3, 102, 110, 203
    atomic 73
    as a tensor3
    of a sphere 62
    of an ellipsoid 66
Polarization charges 7, 8
Polarization
    magnetic, definition 6
    of scattered light 39-45, 93-97
    of matter 2-5
Polarization ellipse 41-42,
Porous grains, see Fluffy grains
Potential
    double well 354-356
    periodic 109-110, 351-353
    retarded 21-22
    U12-6 type 203-204
    van der Waals 202-204
Poynting vector 11, 22, 29, 38-39, 48, 62-63, 165
Poynting-Robertson effect 173-174, 178
Presolar grains 173
Primitive cell 101-102, 106, 148
Principal axis 65, 68, 171
Probability density of temperature 153-155
Protostellar cores
    dust cross section 53-54
    dust emission 291-292
Psycho-physical rule 240
Pyroelectricity 4
Pyroxenes 119
π-bonds 112-115, 117

Radiation const. 135, 266, 272, 340
Radiation damping 23
Radiative transfer
    basic equations 259-265
    in disks 272-278

in galactic nuclei 278-280
in spherical clouds 268-272
Monte Carlo 281-288
Radiation pressure 31-32, 166, 169,
    172, 178, 261, 266, 308, 310-
    312
Random walk 167, 170, 172, 207, 209
Rayleigh limit 55, 60, 66, 68, 80, 81,
    90, 99, 128, 134, 137, 158,
    244, 308
    definition 59
Rayleigh-Gans particles 49, 80
Rayleigh-Jeans limit 121, 138, 140,
    143, 296, 298, 339
Reciprocal lattice 105-107, 109,
Reciprocity theorem 79
Reddening curve 128, 239, 241-246,
    248, 252-253, 314-317
Reflectance of light 75, 76, 88
Reflection efficiency 76
Reflection nebulae 93, 159, 219, 229-
    230, 239, 249, 302-303
Refraction 20, 77
Refractive index, see Optical const.
Retarded potentials 21, 22
Rosseland absorption coefficient 266
R_V (Ratio of total over selective ex-
    tinction) 130, 242-246, 248,
    316-317

Saturation parameter 188, 193-194,
    197-199
Scattering coefficient
    by mass or volume 32, 90, 254
    of ellipsoids 67
    per grain 30
    physical background 63
    at long wavelengths 48
    plots of 85-89, 123-124, 126
    of small spheres 60ff
Scattering diagram 92
Scattering matrix, see Amplitude scat-
    tering matrix
Scattering efficiency
    calculation for spheres 37ff

at long wavelengths 48
of small spheres 60, 63
Scattering plane 40, 44-45, 64, 93
Schrödinger equation 109, 343-345,
    348, 354
Serkowski curve 96
Shape factor 66-67, 100
Shooting stars 176ff
Silicate bands 121,
    at 10 and 18μm 230-232
    of crystalline grains 233-237
Silicate grains 111, 117-123
Silicon bonding 117
Silicon carbide (SiC) 100, 112, 173
Small spheres
    definition 59
    absorption efficiency 60ff
    scattering efficiency 60ff
Snell's law 76
Solar abundances 120
Sound velocity 149
Source function 263, 267, 269, 270,
    279, 280
Specific heat 65, 148, 150-152, 177,
    197, 235
Spectral luminosity 242, 281, 311
Spectral indices 246-247
Spheroids 66-69, 80, 100
Sputtering 130, 182-184
standing waves 146, 149, 326
Stefan-Boltzmann law 339, 340
Stefan-Boltzmann constant, see Ra-
    diation constant
Stellar magnitude 240-241, 362
    conversion to Jansky 239, 240
Sticking probability 162, 163, 166, 191,
    201ff, 209, 214
Stirling formula 320
Stokes parameters 39ff
Supernovae (SN) 120, 130, 172-173,
    181-184
Supernovae rate 182
Surface modes 99, 100
Surface tension 187, 192, 199
Symmetry of crystals 101ff

SiO$_4$-tetrahedron 117-119, 235
σ-bonds 112-115

Thermal hopping 206-208, 212, 233
Thermodynamics, basics 319-340
Thomson scattering cross section 24,
        82
Transition probability, see Einstein
        coefficient
Translational symmetry 101
T Tauri stars 296-298
Tunguska meteorite 181
Tunneling 206, 208-209, 212, 214, 350-
        356
Two-color-diagrams (TCD) 245, 246

UBV diagram 245, 246
Unit cell 101

van der Waals forces 110-111, 115,
        202-203
Vapor pressure 185ff
Vega 241
Very small grains (vsg) 249-254, 192,
        299-303, 305-307
    definition 152

Water (gaseous or liquid) 110, 120,
        185-187, 184, 203-204, 210
Water ice 122, 127, 185-186, 207, 215-
        217, 237
Wave equation 9-10, 37,-38, 78
Wavenumber 10, 12, 26, 28-29, 33,
        38, 41, 45, 75, 107, 109, 146-
        147, 222, 351-352
Werner band 214, 248
White dwarfs 307
Wien's displacement law 141
Wien limit 339

X-ray absorption 82, 83, 122
X-ray scattering 81, 82, 104

2175Å (4.6μm^{-1}) bump 128, 244, 252,
        253, 316
10μm silicate feature 121, 142, 159,
        230-232, 238, 301

Printed and bound by CPI Group (UK) Ltd, Croydon, CR0 4YY

23/10/2024

01778263-0016